Number Theory

New York Seminar 2003

Springer Science+Business Media, LLC

David Chudnovsky
Gregory Chudnovsky
Melvyn Nathanson

Editors

Number Theory

New York Seminar 2003

 Springer

David Chudnovsky
Gregory Chudnovsky
Department of Mathematics
Polytechnic University
6 Metrotech Center
Brooklyn, NY 11201
USA

Melvyn Nathanson
Department of Mathematics & Computer Science
Lehman College
250 Bedford Park Blvd. West
Bronx, NY 10468
USA

Library of Congress Cataloging-in-Publication Data
New York Number Theory Seminar (2003)
    Number theory: New York seminar 2003 / David Chudnovsky, Gregory Chudnovsky, Melvyn
Nathanson.
        p.   cm. —
        ISBN 978-1-4612-6490-3   ISBN 978-1-4419-9060-0 (eBook)
        DOI 10.1007/978-1-4419-9060-0
    1. Number theory—Congresses. I. Chudnovsky, D. (David), 1947– II. Chudnovsky, G.
(Gregory), 1952– III. Nathanson, Melvyn B. (Melvyn Bernard), 1944– IV. Title.
    QA241.N475 2003
    512.7—dc22   2003058458

ISBN 978-1-4612-6490-3   Printed on acid-free paper.

9 8 7 6 5 4 3 2 1   SPIN 10944363

springeronline.com

Preface

This volume marks the 20th anniversary of the New York Number Theory Seminar (NYNTS). The seminar began to meet in the Spring, 1982 semester at the CUNY Graduate Center in midtown Manhattan, and has been meeting continuously at the Graduate Center for two decades, even as the Graduate Center moved from its original location on 42nd Street near Fifth Avenue to temporary quarters in an office building next to Grand Central Station to a new and elegant building in the former B. Altman department store on Fifth Avenue betwen 34th and 35th Streets.

The seminar was originally organized by Harvey Cohn, David and Gregory Chudnovsky, and Melvyn B. Nathanson. In 1982, Harvey Cohn was at City College (CUNY) and the Graduate Center, the Chudnovskys were at Columbia, and Mel Nathanson was at Rutgers. Today, Harvey has retired to California, the Chudnovskys are at Polytechic University of New York, and Nathanson is at Lehman College (CUNY) and the Graduate Center.

For 20 years the NYNTS has tried to present the broad spectrum of number theory and related fields of mathematics, from physics to geometry to computer science. Mathematics, like other sciences, is a mafia-run enterprise, where the local mafias represent currently fashionable fields of research, almost always important, but never nearly as important as the reigning dons believe. We have always tried to invite not only Fields metalists and other standard speakers, but also mathematicians, especially younger and relatively unknown researchers, whose theorems are significant and whose work might become the next big thing in number theory. We do not attempt to predict the future, but we consciously strive to include speakers who are out of the mainstream.

Since its inception, the proceedings of the New York Number Theory Seminar have been published by Springer-Verlag, and we are grateful to Springer and its mathematics editors for their support of the Seminar. We thank Ina Lindemann, the current mathematics editor of Springer, for her continuing interest in NYNTS.

At various times in the past 20 years the seminar has been supported by grants from the NSA Mathematical Sciences Program, the Number Theory Foundation, and the Office of the Provost of Lehman College, and we appreciate their support.

Contents

The spanning number and the independence number of a subset of an abelian group

Béla Bajnok

Department of Mathematics, Gettysburg College
Gettysburg, PA 17325-1486 USA
E-mail: bbajnok@gettysburg.edu

April 29, 2003

Abstract

Let $A = \{a_1, a_2, \ldots, a_m\}$ be a subset of a finite abelian group G. We call A *t-independent* in G, if whenever

$$\lambda_1 a_1 + \lambda_2 a_2 + \cdots + \lambda_m a_m = 0$$

for some integers $\lambda_1, \lambda_2, \ldots, \lambda_m$ with

$$|\lambda_1| + |\lambda_2| + \cdots + |\lambda_m| \leq t,$$

we have $\lambda_1 = \lambda_2 = \cdots = \lambda_m = 0$, and we say that A is *s-spanning* in G, if every element g of G can be written as

$$g = \lambda_1 a_1 + \lambda_2 a_2 + \cdots + \lambda_m a_m$$

for some integers $\lambda_1, \lambda_2, \ldots, \lambda_m$ with

$$|\lambda_1| + |\lambda_2| + \cdots + |\lambda_m| \leq s.$$

In this paper we give an upper bound for the size of a t-independent set and a lower bound for the size of an s-spanning set in G, and determine some cases when this extremal size occurs. We also discuss an interesting connection to spherical combinatorics.

1 Introduction

We illuminate our concepts by the following examples.

Example 1 Consider the set $A = \{1, 4, 6, 9, 11\}$ in the cyclic group $G = \mathbb{Z}_{25}$. We are interested in the degree to which this set is *independent* in G. We find, for example, that $1 + 4 + 4 - 9 = 0$ and $11 + 11 + 9 - 6 = 0$, but that such an equation with only three terms from A cannot be found. We therefore say that A is *3-independent* in G and write $\mathrm{ind}(A) = 3$. It can be shown that A is optimal in each of the following regards:

- no subset of G of size $m > 5$ is 3-independent in G (furthermore, A is essentially the unique 3-independent set in G of size 5);

- no subset of G of size 5 is t-independent for $t > 3$ (that is, for $t > 3$, there will always be t, not necessarily distinct, elements with a signed sum of 0); and

- $n = 25$ is the smallest odd number for which a 3-independent set of size 5 in $\mathbb{Z}_n$ exists. (In fact, it can be shown that $\mathbb{Z}_n$ has a 3-independent set of size 5, if and only if, $n = 20, 22, 24, 25, 26$, or $n \geq 28$.)

The fact that $G = \mathbb{Z}_{25}$ has this relatively large 3-independent subset is due, as explained later, to the fact that 25 has a prime divisor which is congruent to 5 mod 6.

Example 2 How can one place a finite number of points on the d-dimensional sphere $S^d \subset \mathbb{R}^{d+1}$ with the highest *momentum balance*? For the circle S^1, the answer is given by the vertices of a regular polygon, but the issue is far more difficult for $d > 1$. For a positive integer n and a set of integers $A = \{a_1, a_2, \ldots, a_m\}$, define the set of n points $X(A) = \{x_1, x_2, \ldots, x_n\}$ with

$$x_i = \frac{1}{\sqrt{m}} \cdot \left(\cos(\frac{2\pi i a_1}{n}), \sin(\frac{2\pi i a_1}{n}), \ldots, \cos(\frac{2\pi i a_m}{n}), \sin(\frac{2\pi i a_m}{n}) \right)$$

($i = 1, 2, \ldots, n$); thus, for example, for $n = 25$ and $A = \{1, 4, 6, 9, 11\}$, $X(A)$ is a set of 25 points on the unit sphere S^9. It can be shown that this $X(A)$ is a *spherical 3-design*, that is, for every polynomial $f : S^9 \to \mathbb{R}$ of total degree at most 3, the average value of f on S^9 equals the arithmetic average of f on $X(A)$. We can also verify that $X(A)$ is optimal in that

- no set of 25 points is a t-design on S^9 for $t > 3$;

- no set of 25 points is a 3-design on S^d for $d > 9$;

- $n = 25$ is the minimum odd size for which a 3-design on S^9 exists. (It was recently proved that an n-point 3-design on S^9 exists, if and only if, $n = 20, 22, 24$, or $n \geq 25$.)

Example 3 Finally, consider $A = \{3, 4\}$ in $G = \mathbb{Z}_{25}$. Note that every element of G can be generated by a signed sum of at most three terms of A: $1 = 4 - 3, 2 = 3 + 3 - 4, \ldots, 24 = 3 - 4$. We therefore call $A = \{3, 4\}$ a *3-spanning set* in $G = \mathbb{Z}_{25}$, and write span$(A) = 3$. Again, our example is extremal; it can be shown that

- no subset of G of size $m < 2$ is 3-spanning in G;

- no subset of G of size 2 is s-spanning for $s < 3$; and

- $n = 25$ is the largest number for which a 3-spanning set of size 2 in $\mathbb{Z}_n$ exists. (Furthermore, as we will see, $\mathbb{Z}_n$ has a 3-spanning set of size at most 2 for every $n \leq 25$.)

In fact, this example has an even more distinguished property: every element of G can be written *uniquely* as a signed sum of at most 3 elements of A; we call such a set *perfect*. As a consequence of being a perfect 3-spanning set, A is also a maximum size 6-independent set in G.

The fact that $G = \mathbb{Z}_{25}$ has a perfect spanning subset of size 2 is due to the fact that 25 is the sum of two consecutive squares, as explained later.

In the subsequent sections of this paper we define and investigate the afore-mentioned concepts and statements. Topics similar to spanning numbers (e.g. h-bases) and independence numbers (e.g. sum-free sets, Sidon sets, and B_h sequences) have been studied vigorously for a long time, see, for example, [9], [13], [15], [20], [22], [23], [27], and various sections of Guy's book [14]. For general references on spherical designs, see [4], [8], [10], [11], [12], [17], [21], and [25].

2 Spanning numbers

Let G be a finite abelian group of order $|G| = n$, written in additive notation. We are interested in the degree to which a given subset of G spans G. More precisely, we introduce the following definition.

Definition 1 *Let s be a non-negative integer and $A = \{a_1, a_2, \ldots, a_m\}$. We say that A is an s-spanning set in G, if every $g \in G$ can be written as*

$$g = \lambda_1 a_1 + \lambda_2 a_2 + \cdots + \lambda_m a_m$$

for some integers $\lambda_1, \lambda_2, \ldots, \lambda_m$ with

$$|\lambda_1| + |\lambda_2| + \cdots + |\lambda_m| \leq s.$$

*We call the smallest s for which A is s-spanning the **spanning number** of A in G, and denote it by $\mathrm{span}(A)$.*

Equivalently, A is an s-spanning subset of G if for every element $g \in G$, we can find non-negative integers h and k and elements x and y in G, so that x is the sum of h (not necessarily distinct) elements of A, y is the sum of k (not necessarily distinct) elements of A, $h + k \leq s$, and $g = x - y$.

The case $s = 0$ is trivial: the only group G which has a 0-span is the one with a single element; therefore, we may assume that $s \geq 1$ and $n \geq 2$. Obviously, $A = G$ is an s-spanning subset of G for every $s \geq 1$. Here we are interested in small s-spanning sets in G; we denote the size of a minimum s-spanning set of G by $p(G, s)$.

For $s = 1$, it is clear that $\mathrm{span}(A) = 1$, if and only if, for each $g \in G$, A contains at least one of g or $-g$; in particular, A must contain every element of order 2. Let $O(G, 2)$ denote the set of order 2 elements of G; with this notation we have

$$p(G, 1) = |O(G, 2)| + \frac{|G \setminus O(G, 2) \setminus \{0\}|}{2} = \frac{n + |O(G, 2)| - 1}{2}. \tag{1}$$

As a special case, for the cyclic group of order n we have

$$p(\mathbb{Z}_n, 1) = \lfloor n/2 \rfloor. \tag{2}$$

For $s \geq 2$, values of $p(G, s)$ seem difficult to establish, even in the case of the cyclic groups. Computational data shows that

$$p(\mathbb{Z}_n, 2) = \begin{cases} 0 & \text{if } n = 1; \\ 1 & \text{if } n = 2, 3, 4, \mathbf{5}; \\ 2 & \text{if } n = 6, 7, \ldots, 12, \mathbf{13}; \\ 3 & \text{if } n = 14, 15, \ldots, 21; \\ 4 & \text{if } n = 22, 23, \ldots, 33, \text{ and } n = 35; \\ 5 & \text{if } n = 34, \, n = 36, 37, \ldots, 49, \text{ and } n = 51; \end{cases} \tag{3}$$

and

$$p(\mathbb{Z}_n, 3) = \begin{cases} 0 & \text{if } n = 1; \\ 1 & \text{if } n = 2, 3, \ldots, 6, \mathbf{7}; \\ 2 & \text{if } n = 8, 9 \ldots, 24, \mathbf{25}; \\ 3 & \text{if } n = 26, 27, \ldots, 50, \, n = 52, \text{ and } n = 55; \\ 4 & \text{if } n = 51, 53, 54, \, n = 56, 57, \ldots, 100, \text{ and } n = 104. \end{cases} \tag{4}$$

(Values marked in bold-face will be discussed later.)

As these values indicate, $p(\mathbb{Z}_n, s)$ is, in general, not a monotone function of n, though we believe that

$$P(s) := \lim_{n \to \infty} \frac{p(\mathbb{Z}_n, s)^s}{n}$$

exists for every s. The following theorem provides a lower bound for $p(G, s)$ which is of the order $n^{1/s}$ as n goes to infinity.

Theorem 2 *Let m and s be positive integers, and define $a(m, s)$ recursively by $a(m, 0) = a(0, s) = 1$ and*

$$a(m, s) = a(m - 1, s) + a(m, s - 1) + a(m - 1, s - 1).$$

1. We have

$$a(m, s) = \sum_{k=0}^{s} \binom{s}{k} \binom{m}{k} 2^k.$$

2. If G has order n and contains an s-spanning set of size m, then $n \leq a(m, s)$.

Proof. 1. Let us define

$$a'(m, s) := \sum_{k=0}^{s} \binom{s}{k} \binom{m}{k} 2^k.$$

Clearly, $a'(m, 0) = a'(0, s) = 1$; below we prove that $a'(m, s)$ also satisfies the recursion.

We have

$$\begin{aligned} a'(m - 1, s - 1) &= \sum_{k=0}^{s-1} \binom{s-1}{k} \binom{m-1}{k} 2^k \\ &= \sum_{k=0}^{s-2} \binom{s-1}{k} \binom{m-1}{k} 2^k + \binom{m-1}{s-1} 2^{s-1}, \end{aligned}$$

and

$$a'(m-1,s) = \sum_{k=0}^{s}\binom{s}{k}\binom{m-1}{k}2^k$$

$$= \sum_{k=0}^{s-1}\binom{s}{k}\binom{m-1}{k}2^k + \binom{m-1}{s}2^s$$

$$= \sum_{k=0}^{s-1}\binom{s-1}{k-1}\binom{m-1}{k}2^k + \sum_{k=0}^{s-2}\binom{s-1}{k}\binom{m-1}{k}2^k + \binom{m-1}{s-1}2^{s-1} + \binom{m-1}{s}2^s.$$

Next, we add $a'(m-1,s)$ and $a'(m-1,s-1)$. Note that

$$\binom{m-1}{s-1}2^{s-1} + \binom{m-1}{s-1}2^{s-1} + \binom{m-1}{s}2^s = \binom{m}{s}2^s,$$

and

$$\sum_{k=0}^{s-2}\binom{s-1}{k}\binom{m-1}{k}2^k + \sum_{k=0}^{s-2}\binom{s-1}{k}\binom{m-1}{k}2^k = \sum_{k=0}^{s-2}\binom{s-1}{k}\binom{m-1}{k}2^{k+1},$$

and by replacing k by $k-1$, this sum becomes

$$\sum_{k=0}^{s-1}\binom{s-1}{k-1}\binom{m-1}{k-1}2^k.$$

Therefore,

$$a'(m-1,s) + a'(m-1,s-1) = \sum_{k=0}^{s-1}\binom{s-1}{k-1}\binom{m-1}{k}2^k + \sum_{k=0}^{s-1}\binom{s-1}{k-1}\binom{m-1}{k-1}2^k + \binom{m}{s}2^s$$

$$= \sum_{k=0}^{s-1}\binom{s-1}{k-1}\binom{m}{k}2^k + \binom{m}{s}2^s$$

$$= \sum_{k=0}^{s}\binom{s-1}{k-1}\binom{m}{k}2^k$$

$$= \sum_{k=0}^{s}\binom{s}{k}\binom{m}{k}2^k - \sum_{k=0}^{s}\binom{s-1}{k}\binom{m}{k}2^k$$

$$= a'(m,s) - a'(m,s-1).$$

2. Assume that $A = \{a_1,\ldots,a_m\}$ is an s-spanning set in G of size m, and let

$$\Sigma = \{\lambda_1 a_1 + \cdots + \lambda_m a_m \mid \lambda_1,\ldots,\lambda_m \in \mathbb{Z}, |\lambda_1| + \cdots + |\lambda_m| \leq s\}.$$

We will count the elements in the index set

$$I = \{(\lambda_1,\cdots,\lambda_m) \mid \lambda_1,\ldots,\lambda_m \in \mathbb{Z}, |\lambda_1| + \cdots + |\lambda_m| \leq s\},$$

as follows. For $k = 0, 1, 2, \ldots, m$, let I_k be the set of those elements of I where exactly k of the m coördinates are non-zero. How many elements are in I_k? We can choose which k of the m coördinates are non-zero in $\binom{m}{k}$ ways; w.l.o.g. let these coördinates be $\lambda_1, \lambda_2, \ldots, \lambda_k$. Next, we choose the values of $|\lambda_1|, |\lambda_2|, \ldots, |\lambda_k|$: since the sum of these k positive integers is at most s, we have $\binom{s}{k}$ choices. Finally, each of these coördinates can be positive or negative, and therefore

$$|I_k| = \binom{s}{k}\binom{m}{k}2^k,$$

and

$$|I| = \sum_{k=0}^{m}\binom{s}{k}\binom{m}{k}2^k = \sum_{k=0}^{s}\binom{s}{k}\binom{m}{k}2^k = a(m, s).$$

Since A is s-spanning in G, we must have $n = |\Sigma| \leq |I| = a(m, s)$. $\quad\square$

Theorem 2 thus provides a lower bound for the size of an s-spanning set in G which is of the order $n^{1/s}$ as n goes to infinity.

For exact values, we establish the following results.

Proposition 3 *Let $s \geq 1$ be an integer.*

1. *If $2 \leq n \leq 2s + 1$, then the set $\{1\}$ is s-generating in $\mathbb{Z}_n$ and $p(\mathbb{Z}_n, s) = 1$.*

2. *If $2s + 2 \leq n \leq 2s^2 + 2s + 1$, then the set $\{s, s + 1\}$ is s-generating in $\mathbb{Z}_n$ and $p(\mathbb{Z}_n, s) = 2$.*

3. *If $n \geq 2s^2 + 2s + 2$, then $p(\mathbb{Z}_n, s) \geq 3$.*

Proof. 1 is trivial. To prove 2, let

$$\Sigma = \{\lambda_1 s + \lambda_2(s + 1) \mid \lambda_1, \lambda_2 \in \mathbb{Z}, |\lambda_1| + |\lambda_2| \leq s\}.$$

The elements of Σ lie in the interval $[-(s^2 + s), (s^2 + s)]$ and, since the index set

$$I = \{(\lambda_1, \lambda_2) \mid \lambda_1, \lambda_2 \in \mathbb{Z}, |\lambda_1| + |\lambda_2| \leq s\}$$

contains exactly $2s^2 + 2s + 1$ elements, it suffices to prove that no integer in $[-(s^2 + s), (s^2 + s)]$ can be written as an element of Σ in two different ways. Indeed, it is an easy exercise to show that

$$\lambda_1 s + \lambda_2(s + 1) = \lambda_1' s + \lambda_2'(s + 1) \in \Sigma$$

implies $\lambda_1 = \lambda_1'$ and $\lambda_2 = \lambda_2'$; therefore, the set $\{s, s + 1\}$ is s-generating in $\mathbb{Z}_n$. As the s-span of a single element can contain at most $2s + 1$ elements, for values $n \geq 2s + 2$ we must have $p(\mathbb{Z}_n, s) = 2$. Statement 3 follows from Theorem 2 by noting that $a(2, s) = 2s^2 + 2s + 1$. $\quad\square$

Let us now examine the extremal cases of Theorem 2.

Definition 4 *Suppose that A is an s-spanning set of size m in G and that $a(m, s)$ is defined as in Theorem 2. If $|G| = n = a(m, s)$, then we say that A is a **perfect s-spanning set** in G.*

Cases where $\mathbb{Z}_n$ has a perfect s-spanning set for $s = 2$ and $s = 3$ are marked with bold-face in (3) and (4). Trivially, the empty-set is a perfect s-spanning set in $\mathbb{Z}_1$ for every s. With (2) and Proposition 3, we can exhibit some other perfect spanning sets in the cyclic group.

Proposition 5 *Let m, n, and s be positive integers, and let $G = \mathbb{Z}_n$.*

1. *If $n = 2m + 1$, then the set $\{1, 2, \ldots, m\}$ is a perfect 1-spanning set in G.*

2. *If $n = 2s + 1$, then the set $\{1\}$ is a perfect s-spanning set in G.*

3. *If $n = 2s^2 + 2s + 1$, then the set $\{s, s + 1\}$ is a perfect s-spanning set in G.*

Note that the sets given in Proposition 5 are not unique: any element of the set in 1 can be replaced by its negative; in 2, the set $\{a\}$ is perfect for every a which is relatively prime to n; it is not difficult to show that another example in 3 is provided by $A = \{1, 2s + 1\}$ (however, the set $\{s, s + 1\}$ in Proposition 3 cannot be replaced by $\{1, 2s + 1\}$). We could not find perfect spanning sets for $s \geq 2$ and $m \geq 3$. It might be an interesting problem to find and classify all perfect spanning sets.

3 Independence numbers

As in the previous section, we let G be a finite abelian group of order $|G| = n$, written in additive notation, and suppose that A is a subset of G. Here we are interested in the degree to which A is independent in G. More precisely, we introduce the following definition.

Definition 6 *Let t be a non-negative integer and $A = \{a_1, a_2, \ldots, a_m\}$. We say that A is a t-independent set in G, if whenever*

$$\lambda_1 a_1 + \lambda_2 a_2 + \cdots + \lambda_m a_m = 0$$

for some integers $\lambda_1, \lambda_2, \ldots, \lambda_m$ with

$$|\lambda_1| + |\lambda_2| + \cdots + |\lambda_m| \leq t,$$

*we have $\lambda_1 = \lambda_2 = \cdots = \lambda_m = 0$. We call the largest t for which A is t-independent the **independence number** of A in G, and denote it by $\mathrm{ind}(A)$.*

Equivalently, A is a t-independent set in G, if for all non-negative integers h and k with $h + k \leq t$, the sum of h (not necessarily distinct) elements of A can only equal the sum of k (not necessarily distinct) elements of A in a *trivial* way, that is, $h = k$ and the two sums contain the same terms in some order.

Here we are interested in the size of a maximum t-independent set in G; we denote this by $q(G, t)$.

Since $0 \leq \mathrm{ind}(A) \leq n - 1$ holds for every subset A of G (so no subset is "completely" independent), we see that $q(G, 0) = n$ and $q(G, n) = 0$. It is also clear that $\mathrm{ind}(A) = 0$, if and only if, $0 \in A$, hence

$$q(G, 1) = n - 1. \tag{5}$$

For the rest of this section we assume that $t \geq 2$.

We can easily determine the value of $q(G, 2)$ as well. First, note that A cannot contain any element of $\{0\} \cup \mathrm{Ord}(G, 2)$ (the elements of order at most 2); to get a maximum 2-independent set in G, take exactly one of each element or its negative in $G \setminus \mathrm{Ord}(G, 2) \setminus \{0\}$, hence we have

$$q(G, 2) = \frac{n - |\mathrm{Ord}(G, 2)| - 1}{2}. \tag{6}$$

As a special case, for the cyclic group of order n we have

$$q(\mathbb{Z}_n, 2) = \lfloor (n - 1)/2 \rfloor. \tag{7}$$

Note that if $\mathrm{Ord}(G, 2) \cup \{0\} = G$ then $q(G, 2) = 0$; for $n \geq 2$ this occurs only for the elementary abelian 2-group. If $\mathrm{Ord}(G, 2) \cup \{0\} \neq G$ then, since $\mathrm{Ord}(G, 2) \cup \{0\}$ is a subgroup of G, we have $1 \leq |\mathrm{Ord}(G, 2)| + 1 \leq n/2$, and therefore we get the following.

Proposition 7 *If G is isomorphic to the elementary abelian 2-group, then $q(G, 2) = 0$. Otherwise*

$$\frac{1}{4}n \leq q(G, 2) \leq \frac{1}{2}n.$$

Let us now consider $t = 3$. As before, if G does not contain elements of order at least 4, then $q(G, 3) = 0$; this occurs if and only if G is isomorphic to the elementary abelian p-group for $p = 2$ or $p = 3$. In [3] we proved the following.

Theorem 8 ([3]) *If G is isomorphic to the elementary abelian p-group for $p = 2$ or $p = 3$, then $q(G, 3) = 0$. Otherwise*

$$\frac{1}{9}n \leq q(G, 3) \leq \frac{1}{4}n.$$

These bounds can be attained since $q(\mathbb{Z}_9, 3) = 1$ and $q(\mathbb{Z}_4, 3) = 1$.

For the cyclic group $\mathbb{Z}_n$, we can find explicit 3-independent sets as follows. For every n, the odd integers which are less than $n/3$ form a 3-independent set; if n is even, we can go up to (but not including) $n/2$ as then the sum of two odd integers cannot equal n. We can do better in one special case when n is odd; namely, when n has a prime divisor p which is congruent to 5 mod 6, one can show that the set

$$\left\{ pi_1 + 2i_2 + 1 \mid i_1 = 0, 1, \ldots, \frac{n}{p} - 1, \ i_2 = 0, 1, \ldots, \frac{p - 5}{6} \right\} \tag{8}$$

is 3-independent. It is surprising that these examples cannot be improved, as we have the following exact values.

Theorem 9 ([3]) *For the cyclic group* $G = \mathbb{Z}_n$ *we have*

$$q(\mathbb{Z}_n, 3) = \begin{cases} \lfloor \frac{n}{4} \rfloor & \text{if } n \text{ is even,} \\ \left(1 + \frac{1}{p}\right) \frac{n}{6} & \text{if } n \text{ is odd, has prime divisors congruent to 5} \quad (\text{mod } 6), \\ & \text{and } p \text{ is the smallest such divisor,} \\ \lfloor \frac{n}{6} \rfloor & \text{otherwise.} \end{cases}$$

For $t \geq 4$, exact results seem more difficult. With the help of a computer, we generated the following values.

$$q(\mathbb{Z}_n, 4) = \begin{cases} 0 & \text{if } n = 1, 2, 3, 4; \\ 1 & \text{if } n = 5, 6, \ldots, 12; \\ 2 & \text{if } n = 13, 14, \ldots, 26; \\ 3 & \text{if } n = 27, 28, \ldots, 45, \text{ and } n = 47; \\ 4 & \text{if } n = 46, \; n = 48, 49, \ldots, 68, \text{ and } n = 72, 73; \\ 5 & \text{if } n = 69, 70, 71, \text{ and } n = 74, 75, \ldots, 102; \end{cases} \qquad (9)$$

$$q(\mathbb{Z}_n, 5) = \begin{cases} 0 & \text{if } n = 1, 2, 3, 4, 5; \\ 1 & \text{if } n = 6, 7, \ldots, 17, \text{ and } n = 19, 20; \\ 2 & \text{if } n = 18, \; n = 21, 22, \ldots, 37, \; n = 39, 40, 41, \; n = 43, 44, 45, 47; \\ 3 & \text{if } n = 38, 42, 46, \; n = 48, 49, \ldots, 69, \; n = 71, 72, 73, 75, 76, 77, 79, 81, 83, 85, 87; \end{cases} \qquad (10)$$

and

$$q(\mathbb{Z}_n, 6) = \begin{cases} 0 & \text{if } n = 1, 2, 3, \ldots, 6; \\ 1 & \text{if } n = 7, 8, 9, \ldots, 24; \\ 2 & \text{if } n = 25, 26, 27, \ldots, 69; \\ 3 & \text{if } n = 70, 71, \ldots, 151, \text{ and } n = 153, 154, 155, 158, 159, 160. \end{cases} \qquad (11)$$

(Values marked in bold-face will be discussed later.)

Again we see that $q(\mathbb{Z}_n, t)$ is not, in general, a monotone function of n; although for even values of t the sequence seems to possess more regularity and we conjecture that

$$Q(t) := \lim_{n \to \infty} \frac{q(\mathbb{Z}_n, t)^{t/2}}{n}$$

exists for every even t. The following theorem establishes an upper bound for $q(G, s)$ which is of the order $n^{1/\lfloor t/2 \rfloor}$ as n goes to infinity.

Theorem 10 *Let m and t be positive integers, $t \geq 2$, and let us denote*

$$q(m, t) = \begin{cases} a(m, t/2) & \text{if } t \text{ is even,} \\ a(m, (t-1)/2) + a(m-1, (t-1)/2) & \text{if } t \text{ is odd,} \end{cases}$$

where $a(m, t)$ is defined in Theorem 2. If G has order n and contains a t-independent set of size m, then $n \geq q(m, t)$.

Proof. Assume that $A = \{a_1, \ldots, a_m\}$ is a t-independent set in G of size m, and define

$$\Sigma = \{\lambda_1 a_1 + \cdots + \lambda_m a_m \mid \lambda_1, \ldots, \lambda_m \in \mathbb{Z}, |\lambda_1| + \cdots + |\lambda_m| \le \lfloor t/2 \rfloor\}$$

and

$$I = \{(\lambda_1, \cdots, \lambda_m) \mid \lambda_1, \ldots, \lambda_m \in \mathbb{Z}, |\lambda_1| + \cdots + |\lambda_m| \le \lfloor t/2 \rfloor\}.$$

As in the proof of Theorem 2, we have $|I| = a(m, \lfloor t/2 \rfloor)$. Since A is a t-independent set in G, the elements listed in Σ must be all distinct, hence $n \ge |\Sigma| = |I| = a(m, \lfloor t/2 \rfloor)$. If t is even, we are done.

Now let

$$\Sigma' = \{\lambda_1 a_1 + \cdots + \lambda_m a_m \mid \lambda_1, \ldots, \lambda_m \in \mathbb{Z}, \lambda_1 \ge 1, \lambda_1 + |\lambda_2| + \cdots + |\lambda_m| = \lfloor t/2 \rfloor + 1\}$$

and

$$I' = \{(\lambda_1, \cdots, \lambda_m) \mid \lambda_1, \ldots, \lambda_m \in \mathbb{Z}, \lambda_1 \ge 1, \lambda_1 + |\lambda_2| + \cdots + |\lambda_m| = \lfloor t/2 \rfloor + 1\}.$$

We will count the elements in the index set $|I'|$ as follows. For $k = 0, 1, 2, \ldots, m - 1$, let I_k be the set of those elements of I' where exactly k of the $m - 1$ coördinates $\lambda_2, \ldots, \lambda_m$ are non-zero. An argument similar to that in the proof of Theorem 2 shows that

$$|I'_k| = \binom{m-1}{k} \binom{\lfloor t/2 \rfloor}{k} 2^k,$$

hence

$$|I'| = \sum_{k=0}^{m-1} \binom{m-1}{k} \binom{\lfloor t/2 \rfloor}{k} 2^k = a(m-1, \lfloor t/2 \rfloor).$$

If t is odd, then the elements listed in Σ' must be distinct from each other and from those in Σ as well, thus $n \ge |\Sigma| + |\Sigma'| = |I| + |I'| = a(m, \lfloor t/2 \rfloor) + a(m-1, \lfloor t/2 \rfloor)$. $\square$

Theorem 10 thus provides an upper bound for the size of a t-independent set in G which is of the order $n^{1/\lfloor t/2 \rfloor}$ as n goes to infinity.

For exact values, we establish the following results.

Proposition 11 *Let $t \ge 2$ be an integer.*

1. *If $1 \le n \le t$, then $q(\mathbb{Z}_n, t) = 0$.*

2. *If $t + 1 \le n \le \lfloor t^2/2 \rfloor + t$, then the set $\{1\}$ is t-independent in $\mathbb{Z}_n$ and $q(\mathbb{Z}_n, t) = 1$.*

3. (a) *Suppose that t is even. If $n \ge t^2/2 + t + 1$, then the set $\{t/2, t/2 + 1\}$ is t-independent in $\mathbb{Z}_n$ and $q(\mathbb{Z}_n, t) \ge 2$.*

   (b) *Suppose that t is odd. If $n = (t^2 - 1)/2 + t + 1$, then the set $\{1, t\}$ is t-independent in $\mathbb{Z}_n$ and $q(\mathbb{Z}_n, t) = 2$.*

Proof. Let $q(m,t)$ be defined as in Theorem 10. Since $q(1,t) = t+1$, our first claim follows from Theorem 10. To prove 2, note that if $n \geq t+1$, then $\{1\}$ is t-independent in $\mathbb{Z}_n$; furthermore, $q(2,t) = \lfloor t^2/2 \rfloor + t + 1$.

Now let t be even, and assume that $n \geq t^2/2 + t + 1$. We define

$$\Sigma = \{\lambda_1 \frac{t}{2} + \lambda_2(\frac{t}{2} + 1) \mid \lambda_1, \lambda_2 \in \mathbb{Z}, |\lambda_1| + |\lambda_2| \leq t\}.$$

The elements of Σ lie in the interval $[-(t^2/2 + t), (t^2/2 + t)]$ and therefore, to prove 3 (a), it suffices to show that

$$0 = \lambda_1 \frac{t}{2} + \lambda_2(\frac{t}{2} + 1) \in \Sigma$$

implies $\lambda_1 = \lambda_2 = 0$, which is an easy exercise. Statement 3 (b) is essentially similar. □

We now turn to the extremal cases of Theorem 10.

Definition 12 *Suppose that A is a t-independent set of size m in G and that $q(m,t)$ is defined as in Theorem 10. If $|G| = n = q(m,t)$, then we say that A is a **tight t-independent set** in G.*

Cases where $\mathbb{Z}_n$ has a tight t-independent set for $t = 4$, $t = 5$, and $t = 6$ are marked with bold-face in (9), (10), and (11). Trivially, the empty-set is a perfect t-independent set in $\mathbb{Z}_1$ for every t. With (5), (7), Theorem 9, Proposition 11, and one other (sporadic) example, we have the following tight t-independent sets in the cyclic group.

Proposition 13 *Let m, n, and t be positive integers, and let $G = \mathbb{Z}_n$.*

1. *If $n = 2$, then the set $\{1\}$ is a tight 1-independent set in G.*

2. *If $n = 2m + 1$, then the set $\{1, 2, \ldots, m\}$ is a tight 2-independent set in G.*

3. *If $n = 4m$, then the set $\{1, 3, \ldots, 2m - 1\}$ is a tight 3-independent set in G.*

4. *If $n = t + 1$, then the set $\{1\}$ is a tight t-independent set in G.*

5. *Let $n = \lfloor t^2/2 \rfloor + t + 1$. If t is even, then the set $\{t/2, t/2 + 1\}$ is a tight t-independent set in G; if t is odd, then the set $\{1, t\}$ is a tight t-independent set in G.*

6. *If $n = 38$, then the set $\{1, 7, 11\}$ is a tight 5-independent set in G.*

Proposition 13 contains every tight (non-empty) t-independent set that we could find so far; in particular, we could not find tight t-independent sets for $t \geq 4$ and $m \geq 3$ other than the seemingly sporadic example listed last. The problem of finding and classifying all tight t-independent sets remains open.

As it is clear from our exposition, there is a strong relationship between s-spanning sets and t-independent sets when t is even. Namely, we have the following.

Theorem 14 *Let s and t positive integers, t even. Let A be a subset of G, and suppose that $\operatorname{span}(A) = s$ and $\operatorname{ind}(A) = t$.*

1. *The order n of G satisfies $a(m, t/2) \leq n \leq a(m, s)$.*

2. *We have $t \leq 2s$.*

3. *The following three statements are equivalent.*

   (i) *$t = 2s$;*

   (ii) *A is a perfect s-spanning set in G; and*

   (iii) *A is a tight t-independent set in G.*

The analogous relationship when t is odd is considerably more complicated and will be the subject of future study.

4   Spherical designs

Here we discuss an application of the previous section to spherical combinatorics. We are interested in placing a finite number of points on the d-dimensional sphere $S^d \subset \mathbb{R}^{d+1}$ with the highest *momentum balance*. The following definition was introduced by Delsarte, Goethals, and Seidel in 1977 [8].

Definition 15 *Let t be a non-negative integer. A finite set X of points on the d-sphere $S^d \subset \mathbb{R}^{d+1}$ is a **spherical t-design**, if for every polynomial f of total degree t or less, the average value of f over the whole sphere is equal to the arithmetic average of its values on X.*

In other words, X is a spherical t-design if the Chebyshev-type quadrature formula

$$\frac{1}{\sigma_d(S^d)} \int_{S^d} f(\mathbf{x}) d\sigma_d(\mathbf{x}) \approx \frac{1}{|X|} \sum_{\mathbf{x} \in X} f(\mathbf{x}) \tag{12}$$

is exact for all polynomials $f : S^d \to \mathbb{R}$ of total degree at most t (σ_d denotes the surface measure on S^d).

The concept of t-designs on the sphere is analogous to $t - (v, k, \lambda)$ designs in combinatorics (see [24]), and has been studied in various contexts, including representation theory, spherical geometry, and approximation theory. For general references see [4], [8], [10], [11], [12], [17], [21], and [25]. The existence of spherical designs for every t and d and large enough $n = |X|$ was first proved by Seymour and Zaslavsky in 1984 [26].

A central question in the field is to find all integer triples (t, d, n) for which a spherical t-design on S^d exists consisting of n points, and to provide explicit constructions for these parameters.

Clearly, to achieve high momentum balance on the sphere, one needs to take a large number of points. Delsarte, Goethals, and Seidel [8] provide the tight lower bound

$$n \geq N_t^d := \binom{d + \lfloor t/2 \rfloor}{\lfloor t/2 \rfloor} + \binom{d + \lfloor (t-1)/2 \rfloor}{\lfloor (t-1)/2 \rfloor}. \tag{13}$$

We shall refer to the bound N_t^d in (13) as the DGS bound. Spherical designs of this minimum size are called **tight**. Bannai and Damerell [5], [6] proved that tight spherical designs for $d \geq 2$ exist only for $t = 1, 2, 3, 4, 5, 7$, or 11. All tight t-designs are known, except possibly for $t = 4, 5$, or 7. In particular, there is a unique 11-design (d=23 and $n = 196560$).

Let us now attempt to construct spherical designs. One's intuition that the vertices of a regular polygon provide spherical designs on the circle S^1 is indeed correct; more precisely, we have the following.

Proposition 16 *Let t and n be positive integers.*

1. *If $n \leq t$, then there is no n-point spherical t-design on S^1.*

2. *Suppose that $n \geq t + 1$. For a positive integer j, define*

$$\mathbf{z}_n^j := \left(\cos(\frac{2\pi j}{n}), \sin(\frac{2\pi j}{n}) \right). \tag{14}$$

Then the set $X_n := \{ \mathbf{z}_n^j \mid j = 1, 2, \ldots, n \}$ is a t-design on S^1.

Proof. 1 follows from the DGS bound as $N_t^1 = t + 1$. To prove 2, we first note that, using spherical harmonics, one can prove (see [8]) that, in general, a finite set X is a spherical t-design, if and only if, for every integer $1 \leq k \leq t$ and every homogeneous *harmonic* polynomial f of total degree k,

$$\sum_{\mathbf{x} \in X} f(\mathbf{x}) = 0.$$

(A polynomial is harmonic if it is in the kernel of the Laplace operator.) The set of homogeneous harmonic polynomials of total degree k on the circle, $\text{Harm}_k(S^1)$, is a 2-dimensional vector space over the reals and is spanned by the polynomials $\text{Re}(z^k)$ and $\text{Im}(z^k)$ where $z = x + \sqrt{-1}y$ (we can think of the elements of X and S^1 as complex numbers). Therefore, we see that X is a t-design on S^1, if and only if,

$$\sum_{\mathbf{z} \in X} \mathbf{z}^k = 0$$

for $k = 1, 2, \ldots, t$. With X_n as defined above, one finds that

$$\sum_{j=1}^n (\mathbf{z}_n^j)^k = \begin{cases} 0 & \text{if } k \not\equiv 0 \bmod n, \\ n & \text{if } k \equiv 0 \bmod n. \end{cases}$$

Therefore, X_n is a t-design on S^1, if and only if, $k \not\equiv 0 \bmod n$ for $k = 1, 2, \ldots, t$ (using the terminology of our last section, if and only if, $\{1\}$ is a t-independent set in $\mathbb{Z}_n$), or $n \geq t + 1$. $\square$

A further classification of t-designs on the circle can be found in Hong's paper [18]; he proved, for example, that if $n \geq 2t + 3$, then there are infinitely many t-designs on S^1 which do not come from regular polygons.

We now attempt to generalize Proposition 16 for higher dimensions. For simplicity, we assume that d is odd, and let $m = (d + 1)/2$. (The case when d is even can be reduced to this case by a simple technique, see [2] or [19].)

Let $a_1, a_2, \ldots, a_m$ be integers, and set $A := \{a_1, a_2, \ldots, a_m\}$. For a positive integers n, define

$$X_n(A) := \left\{ \frac{1}{\sqrt{m}} \left(\mathbf{z}_n^j(a_1), \mathbf{z}_n^j(a_2), \ldots, \mathbf{z}_n^j(a_m) \right) \mid j = 1, 2, \ldots, n \right\}, \tag{15}$$

where, like in (14),

$$\mathbf{z}_n^j(a_i) := \left(\cos(\frac{2\pi j}{n} a_i), \sin(\frac{2\pi j}{n} a_i) \right).$$

Note that $X_n(A) \subset S^d$. In [2] we proved the following.

Theorem 17 ([2]) *Let t, d, and n be positive integers with $t \leq 3$, d odd, and set $m = (d+1)/2$. For integers $a_1, a_2, \ldots, a_m$, define $X_n(A)$ as in (15). If A is a t-independent set in $\mathbb{Z}_n$, then $X_n(A)$ is a spherical t-desian on S^d.*

Theorem 17 yields the following results.

Corollary 18 *Let n and d be positive integers, d odd, and set $m = (d+1)/2$.*

1. (a) *If $n = 1$, then there is no n-point spherical 1-design on S^d.*

   (b) *If $n \geq 2$, define $a_i = 1$ for $1 \leq i \leq m$. Then the set $X_n(A)$, as defined in (15), is a spherical 1-design on S^d.*

2. (a) *If $n \leq d + 1$, then there is no n-point spherical 2-design on S^d.*

   (b) *If $n \geq d + 2$, define $a_i = i$ for $1 \leq i \leq m$. Then the set $X_n(A)$, as defined in (15), is a spherical 2-design on S^d.*

3. (a) *If $n \leq 2d + 1$, then there is no n-point spherical 3-design on S^d.*

   (b) *If $n \geq 2d + 2$ is even or if $n \geq 3d + 3$ is odd, define $a_i = 2i + 1$ for $1 \leq i \leq m$; if*

   $$n \geq \frac{p}{p+1}(3d + 3)$$

   where p is a divisor of n which is congruent to 5 mod 6, choose A to be any m elements of the set in (8). In each case the set $X_n(A)$, as defined in (15), is a spherical 3-design on S^d.

Proof. Parts (a) are from the DGS bounds N_t^d for $t \leq 3$; parts (b) follow from Theorem 17 since, by (5), (7), and the paragraph before Theorem 9, the sets specified are t-independent for $t = 1, 2$, and 3, respectively (note that in all cases of 2 and 3, $m = (d+1)/2 \leq q(\mathbb{Z}_n, t)$). □

Part 3 of Corollary 18 leaves the question of existence of 3-designs open for some odd values of n. Note that the minimum value of

$$\frac{p}{p+1}(3d+3)$$

is $5(d+1)/2$ (when n is divisible by 5). In [2] we proved that a spherical 3-design on S^d (d odd) exists for *every* odd value of $n \geq 5(d+1)/2$, and conjectured that 3-designs do not exist with $2(d+1) < n < 5(d+1)/2$ and n odd. This conjecture is supported by the numerical evidence of Hardin and Sloane [16]. A recent result of Boumova, Boyvalenkov, and Danev [7] proves that no 3-design exists of odd size n with $n < \approx 2.32(d+1)$. In particular, the case $d = 9$ of Example 2 in our Introduction is completely settled: 3-designs on n points on S^9 exist, if and only if, $n \geq 20$ even, or $n \geq 25$ odd.

The application of t-independent sets to spherical t-designs seems more complicated when $t \geq 4$, and will be the subject of an upcoming paper.

Acknowledgments. The author expresses his gratitude to his students Nicolae Laza for valuable computations and Nikolay Doskov for an improvement of Proposition 3.

References

[1] B. Bajnok. Construction of spherical t-designs. *Geom. Dedicata*, 43:167–179, 1992.

[2] B. Bajnok. Constructions of spherical 3-designs. *Graphs Combin.*, 14/2:97–107, 1998.

[3] B. Bajnok and I. Ruzsa. The independence number of a subset of an abelian group. *Integers*, 3/Paper A2, 23 pp. (electronic), 2003.

[4] E. Bannai. On extremal finite sets in the sphere and other metric spaces. *London Math. Soc. Lecture Note Ser.*, 131:13–38, 1988.

[5] E. Bannai and R. M. Damerell. Tight spherical designs I. *J. Math. Soc. Japan*, 31:199–207, 1979.

[6] E. Bannai and R. M. Damerell. Tight spherical designs II. *J. London Math. Soc. (2)*, 21:13–30, 1980.

[7] S. Boumova, P. Boyvalenkov, and D. Danev. New nonexistence results for spherical designs. In B. Bojanov, editor, *Constructive Theory of Functions*, pages 225–232, Varna, 2002.

[8] P. Delsarte, J. M. Goethals, and J. J. Seidel. Spherical codes and designs. *Geom. Dedicata*, 6:363–388, 1977.

[9] P. Erdős and R. Freud. A Sidon problémakör. *Mat. Lapok*, 1991/2:1–44, 1991.

[10] C. D. Godsil. *Algebraic Combinatorics*. Chapman and Hall, Inc., 1993.

[11] J. M. Goethals and J. J. Seidel. Spherical designs. In D. K. Ray-Chaudhuri, editor, *Relations between combinatorics and other parts of mathematics*, volume 34 of *Proc. Sympos. Pure Math.*, pages 255–272. American Mathematical Society, 1979.

[12] J. M. Goethals and J. J. Seidel. Cubature formulae, polytopes and spherical designs. In C. Davis, B. Grünbaum, and F. A. Sher, editors, *The Geometric Vein: The Coveter Festschrift*, pages 203–218. Springer-Verlag New York, Inc., 1981.

[13] S. W. Graham. B_h sequences. In B. C. Berndt, H. G. Diamond, and A. J. Hildebrand, editors, *Analytic number theory, Vol.1. (Allerton Park, IL, 1995)*, pages 431-449, Progr. Math. 138, Birkhäuser Boston, Boston, MA, 1996.

[14] R. K. Guy. *Unsolved Problems in Number Theory*. Second edition. Springer-Verlag New York, 1994.

[15] H. Halberstam and K. F. Roth. *Sequences*. Second edition. Springer-Verlag New York – Berlin, 1983.

[16] R. H. Hardin and N. J. A. Sloane. McLaren's improved snub cube and other new spherical designs in three dimensions. *Discrete Comput. Geom.*, 15:429–441, 1996.

[17] S. G. Hoggar. Spherical t-designs. In C. J. Colbourn and J. H. Dinitz, editors, *The CRC handbook of combinatorial designs*, pages 462–466. CRC Press, Inc., 1996.

[18] Y. Hong. On Spherical t-designs in $\mathbb{R}^2$. *Europ. J. Combinatorics*, 3:255–258, 1982.

[19] J. Mimura. A construction of spherical 2-design. *Graphs Combin.*, 6:369–372, 1990.

[20] M. B. Nathanson. *Additive Number Theory: Inverse Problems and the Geometry of Sumsets*. Springer-Verlag New York, 1996.

[21] B. Reznick. Sums of even powers of real linear forms. *Mem. Amer. Math. Soc.*, 463, 1992.

[22] I. Ruzsa. Solving linear equations in a set of integers I. *Acta Arith.*, 65/3:259–282, 1993.

[23] I. Ruzsa. Solving linear equations in a set of integers II. *Acta Arith.*, 72/4:385–397, 1995.

[24] J. J. Seidel. Designs and approximation. *Contemp. Math.*, 111:179–186, 1990.

[25] J. J. Seidel. Spherical designs and tensors. In E. Bannai and A. Munemasa, editors, *Progress in algebraic combinatorics*, volume 24 of *Adv. Stud. Pure Math.*, pages 309–321. Mathematical Society of Japan, 1996.

[26] P. ᗺ. Seymour and T. Zaslavsky. Averaging sets: A generalization of mean values and spherical designs. *Adv. Math.*, 52:213–240, 1984.

[27] W. D. Wallis, A. P. Street, and J. S. Wallis. *Combinatorics: Room Squares, Sum-free Sets, Hadamard Matrices, Lecture Notes in Mathematics*, Vol. 292, Part 3. Springer-Verlag, Berlin-New York, 1972.

A formula related to the Frobenius problem in two dimensions [1]

MATTHIAS BECK AND SINAI ROBINS [2]

1   Introduction

Given positive integers $a_1, \ldots, a_n$ with $\gcd(a_1, \ldots, a_n) = 1$, we call an integer t **representable** if there exist nonnegative integers $m_1, \ldots, m_n$ such that

$$t = \sum_{j=1}^{n} m_j a_j \ .$$

In this paper, we discuss the *linear diophantine problem of Frobenius*: namely, find the largest integer which is not representable. We call this largest integer the **Frobenius number** $g(a_1, \ldots, a_n)$. We study a more general problem: namely, we consider $N(t)$, the number of nonnegative integer solutions $(m_1, \ldots, m_n)$ to $\sum_{j=1}^{n} m_j a_j = t$ for any positive integral t. In this paper, we concentrate on two dimensions and obtain the formula

$$N(t) = \frac{t}{a_1 a_2} - \left(\left(\frac{a_2^{-1} t}{a_1} \right) \right) - \left(\left(\frac{a_1^{-1} t}{a_2} \right) \right) \ .$$

Here

$$((x)) = x - [x] - 1/2$$

is called a sawtooth function, $a_1^{-1} a_1 \equiv 1 \pmod{a_2}$, and $a_2^{-1} a_2 \equiv 1 \pmod{a_1}$. From this apparently new identity, we immediately recover and extend two well-known results: $g(a_1, a_2) = a_1 a_2 - a_1 - a_2$ ([Sy]), and exactly half of the integers between 1 and $(a_1 - 1)(a_2 - 1)$ are representable ([Ni-Wi]). Note that we can rephrase the definition of $g(a_1, a_2)$ to be the largest zero of $N(t)$.

More generally, we say that an integer t is **k-representable** if $N(t) = k$; that is, t can be represented in exactly k ways. Define $g_k = g_k(a_1, \ldots, a_n)$ to be the largest k-representable integer. It is fairly easy to see that for each k, eventually all integers are k-representable. Hence g_k is well-defined, and every integer greater than g_k is representable in at least $k + 1$ ways. In particular $g_0(a_1, \ldots, a_n) = g(a_1, \ldots, a_n)$. We prove statements about g_k similar in spirit to the two classical results mentioned above. Finally, we define $S_k = \{t \in \mathbb{N} : N(t) \geq k\}$. It is easy to see that S_k is a semigroup for all positive integers k. Note that $S_0 \supset S_1 \supset S_2 \supset \ldots$ and hence provides an interesting filtration of the positive integers. This filtration forms an arithmetic progression in two dimensions but becomes nonlinear in dimension greater than two.

---

[1] Preprint, December 13, 1999. *Keywords*: the linear diophantine problem of Frobenius, rational polytopes, lattice points.

[2] Dept. of Mathematics, Temple University, Philadelphia, PA 19122
matthias@math.temple.edu, srobins@math.temple.edu

2 Proof of the formula

The Frobenius setting can be translated into a lattice point problem: we look at integer dilates of the rational polytope

$$\mathcal{P} = \left\{ (x_1, \ldots, x_n) \in \mathbb{R}^n \ : \ x_j \geq 0, \ \sum_{j=1}^{n} a_j x_j = 1 \right\}$$

and ask for the largest such dilate which contains no lattice point. In the two-dimensional case, we want to study

$$N(t) = \# \left\{ (m, n) \in \mathbb{Z}^2 \ : \ m, n \geq 0, \ am + bn = t \right\} ,$$

which is the lattice point count on the hypothenuse of the triangle $t\mathcal{T}$, where

$$\mathcal{T} = \left\{ (x, y) \in \mathbb{R}^2 \ : \ x, y \geq 0, \ ax + by \leq 1 \right\} .$$

We compute the quantity $N(t)$ by partial fractions, inspired by their applications to Dedekind sums in [Ge]. We note that one does not have to think of $N(t)$ as the lattice point count of a polytope to understand the proof of the following theorem; however, this geometric interpretation was the motivation for our proof.

Theorem 2.1

$$N(t) = \frac{t}{ab} - \left(\left(\frac{b^{-1}t}{a} \right) \right) - \left(\left(\frac{a^{-1}t}{b} \right) \right) .$$

Proof. $N(t)$ is the constant coefficient of the generating function

$$f(z) := \left(\sum_{m \geq 0} z^{am} \right) \left(\sum_{n \geq 0} z^{bn} \right) z^{-t} = \frac{1}{(1 - z^a)(1 - z^b) z^t} .$$

We will expand this function into partial fractions:

$$f(z) = \sum_{\lambda^a = 1}' \frac{A_\lambda}{z - \lambda} + \sum_{\lambda^b = 1}' \frac{B_\lambda}{z - \lambda} + \frac{C_1}{z - 1} + \frac{C_2}{(z - 1)^2} + \sum_{k=1}^{t} \frac{D_k}{z^k} .$$

(Alternatively, we could shift the constant coefficient to the residue and use the residue theorem.) Here, $'$ means we omit the trivial root of unity $\lambda = 1$. Note that the nontrivial roots of unity yield simple poles, since a and b are relatively prime. The reason for doing this is the following: the constant coefficient of the right-hand side, and thus $N_t(p, q)$, equals

$$-\sum_{\lambda^a=1}{}' \frac{A_\lambda}{\lambda} - \sum_{\lambda^b=1}{}' \frac{B_\lambda}{\lambda} - C_1 + C_2 \ . \tag{1}$$

The computation of the coefficients A_λ for a nontrivial a'th root of unity λ is straightforward:

$$A_\lambda = \lim_{z\to\lambda} \frac{z-\lambda}{(1-z^a)(1-z^b)z^t} = -\frac{1}{a(1-\lambda^b)\lambda^{t-1}} \ .$$

Similarly, we obtain for a nontrivial b'th root of unity λ

$$B_\lambda = -\frac{1}{b(1-\lambda^a)\lambda^{t-1}} \ .$$

The coefficients C_2 and C_1 are simply the two leading coefficients of the Laurent series of f. Using the expansion

$$\frac{1}{1-z^a} = -\frac{1}{ab}(z-1)^{-1} + \frac{a-1}{2a} + O(z-1) \ ,$$

it is easy to see that

$$C_2 = \frac{1}{ab} \qquad \text{and} \qquad C_1 = \frac{1}{ab} - \frac{1}{2a} - \frac{1}{2b} - \frac{t}{ab} \ .$$

Hence we can rewrite (1) as

$$N(t) = \sum_{\lambda^a=1}{}' \frac{1}{a(1-\lambda^b)\lambda^t} + \sum_{\lambda^b=1}{}' \frac{1}{b(1-\lambda^a)\lambda^t} + \frac{1}{2a} + \frac{1}{2b} + \frac{t}{ab} \ . \tag{2}$$

We claim that

$$\frac{1}{a}\sum_{\lambda^a=1}{}' \frac{1}{(1-\lambda)\lambda^t} = -\left(\!\left(\frac{t}{a}\right)\!\right) - \frac{1}{2a} \ , \tag{3}$$

from which the statement of the theorem follows from (2) by

$$\frac{1}{a}\sum_{\lambda^a=1}{}' \frac{1}{(1-\lambda^b)\lambda^t} = \frac{1}{a}\sum_{\lambda^a=1}{}' \frac{1}{(1-\lambda)\lambda^{b^{-1}t}} \overset{(3)}{=} -\left(\!\left(\frac{b^{-1}t}{a}\right)\!\right) - \frac{1}{2a} \ .$$

It remains to prove (3), which is equivalent to the well-known finite Fourier series for the sawtooth function (see, e.g., [Ra-Gr]). For sake of completeness we give a short proof of (3), again based on lattice point considerations: consider the interval

$$\mathcal{I} := \left[0, \frac{t}{a}\right] = \{x \in \mathbb{R} : x \geq 0, \ ax \leq t\} \ ,$$

viewed as a one-dimensional polytope. The lattice point count in $\mathcal{I}$ is clearly

$$\#(\mathcal{I} \cap \mathbb{Z}) = \left[\frac{t}{a}\right] + 1 \ . \tag{4}$$

On the other hand, we can write this number, by applying the above ideas to the interval $\mathcal{I}$, as the constant coefficient of

$$\frac{1}{(1 - z^a)(1 - z) z^t} \, .$$

If we expand this function into partial fractions in the exact same manner as above and compare constant coefficients, which equal the lattice point count, we obtain

$$\#(\mathcal{I} \cap \mathbb{Z}) = \frac{t}{a} + \frac{1}{2a} + \frac{1}{2} + \frac{1}{a} \sum_{\lambda^a = 1}' \frac{1}{(1 - \lambda)\lambda^t} \, . \tag{5}$$

Comparing (4) with (5) yields (3). □

3 Consequences of the formula for $N(t)$

From Theorem 2.1, we can derive two basic results on the Frobenius problem: First, Sylvester's result ([Sy]) is a straightforward consequence:

Corollary 3.1 $g(a, b) = ab - a - b$.

Proof. We have to show that $N(ab - a - b) = 0$ and that $N(t) > 0$ for every $t > ab - a - b$. First, by the periodicity of the sawtooth function,

$$
\begin{aligned}
N(ab - a - b) &= \frac{ab - a - b}{ab} - \left(\!\left(\frac{b^{-1}(ab - a - b)}{a}\right)\!\right) - \left(\!\left(\frac{a^{-1}(ab - a - b)}{b}\right)\!\right) \\
&= 1 - \frac{1}{b} - \frac{1}{a} - \left(\!\left(\frac{-b^{-1}b}{a}\right)\!\right) - \left(\!\left(\frac{-a^{-1}a}{b}\right)\!\right) \\
&= 1 - \frac{1}{b} - \frac{1}{a} - \left(\!\left(\frac{-1}{a}\right)\!\right) - \left(\!\left(\frac{-1}{b}\right)\!\right) \\
&= 1 - \frac{1}{b} - \frac{1}{a} - \left(\frac{1}{2} - \frac{1}{a}\right) - \left(\frac{1}{2} - \frac{1}{b}\right) = 0 \, .
\end{aligned}
$$

For any integer m, $\left(\!\left(\frac{m}{a}\right)\!\right) \leq \frac{1}{2} - \frac{1}{a}$. Hence for any positive integer n,

$$N(ab - a - b + n) \geq \frac{ab - a - b + n}{ab} - \left(\frac{1}{2} - \frac{1}{a}\right) - \left(\frac{1}{2} - \frac{1}{b}\right) = \frac{n}{ab} > 0.$$

□

Corollary 3.2 *Exactly half of the integers between* 1 *and* $(a-1)(b-1)$ *are representable.*

Proof. We first claim that, if $t \in [1, ab-1]$ is not a multiple of a or b,

$$N(t) + N(ab - t) = 1 . \tag{6}$$

This identity follows directly from Theorem 2.1:

$$N(ab - t) = \frac{ab - t}{ab} - \left(\left(\frac{b^{-1}(ab - t)}{a}\right)\right) - \left(\left(\frac{a^{-1}(ab - t)}{b}\right)\right)$$

$$= 1 - \frac{t}{ab} - \left(\left(\frac{-b^{-1}t}{a}\right)\right) - \left(\left(\frac{-a^{-1}t}{b}\right)\right)$$

$$\overset{(\star)}{=} 1 - \frac{t}{ab} + \left(\left(\frac{b^{-1}t}{a}\right)\right) + \left(\left(\frac{a^{-1}t}{b}\right)\right)$$

$$= 1 - N(t) .$$

Here, $(\star)$ follows from the fact that $((-x)) = -((x))$ if $x \notin \mathbb{Z}$. This shows that, for t between 1 and $ab-1$ and not divisible by a or b, exactly one of t and $ab-t$ is not representable. There are

$$ab - a - b + 1 = (a - 1)(b - 1) = g(a, b) + 1$$

integers between 1 and $ab-1$ which are not divisible by a or b. Finally, we note that $N(t) > 0$ if t is a multiple of a or b, by the very definition of $N(t)$. Hence the number of non-representable integers is $\frac{1}{2}(a - 1)(b - 1)$. □

Note that we proved even more. By (6), every positive integer less than ab has at most one representation. Hence, the representable integers in the above corollary are *uniquely* representable. We now study integers that are k-representable.

Corollary 3.3 $N(t + ab) = N(t) + 1.$

Proof. By the periodicity of the sawtooth function,

$$N(t + ab) = \frac{t + ab}{ab} - \left(\left(\frac{b^{-1}(t + ab)}{a}\right)\right) - \left(\left(\frac{a^{-1}(t + ab)}{b}\right)\right)$$

$$= \frac{t}{ab} + 1 - \left(\left(\frac{b^{-1}t}{a}\right)\right) - \left(\left(\frac{a^{-1}t}{b}\right)\right) = N(t) + 1 .$$

□

Corollary 3.4 $g_k(a, b) = (k+1)ab - a - b.$

Proof. By the preceeding corollary, $g_{k+1} = g_k + ab$. The statement follows now inductively. □

Corollary 3.5 *Given* $k \geq 2$*, the smallest k-representable integer is* $ab(k-1)$*.*

Proof. Let n be a nonnegative integer. Then

$$N(ab(k-1) - n) =$$
$$= \frac{ab(k-1) - n}{ab} - \left(\left(\frac{b^{-1}(ab(k-1) - n)}{a}\right)\right) - \left(\left(\frac{a^{-1}(ab(k-1) - n)}{b}\right)\right)$$
$$= k - 1 - \frac{n}{ab} - \left(\left(\frac{-b^{-1}n}{a}\right)\right) - \left(\left(\frac{-a^{-1}n}{b}\right)\right). \tag{7}$$

If $n = 0$, (7) equals k. If n is positive, we use $((x)) \geq -\frac{1}{2}$ to see that

$$N(ab(k-1) - n) \leq k - \frac{n}{ab} < k.$$

□

All nonrepresentable positive integers lie, by definition, in the interval $[1, g(a, b)]$. It is easy to see that the smallest interval containing all uniquely representable integers is $[\min(a, b), g_1]$. For $k \geq 2$, the corresponding interval always has length $2ab - a - b + 1$, and the precise interval is given next.

Corollary 3.6 *Given* $k \geq 2$*, the smallest interval containing all k-representable integers is* $[g_{k-2} + a + b, g_k]$*.*

Proof. By Corollaries 3.4 and 3.5, the smallest integer in the interval is

$$ab(k-1) = g_{k-2} + a + b.$$

The upper bound of the interval follows by definition of g_k. □

Corollary 3.7 *There are exactly* $ab - 1$ *integers which are uniquely representable. Given* $k \geq 2$*, there are exactly* ab *k-representable integers.*

Proof. First, in the interval $[1, ab]$, there are, by Corollaries 3.2 and 3.5,

$$ab - \frac{(a-1)(b-1)}{2} - 1$$

1-representable integers. Using Corollory 3.3, we see that there are

$$\frac{(a-1)(b-1)}{2}$$

1-representable integers above ab. For $k \geq 2$, the statement follows by similar reasoning. $\qquad\square$

4   Final remarks

This paper gives only a glimpse of how the 'polytope-view' of the Frobenius problem can be used to gain new results. In dimensions higher than 2, generalized Dedekind sums ([Ge]) appear in the formulas for $N(t)$, which increases the complexity of the problem. The details will be described in a forthcoming paper ([Be-Di-Ro]).

We conclude with a few remarks regarding extensions of the above corollaries to higher dimensions. Although no 'nice' formula similar to Sylvester's result (Corollary 3.1) is known in dimensions greater than 2, there has been a huge effort devoted to giving bounds and algorithms for the Frobenius number. Secondly, we remark that Corollary 3.2 does not extend in general; however, [Ni-Wi] gives necessary and sufficient conditions on the a_j's under which Corollary 3.2 does extend. Corollary 3.3 extends easily to higher dimensions ([Be-Di-Ro]). We leave the reader with the following

Problem *Extend Corollaries 3.1, 3.4, 3.5, 3.6, and 3.7 to higher dimensions.*

References

[Be-Di-Ro] M. Beck, R. Diaz, S. Robins, The Frobenius problem, rational polytopes, and Fourier-Dedekind sums, *in preparation*.

[Ge] I. Gessel, Generating functions and generalized Dedekind sums, *Electronic J. Comb.* **4** (no. 2) (1997).

[Ni-Wi] A. Nijenhuis, H. S. Wilf, Representations of integers by linear forms in nonnegative integers, *J. Number Theory* **4** (1972), 98-106.

[Ra-Gr] H. Rademacher, E. Grosswald, *Dedekind sums*, Carus Mathematical Monographs, The Mathematical Association of America (1972).

[Sy] J. J. Sylvester, Mathematical questions with their solutions, *Educational Times* **41** (1884), 171-178.

One Bit World

David V. Chudnovsky[1] and Gregory V. Chudnovsky[2]

[1] IMAS, Polytechnic University, Brooklyn, 6 MetroTech Center, NY 11201
david@imas.poly.edu
[2] IMAS, Polytechnic University, Brooklyn, 6 MetroTech Center, NY 11201
gregory@imas.poly.edu

1 Introduction.

We want to acknowledge Michael Gerzon of Oxford who had been an early pioneer of one bit audio.

In the beginning there was a word. Analog audio was the first signal seriously studied and recorded. The first efforts in quantization of analog signals were directed towards audio and are known as PCM (Pulse Code Modulation).

PCM is a technique patented in 1938 by Reeves. It still continues to be a major underpinning of all digital speech, audio and video after over 50 years of the first hardware implementation in 1947 (H.S. Black, Bell Laboratories). Shannon and others analyzed PCM in detail in 1948. It, essentially, consists of three steps of processing of analog signals:

sampling,

quantizing and

binary encoding.

The sampler converts a continuous time analog waveform $x(t)$ into a discrete sequence of samples $x_n = x(n/f_s)$ with a sample frequency f_s, usually after the analog waveform is processed by a lowpass filter with the cutoff frequency $f_s/2$ - Nyquist limit. The samples x_n are quantized and represented in a binary form as binary integers of a given dynamic range, which defines the number of quantization levels.

These days the mathematically sophisticated PCM machinery is pushed under the rug and given a general name of A/D - analog to digital conversion. The dynamic range of A/D converters built by many manufacturers can be impressive, though most of the world is so far relatively happy with 16-bit (or even less than 16-bit) A/D convertors. This essentially means that analog waveforms are only sampled within 16 bit accuracy at uniform time intervals.

The vastness of digital information encoded with the PCM quantization (most often without any compression whatsoever) is staggering. We think that the largest commercial repository of digital information is in the form of the high fidelity audio files. These days anybody can get them. Just download your favorite ripper program from the Web, pop in your favorite music CD and in anywhere between a few minutes and a few hours (depending on how much you paid for that CD drive) you get over 500 Megabyte of PCM data - a stream of 16-bit integers adequately representing music (or voice) sampled at 44.1 KHz.

On the top of PCM files - which are simple and uncompressed - one has a multitude of compression (or compression with encryption) formats. There is even a larger set of formats dealing with compression of PCM- information of video files. The video files can be thought of as two- dimensional PCM files; often represented as a linear PCM file, following the line scan of the video.

Almost all stored digital non-trivial information is in the form of audio or video content. The music information is slightly redundant, but so far no lossless compression on music has achieved even 2 : 1 compression rate.

The PCM represents the first step into the One Bit World

Technically the PCM digitalization of the information can be considered as its conversion into a "one bit" stream if one looks at it's radix 2 positional number representation. It is not what we have in mind, when we speak about one bit world. Here we literally mean sequences of 0 and 1 (in fact, we prefer to call them $+1$ and -1) representing individual bit-words.

The actual reason for the need of one-bit data stream, accurately representing an analog continuous signal, has to do with the hardware. The A to D conversion of high accuracy is a grand idea, but, in reality, the nonlinear effects and manufacturing difficulties make it very hard to achieve A to D converters of high dynamic range. On the other hand, a one-bit A/D convertor is a linear device (in a mathematical sense), and represents, basically, an easily realized comparator. Moreover, once a one bit A to D convertor is built, it is a matter of mathematics and not of circuit practice to turn, in digital domain, a one-bit stream, albeit of a much higher frequency, into a high dynamic range digital sample stream.

This is the principle behind what is known as Delta-Sigma ($\Delta - \Sigma$): differentiate the bandwidth-limited analog signal (Delta) into one-bit signal (and the remaining noise) stream; and then integrate, or, rather, filter (Sigma) the one-bit signal stream into a required digitalization of a high dynamic range.

1.1 The Δ - Σ Solution.

The first account of the Delta-Sigma belongs to C.C. Cutler in his U.S. patent of March 1960, [CU]. H. Inose and Y. Yoshida [IY] introduced the term "Delta - Sigma" and did the first analysis. In this country J.C. Candy in 1974 [CA74] introduced the name Sigma - Delta and popularized its use for A/D and D/A converters.

Starting in 1987 in a series of papers by R.M. Gray and his coworkers the rigorous mathematical treatment of the Delta-Sigma appeared, [CA97], [GR87], [GR97], [GRCW], [GRND], [NST]. Not surprisingly, this is a number-theoretical treatment, relaying heavily on the theory of uniform distribution.

We give just the barest example of a relationship with diophantine approximations, following Gray [GR89]. We do not have to dwell on the modern development of uniform distribution as applied to the Delta-Sigma for different classes of functions. We do want to refer to interesting papers of Gunturk, some of which were presented at this seminar; cf. [GC].

Delta-Sigma modulation is defined by the following (nonlinear) difference equations in the time domain:

$$u_n = \begin{cases} u^* & \text{if } n = 0 \\ x_{n-1} - \epsilon_{n-1} & \text{for } n = 1, 2, \ldots \end{cases}$$

Here ϵ_m (quantization noise) is defined in term of $q(u)$. $q(u)$ is the output from the one bit quantizer, defined in the simplest case as:

$$q(u) = \begin{cases} b & \text{if } u \geq 0 \\ -b & \text{otherwise} \end{cases}$$

and

$$\epsilon_n = q(u_n) - u_n, \quad -b \leq \epsilon_n \leq b$$

is the binary quantization noise.

The intuitive idea of oversampled A/D is to shape the binary quantization noise with the filter, so that the quantization noise is mostly out of the base-band. A low pass digital filter is used as a decoding filter to produce a digital representation of the analog input.

If the input $x_n \in [-b, b]$ and the linear decoding filter H has an impulse response $(h_0, h_1, \ldots, h_n)$ then the Sigma- Delta A/D converter with decoding filter H band sampled at time N will satisfy

$$H(q(u_n)) = \sum_{i=0}^{N} h_{N-i} q(u_i) = \sum_{i=0}^{N} h_{N-i} x_{i-1} + \sum_{i=0}^{N} h_{N-i}(\epsilon_i - \epsilon_{i-1}).$$

Consider a good noise shaping filter. E.g., a $sinc^2$ filter is formed by cascade realization of two $sinc$ (comb) filters. Unlike the $sinc$ filter where each sample has equal weight, the weight of a $sinc^2$ filter is given by

$$h_k = \begin{cases} \frac{k-1}{N^2} & \text{if } k = 2, 3, \ldots, N+1 \\ \frac{2N-k+1}{N^2} & \text{for } k = N+2, \ldots, 2N \end{cases}$$

which has a triangular shape.

The first example of the appearance of diophantine approximation (uniform distribution theory) is the seemingly simple case of the DC signal (constant value): $x_n = x$. In the context of the simplest quantizer $q(x)$ above, this leads to the problem of discrepancy of the sequence of $\{n \cdot \beta\}$ for $\beta = \frac{x+b}{2b}$. With the notation of the discrepancy function $D_N(\{\beta\}, \ldots, \{N\beta\})$, determined by the continued fraction expansion of β, we get Gray's bound on the accuracy of DC-level reconstruction:

The absolute decoding error with $sinc^2$ decoding filter satisfies

$$\mid x - Q(x) \mid \leq \frac{2b}{N} D_N(\{\beta\}, \ldots, \{N\beta\}).$$

Now we can characterize the importance of the Delta-Sigma in the acquisition of Petabytes of audio data. It is the Delta-Sigma that makes current CDs and other HiFi music possible.

In the current CD mastering process an oversampled one-bit Delta - Sigma A/D is used as a front end, but the one-bit output signal must be decimated and re-quantized in real time to form the regular sample rate PCM signal.

Recording a one-bit signal instead of a multi-bit recording format has the following advantages:

1) simple system structure, because of the serial nature without any implied framing (nothing like the classical endian mode incompatibility);

2) because the signal is oversampled, the system characteristics approach that of high quality analog audio;

3) the bit stream for a known sampling rate is completely independent of the noise shaping filter, thus tapes recorded with different noise shapers can be interchanged, or;

4) recorder performance can be optimized by changing the noise shaping filter and not the format;

5) tape recorders are upward and downward compatible in terms of the format and only the quality of replay changes.

These new formats are the underpinning of new audio files: SACD - super audio compact disc and DVD audio. We refer to Gerzon's last paper for the original motivation [CRGE].

The Delta-Sigma relaxes demand on the analog circuitry at the expense of the increased demand placed on digital circuitry. It is the filtering DSP part - commonly known as noise shaping - that becomes paramount. Filters of length up to $2K$ are known to have been built into the Delta-Sigma. The quantization part of the Delta-Sigma is a nonlinear iterative mapping, and thus has wonderful properties of complex dynamical systems (limit cycles, various attractors, ... etc.). Moreover, to get a better noise shape Delta-Sigma is often cascaded to provide even more and more sophisticated dynamical systems, with even less chances of a complete analysis (cf. [NST]).

One of the most difficult problems in $\Delta - \Sigma$ design is the choice of a high frequency - a high multiple of the Nyquist limit - at which to quantize the signal. For a sufficiently high multiple no spectral information is lost but at the expense of much more storage and bandwidth requirements.

One of the advantages and simultaneously a problem with Delta-Sigma is the local character of its one bit quantization. When one can have a buffered stream of analog signals, one can make global decisions on one bit quantization in an asynchronous fashion, rather than providing one new bit every clock tick.

This global optimal one-bit quantization is very appropriate for the Delta-Sigma in D/A case (when the stream of data is known a priory).

There is yet another hardware term describing this approach - Pulse Width Modulation (PWM). PWM is one-bit representation of an analog form as a train of $+1$ or -1 square pulses (often, PWM is even mistaken for Delta part in Delta-Sigma).

The Optimal PWM (OPWM) approach provides the optimal number of one bit samples needed to reconstruct the original signal stream (in contrast with $\Delta - \Sigma$). However, in general the PWM and OPWM approaches do not have any particular frequency at which the edges of the pulses are sampled - in theory the location of the edges is arbitrary.

The OPWM problem for arbitrary harmonic series is solved in this paper using a combination of methods of orthogonal polynomials and nonlinear completely integrable systems.

2 Optimal Pulse Width Modulation (OPWM).

2.1 General Formulation.

The most general Optimal PWM formulation is the following. Given a spectral bandwidth limited (analog) waveform, find a square form of amplitudes ± 1 that most closely approximates this waveform at the low frequency range. More precisely, given a periodic harmonic series

$$\psi(t) = b_o + \sum_{n=1}^{N} (a_n \sin nt + b_n \cos nt)$$

(with the fundamental domain $[0, 2\pi]$), find the periodic pulse train $f(t)$ with $2N+1$ (unknown) edges α_i ($i = 1, \ldots, 2N+1$) in the fundamental domain such that the Fourier expansion of $f(t)$ has the leading $2N+1$ coefficients the same as $\psi(t)$.

The main period part of the pulse train can be expressed as follows

$$0 = \alpha_0 \leq \alpha_1 < \alpha_2 < \ldots < \alpha_{2N} < \alpha_{2N+1} \leq \alpha_{2N+2} = 2\pi.$$

There are, of course, power and other isoperimetric constraints on the harmonic form $\psi(t)$ for it to be accurately approximated by a square form $f(t)$ (with the leading $2N + 1$ harmonic terms). The simplest of them is a consequence of the Parceval identity (and expresses a fact that the amplitudes of individual waves have to be limited in order to be represented by one bit pulses):

$$2b_0^2 + \sum_{n=1}^{N} (a_n^2 + b_n^2) \leq 2.$$

In general, we ask for

$$\max_{t \in [0, 2\pi]} |\psi(t)| \leq 1.$$

If a good pulse approximation $f(t)$ to $\psi(t)$ can be found, this $f(t)$ can be used as a "digital" (or, rather, a one- bit) representation of $\psi(t)$. It can be quantized in the t - domain by rounding the corresponding values of α_i to a fixed frequency. It can be further compressed using additional properties of the distribution of α_i. Jumping a little bit ahead such asymptotic properties of α_i do exist, because α_i are related to the zeroes of orthogonal polynomials. This representation is also excellent for storage and transmission, especially for audio signals.

Of course, in the real world, i.e., in the analog world, the square waveform does not look at all as a harmonic one. Thus to make a square form look (and, most importantly, sound) good, it has to pass through a low pass filter to restore its spectral bandwidth limitation. In audio playback this is the famous problem of psycho-acoustic spectral matching (according to an experimentally observed graph of spectral weights). For instrumentation providing the best quality of reproduction, the one bit nature of the encoded waveform is crucial, because one bit A/D and D/A converters provide the best signal quality.

2.2 The Classical Pulse Width Modulation (PWM) Problem.

The classical PWM - Pulse Width Modulation had been developed in the 60s and 70s as a one-bit representation of waveforms with a very interesting feature: only square forms of amplitudes $+1$ or -1 are allowed, with positive and negative edges at arbitrary sample times. This is in a sharp contrast with PCM/DPCM and Delta-Sigma (Sigma-Delta), where the sampling happens at uniformly spaced times. Such an asynchronicity /non-uniform sampling is an important tool in any attempt to fight the Nyquist limit on the number of sample points vs. spectral support of the signal.

Initial interest in PWM had been almost entirely driven by special requirements of power electronics. There the only important signal to be represented is the primary (lowest) sine harmonics. Typical applications of PWM are in power electronics: digitally controlled electric motors, high voltage power transmission and control of quality of electrical power grids.

In simple mathematical terms the basic PWM problem is that of approximating a sin wave of a given amplitude A:

$$A \sin t \text{ on } [0, 2\pi]$$

by a train of square pulses of $+1$ or -1 amplitude (independent of A) such that spectral characteristic of the periodic pulse train (square waveform) is the same as that of a sin wave $A \sin t$ in the low frequency range.

Because of the quarter period symmetry, it is enough to consider a square form only on $[0, \pi/2]$. If one has N square forms to define on $[0, \pi/2]$ that "spectrally approximating" sine wave, one has N free parameters (edges of the pulses). Patel in 1970 had been the first to derive an explicit set of equations on N pulse edges to eliminate N harmonics in the Fourier expansion of the full periodic pulse train (with $4N$ pulses per full period), see [PH]. It should be noted that in view of the quarter period symmetry, only $\sin(nt)$ terms are left with n odd. The resulting equations have the following form.

For the quarter-period symmetric (periodic) pulse train $f(t)$ with edges $0 = \alpha_0 < \alpha_1 < \ldots < \alpha_N < \alpha_{N+1} = \pi/2$ (and $f(\alpha_0) = -1$, $f(\alpha_{N+1}) = +1$), we have the following harmonic expansion

$$f(t) = \sum_{n=1,n-\text{odd}}^{\infty} V_n \cdot \sin(nt),$$

where

$$V_n = \frac{4}{n\pi} \cdot \left(-1 + 2 \cdot \sum_{i=1}^{N} (-1)^{i+1} \cos(n\alpha_i)\right).$$

Thus for the approximation of the basic waveform

$$a_1 \sin t$$

to within $N + 1$ first harmonics by the pulse train, the following set of N transcendental equations on unknown α_i has to be satisfied

$$\sum_{i=1}^{N} (-1)^{i+1} \cos(n\alpha_i) = c_n : n = 1, 3, \ldots, 2N - 1$$

where

$$c_n = \begin{cases} (\frac{\pi}{4}a_1 + 1/2) & \text{if } n = 1 \\ 1/2 & \text{if } n > 1 \end{cases}$$

In general, for arbitrary quater-period symmetric harmonic polynomial we get the "system of Classical PWM equations" (in the transcendental form) by writing $\beta_i = \alpha_i$ for odd i and $\beta_i = \pi - \alpha_i$ for even i:

$$\sum_{i=1}^{N} \cos(n\beta_i) = c_n : n = 1, 3, \ldots, 2N - 1.$$

The case of an approximation of a single basic waveform (main PWM case) leads to the following system of transcendental equations:

$$\sum_{i=1}^{n} \cos\alpha_i = a, \quad \sum_{i=1}^{n} \cos(2m - 1)\alpha_i = 0, \quad m = 2, \ldots, n$$

3 Orthogonal Polynomials, Solitons and Classical PWM Problem

The classical PWM problem can be reduced (see below) to an algebraic problem of "odd sums of powers" where one has to determine the set $\{x_i\}$ $(i = 1 \ldots n)$ from the equations on "odd sums of powers":

$$\sum_{i=1}^{n} x_i^{2m-1} = s_{2m-1}, \; m = 1 \ldots n.$$

This and more general OPWM problem are analytically solved using Padé approximations and completely integrable isospectral deformation equations. We present here an analytical solution, and we describe fast methods of the numerical solution needed for practical applications.

This work on the Classical PWM has a variety of possible practical applications. It is a part of an effort at Polytechnic University conducted by Dariusz Czarkowski, Ivan Selesnik and authors of this paper. The main emphasis of that effort is on practical implementation of the PWM solution that can be useful in power electronics. See a detailed review in [CCCS].

A very interesting feature of the general solution to the PWM problem lies in its connection to the classical areas of mathematics - symmetric functions, orthogonal polynomials, and the theory of completely integrable systems. The most surprising relationship is with the Korteweg-de Vries (KdV) hierarchy of infinite dimensional Hamiltonians. We show how the complete solution of the PWM problem describes, in fact, the class of all rational solutions of KdV equations.

4 Subsequences of Symmetric Functions and zeros of Orthogonal Polynomials

The problem of finding the set of n elements $\{x_i\}$ with given values of n arbitrary symmetric functions $\{s_j\}$ in $x_i, i = 1 \ldots n$, is in general a very complicated one because of the nontrivial nature of relations between symmetric functions of high degrees. Only in special cases can this problem be reduced to an "exactly solvable" one. For this one looks at the polynomial $P(x)$, whose roots $\{x_i\}$ are:

$$P(x) = \prod_{i=1}^{n} (x - x_i)$$

This polynomial $P(x)$ is uniquely identifies the set $\{x_i\}$. The coefficients of $P(x)$ are elementary symmetric functions in x_i, which have to be determined in order to determine the set $\{x_i\}$. Newton's formulas for sum of powers of x_i show that once one knows all first n symmetric functions

$$s_j = \sum_{i=1}^{n} x_i{}^j, \; j = 1 \ldots n$$

then one easily finds - by means of Newtons's linear recurrences - the elementary symmetric functions in x_i - the coefficients of $P(x)$ - consequently finding the set $\{x_i\}$ as the set of roots of $P(x)$.

The derivation of Newton's recurrences is simple; it requires a look at the logarithmic derivative of $P(x)$:

$$\frac{P'(x)}{P(x)} = \sum_{i=1}^{n} \frac{1}{x - x_i}.$$

Expanding the logarithmic derivative at $x = \infty$, we get:

$$\frac{P'(x)}{P(x)} = \sum_{m=0}^{\infty} \frac{s_m}{x^{m+1}}$$

for $s_m = \sum_{i=1}^{n} x_i{}^m$. Comparing the standard expansion of $P(x)$ at $x = \infty$

$$P(x) = x^n + a_1 x^{n-1} + \ldots + a_{n-1} x + a_n$$

one gets Newton's linear recursion relations between a_i and s_j.

Newton's identities explicitly are:

$$k \, a_k = -\sum_{i=1}^{k} s_i \cdot a_{k-i}$$

Other relations between a_i and s_j can be found using the following identity (that we will use in later generalizations):

$$P(x) = x^n e^{-\sum_{m=1}^{\infty} \frac{s_m}{m x^m}}.$$

Very important combinatorial interpretations of the last identity were used extensively by MacMahon and others ([M], [Li]):

$$a_k = \sum_{\lambda} (-1)^{\sum \lambda_i} \frac{s_1^{\lambda_1} \cdot s_2^{\lambda_2} \cdot s_3^{\lambda_3} \cdots}{1^{\lambda_1} \cdot 2^{\lambda_2} \cdot 3^{\lambda_3} \cdots \lambda_1! \cdot \lambda_2! \cdot \lambda_3! \cdots}$$

for all partitions λ where $1 \cdot \lambda_1 + 2 \cdot \lambda_2 + 3 \cdot \lambda_3 \ldots = k$.

What happens, if instead of classical consecutive sets of symmetric functions one knows only the values of n non-consecutive symmetric functions? The problem poised in the PWM method of power electronics looks first at n odd power sum symmetric functions s_{2m-1} for $m = 1 \ldots n$. The solution of that problem also leads to a sequence of linear recurrences, but this time not among the invariants themselves but the sequences of polynomials associated with them (these are $P_n(x) = P(x)$ as n varies). This solution is based on Padé approximations and orthogonal polynomials.

We start with the general relation between Padé approximations and the generalization of Newton relations between power and elementary symmetric functions. Let us look at the Padé approximation of the order (n, d) to the series $g(x)$ at $x = \infty$. Here the function $g(x)$ is defined via the generating function of an arbitrary sequence S_m:

$$g(x) = e^{-\sum_{m=1}^{\infty} \frac{S_m}{m x^m}}.$$

The definition of the Padé approximation of the order (n, d) to $g(x)$ at (the neighborhood of) $x = \infty$ is the following - it is a rational function $\frac{P_n(x)}{Q_d(x)}$ with $P_n(x)$ a polynomial of degree n and $Q_d(x)$ a polynomial of degree d - such that the expansion of $\frac{P_n(x)}{Q_d(x)}$ matches the expansion of $g(x)$ at $x = \infty$ up to the maximal order. This means that

$$\frac{P_n(x)}{Q_d(x)} - x^{n-d}g(x) = O(x^{-2d-1})$$

or

$$P_n(x) - Q_d(x)x^{n-d}g(x) = O(x^{-d-1}).$$

After taking the logarithmic derivative of this expression, we end up with the following representation of the definition of Padé approximation:

$$\frac{P'_n}{P_n} - \frac{Q'_d}{Q_d} = \frac{d}{dx}\log x^{n-d}g(x) + O(x^{-n-d-2}).$$

Now if we write normalized (with the leading coefficient 1) polynomials $P_n(x)$ and $Q_d(x)$ in terms of their roots:

$$P_n(x) = \prod_{i=1}^{n}(x - x_i); \quad Q_d(x) = \prod_{k=1}^{d}(x - y_k),$$

we get an identification of symmetric functions in x_i and y_k with the sequence of S_m in the definition of $g(x) = e^{-\sum_{m=1}^{\infty} \frac{S_m}{m x^m}}$. Namely, we get:

$$\sum_{i=1}^{n} x_i^j - \sum_{k=1}^{d} y_k^j = S_j$$

for $j = 0 \ldots n + d$.

Of course, in the case $d = 0$ one recovers Newton's identities. The case $d = n$ – the so-called case of the "diagonal" Padé approximations – is the most interesting case. It is also the case that solves the problem of "sums of odd powers". This is how it works. Consider the "anti-symmetric" case when $y_i = -x_i$ for $i = 1 \ldots n$ and $d = n$. In this case $S_{2m} = 0$ for $m \geq 0$ and

$S_{2m-1} = 2s_{2m-1}$. We thus get Padé approximations of the order (n, n) to the following function:

$$G(x) = e^{-2\sum_{m \text{ odd}}^{\infty} \frac{s_m}{m x^m}}.$$

The Padé approximants $\frac{P_n(x)}{Q_n(x)}$ have the property:

$$Q_n(x) = (-1)^n P_n(-x),$$

because $G(x)$ satisfies a functional identity: $G(-x) = 1/G(x)$.

This gives us our main result:

Theorem 1. The solution $\{x_i\}$ $(i = 1 \ldots n)$ to the problem of "odd sums of powers":

$$\sum_{i=1}^{n} x_i^{2m-1} = s_{2m-1}, \ m = 1 \ldots n$$

is given by the roots of the numerator $P_n(x) = \prod_{i=1}^{n}(x - x_i)$ in the Padé approximation problem of order (n, n) to the function

$$G(x) = e^{-2\sum_{m \text{ odd}}^{\infty} \frac{s_m}{m x^m}}.$$

Another way to verify this approximation without specialization from the case of general sequence s_m, is simply to take the expansion of the logarithmic derivative of $\frac{P_n(x)}{P_n(-x)}$ at $x = \infty$. Expanding the logarithmic derivative, and then integrating it (formally) in x, one gets a very simple identity

$$(-1)^n \cdot \frac{P_n(x)}{P_n(-x)} = e^{-2\sum_{m \text{ odd}}^{\infty} \frac{x^{-m}}{m} \sum_{i=1}^{n} x_i^m}.$$

From this identity Theorem 1 follows. This is a new formula that Newton could have found.

Proof of Theorem 1. First of all, the Padé approximation rational function $\frac{P_n(x)}{Q_n(x)}$ of order (n, n) is unique. Then, if $\frac{P_n(x)}{Q_n(x)}$ is a Padé approximation of order (n, n) to $G(x)$, we assume that this representation of the rational function is irreducible (i.e., that $P_n(x)$ and $Q_n(x)$ are relatively prime). Then $\frac{Q_n(x)}{P_n(x)}$ is a Padé approximation of order (n, n) to $1/G(x)$, and $\frac{P_n(-x)}{Q_n(-x)}$ is a Padé approximation of order (n, n) to $G(-x)$. Because of the functional equation $G(-x) = 1/G(x)$, and the uniqueness of the Padé approximations, we get $\frac{Q_n(x)}{P_n(x)} = \frac{P_n(-x)}{Q_n(-x)}$. This equation means that $Q_n(x) = \alpha P_n(-x)$. Moreover, since the expansion of $G(x)$ at $x = \infty$ starts at 1, we have $\frac{P_n(x)}{Q_n(x)} \to 1$ as $x \to \infty$. Thus $Q_n(x) = (-1)^n P_n(-x)$. Taking into account the "main" identity

$$(-1)^n \cdot \frac{P_n(x)}{P_n(-x)} = e^{-2\sum_{m \text{ odd}}^{\infty} \frac{x^{-m}}{m} \sum_{i=1}^{n} x_i^m}$$

we can see that the right hand side of this identity and the expansion of $G(x)$ at $x = \infty$ has to agree up to (but not including) x^{-2n-1}. This means that we have $\sum_{i=1}^{n} x_i^{2m-1} = s_{2m-1}$ for $m = 1 \ldots n$. Note that in the definition of $G(x)$, the values of s_{2m-1} for $m > n$ have no impact on the definition of $P_n(x)$ because they enter the expansion of $G(x)$ only at x^{-k} for $k \geq 2n + 1$. This is so because $e^{\sum_{k=1}^{\infty} \beta_k x^{-k}} = e^{\sum_{k=1}^{m} \beta_k x^{-k}} + O(x^{-m-1})$.

Since we identified the solution to the "sums of odd powers" problem with numerator (or denominator) in the (diagonal) Padé approximation problem, we infer from the standard theory of continued fraction expansion that rational functions $\frac{P_n(x)}{P_n(-x)}$ are partial fractions in the continued fraction expansion of the generating function $G(x)$ at $x = \infty$. This also means (see [S] for these and other facts of the theory of continued fraction expansions and orthogonal polynomials) that the sequence of polynomials $P_n(x)$ is the sequence of orthogonal polynomials (with respect to the weight that is Hilbert transform of $G(z)$), and that the sequence of polynomials $P_n(x)$ satisfies three-term linear recurrence relation. Since the same recurrence is satisfied both by numerators and denominators of the partial fractions, it means that the recurrence is satisfied by two sequences - $P_n(x)$ and $(-1)^n \cdot P_n(-x)$. With the leading coefficient of $P_n(x)$ is 1, one gets a particularly simple three-term recurrence relation among $P_n(x)$:

$$P_{n+1}(x) = x \cdot P_n(x) + C_n \cdot P_{n-1}(x)$$

for $n = 0 \ldots$.

5 Algebraic Problem of Odd Sums of Powers vs the Transcendental Problem of Sums of Odd Cosines

The original definition of PWM problem dealt with transcendental equations $\sum_{i=1}^{n} \cos(2m - 1)\alpha_i = c_{2m-1}$, and not algebraic equations $\sum_{i=1}^{n} x_i^{2m-1} = s_{2m-1}$. A very important contribution to the PWM problem by D. Czarkowski and I. Selesnick is in the explicit reduction of the transcendental PWM problem to the algebraic one for consecutive $m = 1, \ldots, n$. This is what allows us here to use the solution of the algebraic problem of "sums of odd powers" for the solution of the original transcendental PWM problem. We will show now how the transcendental case is explicitly expressed using the introduced notations of $G(x)$ and $P_n(x)$. The basic transformation is $\cos \alpha_i = x_i$ or in the algebraic form: $x = (z + z^{-1})/2$, for $z = e^{\alpha}$. Now, to get from the algebraic "sums of odd powers" solution to the transcendental one, consider the set of roots z_i, z_i^{-1}; $i = 1, \ldots, n$. Then "sums of odd powers" for these roots gives

$$\sum_{i=1}^{n} \cos(2m - 1)\alpha_i \text{ for } \cos \alpha_i = z_i.$$

This allows us to represent the polynomials $P_n(x)$ in z-form. Namely, if we define

$$P_a(z) = \prod_{i=1}^{n}(z - z_i), \quad P_b(z) = \prod_{i=1}^{n}(z - \frac{1}{z_i}),$$

then we can express $P_a(z) \cdot P_b(z)$ as a polynomial in $x = \frac{1}{2}(z + 1/z)$: $P_a(z) \cdot P_b(z) = z^n \cdot P_n(x)$. This allows us to express the Padé approximation to $G(x)$ (and $G(x)$) as a similar function of z. We have:

$$(-1)^n \cdot \frac{P_n(x)}{P_n(-x)} = \frac{P_a(z)P_b(z)}{P_a(-z)P_b(-z)}.$$

Expanding each term on the right as a function of z^{-1}, we get

$$(-1)^n \cdot \frac{P_n(x)}{P_n(-x)} = e^{-2\sum_{m=1}^{\infty} T_m/mz^m}$$

where $T_m = \sum_{i=1}^{n} z_i{}^m + z_i{}^{-m} = 2\sum_{i=1}^{n} \cos m\alpha_i$.

This means that the function $G(x)$ (or its Padé approximation) that determines the solution of the algebraic "sums of odd powers" problem (in x) can be reduced to the transcendental "sums of odd cosines" problem (in z). Specifically, $G(x)$, as a function of z has the following form:

$$G(x) = e^{-4\sum_{m \, \text{odd}}^{\infty} \frac{c_m}{mz^m}},$$

where the sequence $\{c_m\}$ arises from the following general transcendental "sums of odd cosines" problem

$$\sum_{i=1}^{n} \cos(2m - 1)\alpha_i = c_{2m-1}, \quad m = 1, \ldots, n$$

corresponding to the algebraic "sums of odd powers" problem:

$$\sum_{i=1}^{n} x_i{}^{2m-1} = s_{2m-1}, \quad m = 1, \ldots, n$$

with $x_i = \cos\alpha_i$ for $i = 1, \ldots, n$.

Notice that the "explicit expression" for x_i (α_i) or $P_n(x)$ simply means that the continued fraction expansion of $G(x)$ (in x or z at infinity) is known "explicitly", or equivalently that the coefficients C_n in the main three-term recurrence describing $P_n(x)$ are "explicit" (i.e., classical elementary or transcendental) functions of n. From this point of view, the main PWM transcendental problem

$$\sum_{i=1}^{n} \cos\alpha_i = a, \quad \sum_{i=1}^{n} \cos(2m-1)\alpha_i = 0, \quad m = 2, \ldots, n$$

is not "explicitly solvable" because the corresponding function $G(x)$ (see the expression above in terms of z and the corresponding sequence $\{c_m\}$):

$$G_a(x) = e^{-4a(x - \sqrt{x^2-1})}$$

does not have an "explicit" continued fraction expansion at $x = \infty$. On the other hand, a very similar algebraic problem:

$$\sum_{i=1}^{n} x_i = a; \quad \sum_{i=1}^{n} x_i^{2m-1} = 0; \quad m = 2, \ldots, n$$

does have an "explicit" solution, since the corresponding function $e^{-2a/x}$ has a classical continued fraction expansion derived by Euler. The polynomials $P_n(x)$ arising in this special algebraic problem are well-known under the name of Bessel polynomials.

An application of the solution of the transcendental "sums of odd cosines" problem in PWM applications lies in is the ability to construct high quality digital (step-function) approximations to arbitrary harmonic series. Since the proposed algorithm can be executed quite fast on any processor with high-performance DSP capabilities, it opens the possibility of better on-the-fly construction of arbitrary (analog) waveforms using simple digital logic.

6 Completely Integrable Difference-Differential Equations of Korteweg-de Vries type.

In connection with Padé approximation solution to the "sum of odd powers" problem, we can ask what "nonclassical" objects this solution is built from. These objects are "soliton equations" and their solutions - isospectral deformation equations of the full Korteweg-de Vries (KdV) hierarchy.

The only parameter left in our solution is the coefficient factor C_n of the three-term recurrence for polynomials P_n:

$$P_{n+1}(x) = x \cdot P_n(x) + C_n \cdot P_{n-1}(x). \tag{1}$$

This parameter C_n is a function of n and the whole generating sequence $\{s_m : m - \text{odd}\}$ of odd symmetric functions. It is C_n that is a solution of the KdV type hierarchy of completely integrable p.d.e.s in variables s_m. In addition to p.d.e.s in s_m the parameter C_n satisfy difference-differential equations in n and each of s_m.

The formal derivation of the full hierarchy of such equations is based on our paper [Ch3]. To see simply how one can derive them we can start with (1). Think of x as a spectral parameter: $x = \lambda$. Then (1) is an eigenvalue problem for the second order linear difference operator in n. The crucial next step is the realization that polynomials P_n also satisfying linear differential equations in each of the variables s_m. Once such a differential equation in s_m is derived, one looks at the consistency condition of this equation and (1). Such a consistency condition is a classical isospectral deformation condition.

This consistency condition implies a nonlinear difference-differential equation on C_n in n and s_m (for any odd m). The resulting nonlinear equation belongs to a completely integrable class. Eliminating n for s_m and $s_{m'}$, one gets a KdV type p.d.e. on C_n in s_m and $s_{m'}$.

To see how these are derived let us look at the case of $m = 1$ and the variable s_1. The corresponding partial differential equation on P_n is:

$$x \cdot P_{n,s_1} = P_n + E_n \cdot P_{n-1} \tag{2}$$

where the parameter E_n does not depend on x. The consistency condition between (1) and (2) leads to the following two difference-differential equations

$$C_n = E_n \cdot E_{n+1}; \quad C_{n,s_1} = E_{n+1} - E_n \tag{3}$$

We also summarize here all relationships with Toeplitz determinants in the expansion

$$G_a(x) = \sum_{m=0}^{\infty} \frac{c_m}{x^m}$$

of the approximated generating function of "odd sums of powers" $\{s_{2m-1}\}$:

$$G = e^{-2 \sum_{m \text{ odd}}^{\infty} \frac{s_m}{m x^m}}.$$

Specifically we use the standard representation of the main and the auxiliary Toeplitz determinants:

$$D_n = \det(c_{i+j})_{i,j=0}^{n-1}; \quad \Delta_n = \det(c_{i+j+1})_{i,j=0}^{n-1}.$$

The relationships between C_n and the determinants are the following ones:

$$C_n = -\frac{\Delta_{n+1} \cdot \Delta_{n-1}}{\Delta_n^2}; \quad \Delta_n^2 = (-1)^n 2 D_n \cdot D_{n+1}.$$

The expression of E_n in the difference-differential equations above is:

$$E_n^2 = -\frac{D_{n+1} \cdot D_{n-1}}{D_n^2}.$$

The main object D_n is expressed in terms of the τ-function (typical notation for KdV equations):

$$D_n = \tau_n^2,$$

where τ_n is a polynomial in s_{2m-1}. We need, however, to normalize the polynomial τ_n, so that the leading power of s_1 (and it is $s_1^{n(n-1)/2}$) would have a coefficient of 1. In this case we can write more accurately:

$$D_n = r_n \cdot \tau_n^2,$$

The rational numbers r_n are easily determined using the following expressions:

$$E_n = -\frac{1}{2n-1} \cdot \frac{\tau_{n+1} \cdot \tau_{n-1}}{\tau_n^2}; \quad C_n = \frac{1}{4n^2 - 1} \cdot \frac{\tau_{n+2} \cdot \tau_{n-1}}{\tau_n \cdot \tau_{n+1}}.$$

This implies $r_{n+1} = -r_n^2/((2n-1)^2 \cdot r_{n-1})$. Substituting these expressions in equations (3) one can get the difference-differential equation on τ_n in n and s_1:

$$\tau_{n+1,x} \cdot \tau_{n-1} - \tau_{n+1} \cdot \tau_{n-1,x} = (2n-1) \cdot \tau_n^2.$$

7 Korteweg-de Vries Hierarchy and all that

Explicit recursion relations defining all commuting (higher) Korteweg-de Vries (KdV) flows are well-known. These relations connect the infinite sequence of conserved quantities H_m of the original KdV equation - these are the (infinite dimensional) Hamiltonians of the higher KdV equations - with the vector flows X_m of (m -th) KdV equation. The m-th KdV is:

$$u_{t_m} = X_m(u); \quad \text{for } X_m(u) = \partial_x \frac{\delta H_m}{\delta u}.$$

$\delta H_m/\delta u$ is the gradient of the functional H_m of u, e.g., for the first KdV Hamiltonian (the **actual** KdV equation), we gave:

$$\frac{\delta H_2}{\delta u} = \frac{3}{2}u^2 - \frac{1}{2}u'', \quad \text{for } H_2 = \int (\frac{1}{2}u^3 + \frac{1}{4}(u')^2)\, dx.$$

The flows $X_m(u)$ are commuting (as Poisson structures induced by H_m). Recursion connecting successive Hamiltonians are relatively simple:

$$X_m(u) = \partial_x \frac{\delta H_m}{\delta u} = (-\frac{1}{2}\partial_x^3 + 2u\partial_x + u_x)\frac{\delta H_{m-1}}{\delta u}.$$

One can also write all higher KdV flows X_m in the explicit form using this recursion as follows:

$$X_m = N_u^m X_1; \quad X_1(u) = u_x$$

using the operator:

$$N_u = -\frac{1}{2}\partial_x^2 + 2u + u_x\partial_x^{-1}.$$

The first few higher KdV equations thus are:

$$u_{t_1} = u';$$

$$u_{t_2} = 3uu' - \frac{1}{2}u''';$$

$$u_{t_3} = \frac{15}{2}u^2u' - 5u'u'' - \frac{5}{2}uu''' + \frac{1}{4}u^{(5)};$$

One should not confuse x variable in KdV equations with x used in the "sums of odd powers" PWM problem. The variable x in the PWM problem is the **spectral** variable, usually denoted as λ in

$$G = e^{-2\sum_{m\,\text{odd}}^{\infty} \frac{s_m}{m\lambda^m}},$$

and the polynomials, whose roots solve the PWM problem will be denoted as $P_n(\lambda)$.

Further, the identification of odd moments s_{2m-1} with the canonical variables of p.d.e.s in the KdV hierarchy, is the following one:

$$x = s_1;$$

and for the higher flows "time variables" t_m we have

$$t_m = \frac{s_{2m-1}}{(2m-1)2^{m-1}}$$

(so the standard KdV time is $t = s_3/6$).

8 Solution to the PWM Problem via KdV rational Solutions

The relationship between the KdV hierarchy of equations and the general "sums of odd powers" problem is very interesting one. Roughly speaking the parameter C_n in the recurrence relation defining the orthogonal polynomials in the "sums of odd powers" problem, as functions of odd moments s_{2m-1} satisfy all p.d.e.s in the KdV hierarchy, with $x = s_1$ and higher flows "time variables" t_m being $t_m = \frac{s_{2m-1}}{(2m-1)2^{m-1}}$.

Moreover – and this is what distinguishes the "sums of odd powers" case, and completely characterizes it in terms of KdV equations - **rational solutions to the KdV equation and to the full KdV hierarchy are completely described by the solution to "sums of odd powers" problem.**

The class of all rational solutions to KdV, well-studied since 1977, and still under great deal of investigation today, has many interesting and important properties - this class is a limit case of the famous N-soliton solutions corresponding to special rational curves; it has a famous many-particle interpretation in terms of dynamics of poles of these solutions in x- (and t_m-) planes, etc.

The specific KdV relationship is the following. We look, as above, at the orthogonal polynomials $P_n(\lambda)$ representing the Padé approximants to the generating function G of the sequence of odd moments s_{2m-1}:

$$G = e^{-2\sum_{m\,\text{odd}}^{\infty} \frac{s_m}{m\lambda^m}},$$

and satisfying the three-term recurrence

$$P_{n+1}(\lambda) = \lambda \cdot P_n(\lambda) + C_n \cdot P_{n-1}(\lambda).$$

"Explicit" expressions for $P_n(\lambda)$ and C_n involve, as above, Toeplitz determinants in the coefficients c_m of the expansion of G at $\lambda = \infty$:

$$G_a(\lambda) = \sum_{m=0}^{\infty} \frac{c_m}{\lambda^m}.$$

Specifically we use the standard representation of the main Toeplitz determinant (as above):

$$D_n = \det(c_{i+j})_{i,j=0}^{n-1}.$$

There is also the auxiliary Toeplitz determinant Δ_n:

$$\Delta_n = \det(c_{i+j+1})_{i,j=0}^{n-1}.$$

The relationships between C_n and the determinants are the following ones:

$$C_n = -\frac{\Delta_{n+1} \cdot \Delta_{n-1}}{\Delta_n^2};$$

$$\Delta_n^2 = (-1)^n \, 2 \, D_n \cdot D_{n+1}.$$

The main parameter D_n is expressed in terms of the τ-function of KdV-type equations:

$$D_n = \tau_n^2,$$

where τ_n is a polynomial in s_{2m-1}.

The KdV solutions are expressed in terms of the potential u, very simply related to τ as follows:

$$u_n = -2 \partial_x^2(\log \tau_n) = -\partial_x^2(\log D_n).$$

(Here as above $x = s_1$).

The relationship between KdV hierarchy and the "sums of odd powers" recurrences is the following one:

Theorem 2. If the generating sequence of odd moments $\{s_{2m-1}\}$ is considered as a sequence of independent variables, then with the identification $x = s_1$, and $t_m = \frac{s_{2m-1}}{(2m-1)2^{m-1}}$, the τ functions $\tau_n = \sqrt{D_n}$ are **all** rational solutions of KdV equation (and all commuting higher KdV equations). The polynomials τ_n are characterized by the their degree in x - which is $\frac{n(n-1)}{2}$.

An important consequence of this theorem is the characterization for the first time of the full manifold of rational solutions as explicit functions of actual higher KdV natural parameters t_m (see above the identification $t_m = \frac{s_{2m-1}}{(2m-1)2^{m-1}}$ with odd moments variables). This completes the study of rational solutions of KdV that we started more than 20 years ago [Ch4].

We write down explicitly the first few τ_n, normalizing them (for uniqueness) with the coefficient at $x^{n(n-1)/2}$ being 1 (remember that $x = s_1$):

$$\tau_2 = s_1;$$

$$\tau_3 = s_1^3 - s_3;$$

$$\tau_4 = s_1^6 - 5s_1^3 s_3 + 5s_3^2 + 9s_1 s_5;$$

$$\tau_5 = s_1^{10} - 15s_1^7 s_3 - 175s_1 s_3^3 + 63(5s_1^2 s_3 + s_1^5 - 3s_5)s_5 + 225(s_3 - s_1^3);$$

The recurrence that these normalized polynomials τ_n satisfy is a known one (it also follows from the difference-differential equation on C_n in n and s_1), that contains a crucial ambiguity, hiding in constants of integration the explicit dependency on s_i:

$$\tau_{n+1,x} \cdot \tau_{n-1} - \tau_{n+1} \cdot \tau_{n-1,x} = (2n-1) \cdot \tau_n^2.$$

Theorem 2 and the direct relation to the Padé approximation problem to G provides a much simpler theory of rational solutions to the KdV hierarchy than all other descriptions (we refer to [AMM] and [Ch4] for an original exposition, and [Z] for the modern presentation and review).

An interesting corollary of Theorem 2 is that the polynomials D_n (τ_n) and rational functions C_n are derived by means of successive Bäcklund transformations from the initial $n = 0$ $(u = 0)$ case.

9 New Difference - Differential Equations In Nonintegrable Cases - PWM Examples.

Whenever the function G that is expanded into its continued fraction, does not satisfy a Riccatti equation over $\mathbf{C}(x)$, there is no simple Painlevé equation/recurrence on partial quotients C_n. Most G fall into this category and examples of G from PWM problems are not "integrable" either. The most interesting example of G depends on the parameter a (voltage level):

$$G = e^{-4a(x - \sqrt{x^2 - 1})}$$

expanded at $x = \infty$. The corresponding partial quotient $C_n = C_n(a)$ has to be determined as a function of a in order to compute solutions $P_n(x)$ of the main PWM problem. We show that $C_n(a)$, though not a Painlevé function, satisfies an algebraic difference -differential equation in n and a.

We start with the definition of Padé approximation to G:

$$q_n \cdot G - p_n = O(x^{-n-1}),$$

where $q_n = q_n(x), p_n = p_n(x)$ are polynomials of degree n in x, and $p_n(x) = (-1)^n q_n(-x)$. From the expansion of G at $x = \infty$ one gets leading coefficients of q_n: $q_n = x^n + ax^{n-1} + q_{2,n}x^{n-2} + \ldots$, where $q_{2,n+1} = q_{2,n} + C_i$, or $q_{2,n} = \sum_{i=1}^{n-1} C_i$. G satisfies a linear p.d.e. over $\mathbf{Q}(x, a)$:

$$L \cdot G_x = 4aG \quad \text{for} \quad L = (x^2 - 1) \cdot d_x - a \cdot x \cdot d_a.$$

We can differentiate the definition of the Padé approximation, and get:

$$Q_n^{(1)} \cdot G - P_n^1 = O(x^{-n});$$

$$Q_n^{(1)} = (x^2 - 1)q_{n,x} - axq_{n,a} + 4aq_n; \quad P_n^{(1)} = (x^2 - 1)p_{n,x} - axp_{n,a}.$$

Because of orthogonality properties of $q_n(x)$, we can express $Q_n^{(1)}$ as a linear combination of only a few of q_m's:

$$Q_n^{(1)} = \alpha_n \cdot q_{n-1} + 2a \cdot q_n + n \cdot x \cdot q_n.$$

Here we define an expression built from previous C_m:

$$\alpha_n = -2 \cdot \sum_{i=1}^{n-1} C_i - a \cdot \sum_{i=1}^{n-1} C_{i,a} + 2a^2 - n.$$

As a result we have a linear partial difference-differential equation on q_n in n, a and x. This equation is compatible with the original three-term linear recurrence on q_n. The consistency condition becomes our new equation on $C_n(a)$. To represent this consistency condition in a transparent form we use 2×2 matrices and a vector of two consecutive polynomials and the corresponding shift operator $+$:

$$\bar{q} = \begin{pmatrix} q_n \\ q_{n-1} \end{pmatrix}; \quad \bar{q}^+ = \begin{pmatrix} q_{n+1} \\ q_n \end{pmatrix}.$$

Then we have the original recurrence written as

$$\bar{q}^+ = M_n \cdot \bar{q}; \quad M_n = \begin{pmatrix} x & C_n \\ 1 & 0 \end{pmatrix}.$$

We can write the expression of $Q_n^{(1)}$ as follows:

$$L(\bar{q}) = K_n \cdot \bar{q}; \quad K_n = \begin{pmatrix} nx - 2a & \alpha_n \\ \alpha_{n-1}/C_{n-1} & (n-1)x - 2a - \alpha_{n-1}/C_{n-1} \cdot x \end{pmatrix}.$$

The compatibility condition now becomes

$$L(M_n) + M_n \cdot K_n = (K_n)^+ \cdot M_n$$

Using this definition of α_n in the first equation we can write a single differential equation in a on C_n that involves only C_{n-1} and C_m for $m < n-1$:

$$C_{n,a} = -\xi_{n-1} \cdot C_n + \frac{\alpha_n}{a}$$

for

$$\xi_{n-1} = \frac{1}{a} \cdot \left(\frac{\alpha_{n-1}}{C_{n-1}} + 2\right).$$

Its solution can be written in quadratures as follows:

$$C_n = \left(e^{-\int \xi_{n-1}}\right) \cdot \left(\int \frac{\alpha_n}{a} \cdot e^{\int \xi_{n-1}}\right).$$

The initial condition on C_n as a function of a is: $C_n \mid_{a=0} = -1/4$.

A more detailed algebraic analysis of C_n as a rational function of a reveals more algebraic structure not dissimilar to that of solutions of Painlevé equations. The canonical determinant is

$$\Delta_n = \det(c_{i+j+1})_{i,j=0}^{n-1}$$

in $G = \sum_{n=0}^{\infty} c_n/x^n$. In these notations we have

$$C_n = -\frac{\Delta_{n+1}\Delta_{n-1}}{\Delta_n^2}.$$

Canonical polynomials ω_n in a:

$$\deg \omega_n = (n+1)^2/2$$

for odd n,

$$\deg \omega_n = \frac{n(n+2)}{2}$$

for even n.

In terms of polynomials ω_k we get the following expressions

$$C_k = n_k \cdot \frac{\omega_{k-3} \cdot \omega_k}{\omega_{k-2} \cdot \omega_{k-1}}$$

$$\Delta_n = -\text{const}_n \cdot \omega_{n-1} \cdot \omega_{n-2} \cdot a^n.$$

$$\frac{n_k}{n_{k+1}} = \frac{2k+3}{2k-1}$$

We can write then

$$e^{\int \xi_n} = a^{2n+4} \cdot \frac{\omega_n}{\omega_{n-2}}.$$

We can get the following expression for ω_n (and C_n) in terms of ω_m for $m < n$:

$$\int \frac{\alpha_n}{a} \cdot a^{2n+2} \cdot \frac{\omega_{n-1}}{\omega_{n-3}} da + \text{Const} = n_k \cdot a^{2n+2} \frac{\omega_n}{\omega_{n-2}}.$$

These differential equations or integral representations allow for a simple recursive computation of $C_n(a)$ as a rational function from $\mathbf{Q}(a)$.

10 Complexity of Computations of PWM polynomials and solutions

What is the complexity of computations of PWM polynomials and their roots - i.e., of the solution to the classical PWM problem? One can ask the same question about all n first sums of powers. If one uses just Newton identities, the complexity is $O(n^2)$, but a much faster scheme can be found. The key to this is:

$$P(x) = x^n e^{-\sum_{m=1}^{\infty} \frac{s_m}{m x^m}}.$$

Indeed, according to Brent's theorem N terms of the power series expansion of $e^{V(x)}$ can be computed in only $O(N \log N)$ steps from the power series expansion of $V(x)$ (see [K1], Sec. 4.7, Ex. 4.) This algorithm requires only use of the FFT-technique for computation of fast convolution.

Similar complexity considerations can be applied to the problem of fast computation of polynomials $P_n(x)$ that give the solution to the classical PWM problem of consecutive odd power sums. In fact, the arguments presented in this and the following sections are applied without significant changes to the solution of the general OPWM problem that uses two-point Padé approximations. A classical Levinson algorithm of solving systems of Toeplitz linear equations, or algorithms based on three-term recurrence relations satisfied by $P_m(x)$ give the complexity of computations of (all coefficients of) $P_n(x)$ as $O(n^2)$. These algorithms are, perhaps, the best in the range of moderate n because of their simple nature and the fact that they use almost no additional space. Also the $O(n^2)$ complexity algorithm provides with the determination of not only the single $P_n(x)$ but all $P_m(x)$ for $m \leq n$. For large n these algorithms became impractical. Thus one needs to use fast algorithms.

This is how a fast algorithm of computations of (all coefficients of) $P_n(x)$ of the total complexity of $O(n \log^2 n)$ works. First, one has to apply Brent's theorem to compute $O(n)$ terms of the power series expansion (at infinity $x = \infty$) of

$$G(x) = e^{-2 \sum_{m \text{ odd}}^{\infty} \frac{s_m}{m x^m}}.$$

from the first $O(n)$ known coefficients s_m with the complexity of only $O(n \log n)$. Then one has to use fast Padé approximation algorithms (or, equivalent fast polynomial gcd algorithms). There are a variety of these algorithms, with the most popular from [BGY]. Its complexity is $O(n \log^2 n)$. Thus we can compute $P_n(x)$ in at most $O(n \log^2 n)$ operations. Of course, this method should be used only for a large n (with additional precision of calculations) since it relies on a variety of extensions of FFT methods.

11 A Simple Algorithm for Orthogonal Polynomials Computations

A simple algorithm of computation of all $P_m(x)$ for all $m = 0, \ldots, n$, having the complexity of $O(n^2)$, that we mentioned above, is easy to describe. Let us look at the expansion of $G(x)$ at $x = \infty$, written in the following form:

$$G(x) = \sum_{k=0}^{\infty} (-1)^k c_k \cdot x^{-k}.$$

By Theorem 1 polynomials $P_n(x)$ are defined from the Padé approximation problem to $G(x)$, i.e., the remainder function

$$R_n(x) = P_n(-x) \cdot G(x) - (-1)^n \cdot P_n(x)$$

has the following expansion at $x = \infty$: $R_n(x) = O(\frac{1}{x^{n+1}})$. Also, since $P_n(x)$ are orthogonal polynomials, they satisfy the three-term recurrence

$$P_{n+1}(x) = x \cdot P_n(x) + C_n \cdot P_{n-1}(x)$$

for $n = 0 \ldots$. The initial conditions can be chosen here as: $P_{-1} = 1$, $P_0 = 1$. Let us write $P_n(x)$ in terms of its coefficients: $P_n(x) = \sum_{i=0}^{n} P_{n,i} \cdot x^i$.

If $P_{n-1}(x)$ and $P_n(x)$ are known, then in order to determine the single unknown C_n, and thus $P_{n+1}(x)$ (via the recurrence), we have to look at the coefficient at x^{-n} of $R_n(x)$. Assume as before, that the leading coefficient of $P_n(x)$ is 1, i.e., $P_{n,n} = 1$. Looking at the coefficient at x^{-n} in the expansion of $R_n(x)$ we get the following expression for C_n:

$$C_n = -\frac{\sum_{j=0}^{n} P_{n,j} \cdot c_{n+j+1}}{\sum_{i=0}^{n-1} P_{n-1,i} \cdot c_{n+i}}.$$

Once C_n is determined, the coefficients $P_{n+1,i}$ are easily determined recursively.

In order for this algorithm to work, one needs coefficients c_i in the expansion of $G(x)$. As we mentioned above, the complexity of Brent's algorithm of computations of c_i up to $O(n)$ that uses FFT is $O(n \log n)$.

In fact often the complexity is bounded only by $O(n)$. For example, in the most interesting case of the PWM problem:

$$G(x) = G_a(x) = e^{-4a(x - \sqrt{x^2 - 1})}$$

the algorithm for computing c_i is a very simple one of complexity $O(n)$ only. This algorithm follows the authors' general power series algorithms [Ch5]. The key to this algorithm is to notice that $G_a(x)$ satisfies second order linear differential equation (with singularities at $x = -1, 1, \infty$ and an apparent singularity at $x = 0$):

$$y''x(x^2 - 1) + y'(8ax(x^2 - 1) + 1) + y(4a(1 - 4ax)) = 0.$$

If we look at the expansion of $G_a(x)$ at $x = \infty$: $G_a(x) = \sum_{n=0}^{\infty} \frac{(-1)^n c_n}{x^n}$, then we get the 4-term recurrence on c_n:

$$c_{n+2} = \frac{1}{8a(n+2)} \cdot (c_{n+1}((n+1)^2 + (n+1) - 16a^2) + c_n(8an + 4a) + c_{n-1}(-(n-1)^2 - 2(n-1))).$$

The initial conditions are $c_n = 0$ for $n < 0$ and

$$c_0 = 1, \; c_1 = 2a, \; c_2 = 2a^2, \ldots.$$

From these values and the recurrence for c_n one derives the C_n factors in the three-term recurrence for orthogonal polynomials $P_n(x)$. Here are a few initial C_n:

$$C_0 = -a, \; C_1 = \frac{4a^2 - 3}{12}, \; C_2 = \frac{45 - 60a^2 + 16a^4}{60(4a^2 - 3)}.$$

Here all C_n are rational functions in a of rather special structure. Since the case of the continued fraction expansion of $G_a(x)$ is not "explicitly solvable", C_n is not a "known" function of a and n. An important consequence of the "unsolvability" of C_n is the growth of coefficients of C_n as rational functions in a with integer coefficients. According to standard conjectures about explicit and non-explicit continued fraction expansions (see [Ch1]), the coefficients of C_n as the rational function in a over $\mathbf{Z}$ are growing as $e^{O(n^2)}$ for large n. In fact, for $n \geq 12$, the coefficients of C_n in a are large integers. This makes it impractical to precompute with full accuracy C_n for large n. It is also unnecessary to analytically determine $C_n = C_n(a)$ explicitly, since we need to know $C_n(a)$ only in the range of a that is significant for applications - this is the range where the weight of orthogonal polynomials $P_n(x)$ is real.

Let us show explicitly $P_9(x) = P_9(x, a)$ the polynomial:

$(-1099511627776\, a^{41}$ $+$ $49478023249920\, a^{40} x$ $+$
$3809807790243840\, a^{38} x\, (-3 + 4\, x^2) - 272129127874560\, a^{39}\, (-1 + 4\, x^2) -$
$714338960670720\, a^{37}\, (43 - 328\, x^2 + 208\, x^4) + 714338960670720\, a^{36} x\, (1685 -$
$4280\, x^2 + 1456\, x^4) + 139296097330790400\, a^{34} x\, (-551 + 2005\, x^2 - 1316\, x^4 +$
$128\, x^6) - 11608008110899200\, a^{35}\, (-181 + 1980\, x^2 - 2400\, x^4 + 448\, x^6) -$
$46732801445936511434904375000000\, a^2 x\, (3 - 45\, x^2 + 180\, x^4 - 264\, x^6 + 128\, x^8) +$
$43812001355565479470222851562500\, x\, (5 - 80\, x^2 + 336\, x^4 - 512\, x^6 + 256\, x^8) -$
$43812001355565479470222851562500\, a\, (1 - 40\, x^2 + 240\, x^4 - 448\, x^6 + 256\, x^8) +$
$23366400722968255717452187500000\, a^3\, (1 - 40\, x^2 + 240\, x^4 - 448\, x^6 + 256\, x^8) +$
$34824024332697600\, a^{32} x\, (95319 - 442680\, x^2 + 420084\, x^4 - 82368\, x^6 + 1088\, x^8) -$
$2742391916199936000\, a^{30} x\, (37688 - 209833\, x^2 + 255880\, x^4 - 75296\, x^6 +$
$2176\, x^8) - 17412012166348800\, a^{33}\, (5605 - 78334\, x^2 + 136328\, x^4 - 49952\, x^6 +$
$2176\, x^8) + 31155200963957674289936250000000\, a^4 x\, (123 - 1732\, x^2 + 6544\, x^4 -$
$9120\, x^6 + 4224\, x^8) - 31155200963957674289936250000000\, a^5\, (17 - 672\, x^2 +$

$4000\,x^4 - 7424\,x^6 + 4224\,x^8) - 118686479862695902056900000000\,a^6\,x\,(499 - 6630\,x^2 + 23560\,x^4 - 30816\,x^6 + 13376\,x^8) + 274239191619993600\,a^{31}\,(11921 - 199920\,x^2 + 444760\,x^4 - 238880\,x^6 + 21760\,x^8) + 5934323993134795102845000000000\,a^7\,(115 - 4444\,x^2 + 25984\,x^4 - 47552\,x^6 + 26752\,x^8) + 5656183327162368000\,a^{28}\,x\,(424359 - 2723796\,x^2 + 4003184\,x^4 - 1561088\,x^6 + 73984\,x^8) + 7912431990846393470460000000\,a^8\,x\,(7341 - 92592\,x^2 + 309536\,x^4 - 376640\,x^6 + 150016\,x^8) - 7912431990846393470460000000\,a^9\,(715 - 26788\,x^2 + 152384\,x^4 - 272256\,x^6 + 150016\,x^8) - 1885394442387456000\,a^{29}\,(43413 - 840060\,x^2 + 2243136\,x^4 - 1565056\,x^6 + 221952\,x^8) - 19181653311142772049600000000\,a^{10}\,x\,(20157 - 242368\,x^2 + 763992\,x^4 - 860800\,x^6 + 308096\,x^8) + 95908266555713860248000000000\,a^{11}\,(3375 - 121872\,x^2 + 668960\,x^4 - 1155008\,x^6 + 616192\,x^8) - 299786523055711211520000000\,a^{14}\,x\,(211983 - 2320480\,x^2 + 6524340\,x^4 - 6264792\,x^6 + 1704864\,x^8) - 434288826088685568000\,a^{22}\,x\,(1424718 - 12271325\,x^2 + 25980972\,x^4 - 16934880\,x^6 + 2078080\,x^8) + 3582882815231655936000\,a^{19}\,(48491 - 1443298\,x^2 + 6422140\,x^4 - 8729812\,x^6 + 3490712\,x^8) + 11241994614589170432000000\,a^{15}\,(36587 - 1210992\,x^2 + 6064752\,x^4 - 9487808\,x^6 + 4546304\,x^8) - 2828091663581184000\,a^{26}\,x\,(15025965 - 108265340\,x^2 + 184227288\,x^4 - 88762560\,x^6 + 6125440\,x^8) + 1414045831790592000\,a^{27}\,(1107681 - 24079896\,x^2 + 74194336\,x^4 - 62926080\,x^6 + 12250880\,x^8) + 37118703084503040000\,a^{24}\,x\,(15659431 - 124251072\,x^2 + 238180224\,x^4 - 135547008\,x^6 + 12747008\,x^8) + 4702533694991548416000000\,a^{16}\,x\,(3532305 - 36755470\,x^2 + 97319348\,x^4 - 85924160\,x^6 + 19796160\,x^8) + 36190735507390464000\,a^{23}\,(743683 - 19351944\,x^2 + 73808448\,x^4 - 82706240\,x^6 + 24936960\,x^8) - 4478603519039569920000\,a^{18}\,x\,(7416360 - 72992577\,x^2 + 181023556\,x^4 - 146106480\,x^6 + 27925696\,x^8) + 281049865364729260800000\,a^{12}\,x\,(6507637 - 74700980\,x^2 + 222417104\,x^4 - 231548736\,x^6 + 72980736\,x^8) - 281049865364729260800000\,a^{13}\,(475795 - 16484016\,x^2 + 86701648\,x^4 - 143211712\,x^6 + 72980736\,x^8) - 1175633423747887104000000\,a^{17}\,(818493 - 25762098\,x^2 + 122055168\,x^4 - 178987040\,x^6 + 79184640\,x^8) - 5302671869214720000\,a^{25}\,(4380839 - 104950052\,x^2 + 363264160\,x^4 - 359205952\,x^6 + 89229056\,x^8) + 1357152581527142400000\,a^{20}\,x\,(37857867 - 350070756\,x^2 + 806628912\,x^4 - 589406592\,x^6 + 91422464\,x^8) - 271430516305428480000\,a^{21}\,(9003483 - 251665140\,x^2 + 1042516880\,x^4 - 1296841600\,x^6 + 457112320\,x^8)) \;/$

$(8821612800\,(12714083695698776015625 - 67808446377060138750000\,a^2 + 14917858202953230525000000\,a^4 - 17996146403562627300000000\,a^6 + 1345549192023948912000000\,a^8 - 66992179236769987200000\,a^{10} + 23251106676342504960000\,a^{12} - 5793671320995317760000\,a^{14} + 1055273129097707520000\,a^{16} - 141774665486598144000\,a^{18} + 14064801509670912000\,a^{20} - 1023040734363648000\,a^{22} + 53635589760614400\,a^{24} - 1963734545203200\,a^{26} + 47436571607040\,a^{28} - 676457349120\,a^{30} + 4294967296\,a^{32}))$

12 Fast Algorithms For Computing Solutions to PWM Problems

We already know that while the slow algorithms of computation of $P_n(x)$ can be completed in $O(n^2)$ operations, the fast FFT-algorithms can be completed in $O(n \log^2 n)$ operations. After $P_n(x)$ is determined, we need, in addition, determine the set $\{x_i\}$ of all roots of $P_n(x)$. One can use general methods of computations of roots of univariant polynomials; that would bring the overall minimal complexity higher. We do not need to do it in our case because orthogonality properties of $P_n(x)$ allows one to have fast and numerically stable methods of computing $\{x_i\}$. We present one such algorithm, suitable for moderate and large range of n.

Fast polynomial root finding requires first "root separation" - determination of domains in the complex plane containing only one root. Next, individual roots are "polished", i.e., computed with the high accuracy, usually using Newton-type iterative algorithms, Both parts of this program of polynomial root- finding can be very well accomplished for polynomials $P_n(x)$ in PWM-problems relevant for applications.

In all practically relevant cases of the PWM transcendental "sums of odd cosines" problem, $x_i = \cos \alpha_i$ for $i = 1, \ldots, n$ and real α_i. This means that x_i are real and lie in the interval $[-1, 1]$ for all n. Because of the orthogonality properties of $P_n(x)$, we get the Sturm property: the roots of $P_{n-1}(x)$ separate the roots of $P_n(x)$. This provides a variety of methods of separating the roots of any given $P_n(x)$. The simplest method of the overall complexity of $O(n^2 \log^2 n)$, is to compute roots of all $P_m(x)$ for all $m = 0, \ldots, n$ recursively, using roots of the previous polynomial to separate and then accurately compute the roots of the next $P_m(x)$ polynomial. Using fast algorithms of computations of Padé approximations this method can be improved to the overall complexity of only $O(n \log^2 n)$ operations for computing the set $\{x_i\}$ of all roots of a polynomial $P_n(x)$.

For these fast algorithms we use fast polynomial evaluation: for any given set of N points $\{X_i\}$ and a polynomial $P(X)$ of degree N in X, one needs at most $O(N \log^2 N)$ operations to evaluate $P(X_i)$ for all $i = 1, \ldots, N$. Using this technique one can rapidly find true roots X_i of the polynomial $P(X)$, starting with approximate (but well-separated) values $X_{0,i}$ of these roots. In the case of single roots (and for all physically relevant PWM problems roots are single) one uses a classical Newton-Raphson method.

For orthogonal polynomials $P_n(x)$, arising from the algebraic version of PWM problem, we can use fast algorithms of evaluation at n points of $P_n(x)$ and $P'_n(x)$ to get n Newton-Raphson approximations running at the same time:

$$x_i = x_i - \frac{P_n(x_i)}{P'_n(x_i)}; \; i = 1, \ldots, n.$$

To get rapid (geometric) rate of convergence of this algorithm one needs initial conditions of x_i, corresponding to centers of intervals separating the roots of $P_n(x)$.

A fast algorithm arises when we start directly from the initial approximation to x_i for the given $P_n(x)$. Such approximation can be rigorously derived using the classical properties of orthogonal polynomials (see [S]). Roughly, the leading term in $1/n$-expansion of (properly ordered) roots x_i of $P_n(x)$ is given by asymptotically as follows:

$$x_i = \cos \alpha_i, \ \alpha_i \sim \frac{\pi i}{n}; \ i = 1, \ldots, n.$$

More accurately, α_i are separated by $O(1/n)$; if these angles are ordered, then $A/n \leq \alpha_i - \alpha_{i+1} \leq B/n$. Thus one can chose $O(n/\epsilon)$ total starting points x in the Newton-Raphson iterations with the property that any real root x_i is within the distance ϵ/n from at least one starting point x of the iteration. In the case of a fixed machine precision the number of iteration is constant, providing us with the fast algorithm of computing the set $\{x_i\}$ with at most $O(n \log^2 n)$ operations.

13 Solvable Extensions of the Classical PWM Problem

A variety of extensions of the "odd sums of powers" problem can be solved using Padé approximation techniques. For this one uses methods of generalized graded Padé approximations developed by authors. Some of these problems also arise in practical applications of signal processing (phase unwrapping, channel identification, etc.). A particular example of the problem, generalizing the "odd sums of powers" is the problem where for a given n and $N \geq 1$ one knows n first consecutive sums of powers of $\{x_i\}$ ($i = 1, \ldots, n$) but every N-th one (i.e., $1, \ldots, N - 1, N + 1, \ldots$). This problem is analytically solved using simultaneous Padé approximations to $N - 1$ functions in a way almost identical to the one presented above for $N = 2$.

14 The complete solution to the OPWM Problem

The main ingredients of the solution to the OPWM problem are:

1) Analytic solution of the underlying approximation problem by reduction to the problem of two-point Padé approximation and orthogonal polynomials on the unit circle;

2) Determination of the variational part of the OPWM problem (the dependency on the amplitudes of different frequency bands) in terms of the completely integrable commuting Hamiltonian flows of Sin-Gordon and Korteweg-de Vries type.

14.1 The analytic approach to the OPWM Problem.

Let us look at the periodic ± 1 pulse train $f(t)$ with the $2N$ pulses in the fundamental period $[0, 2\pi]$ and the corresponding pulse edges at α_j:

$$0 = \alpha_0 < \alpha_1 < \alpha_2 < \ldots < \alpha_{2N+1} < \alpha_{2N+2} = 2\pi$$

(normalized at $f(0) = -1, f(2\pi) = 1$). Let us look at the Fourier expansion of $f(t)$:

$$f(t) = b_0 + \sum_{n=1}^{\infty} a_n \sin nt + b_n \cos nt.$$

Let us define now the (complex) Fourier coefficient c_n of $f(t)$:

$$c_n = b_n + i \cdot a_n; \quad c_n = \frac{1}{\pi} \int_0^{2\pi} f(t) e^{int} dt$$

for $n > 0$. Then the Fourier coefficients in the expansion of pulse train $f(t)$ have the following form:

$$\pi b_0 = \sum_{j=1}^{N} \alpha_{2j} - \sum_{j=1}^{N+1} \alpha_{2j-1} + \pi;$$

$$\pi c_n = \frac{-2i}{n} \{ \sum_{j=1}^{N} e^{in\alpha_{2j}} - \sum_{j=1}^{N+1} e^{in\alpha_{2j-1}} + 1 \};$$

for $n > 0$.

To put this expansion in a more symmetric form, we associate points z_j on the unit circle with the phases (angles) α_j:

$$z_j = e^{i\alpha_j} \text{ for } j = 1, \ldots, 2N + 1.$$

This allows us to represent the Fourier coefficients b_0, c_n of $f(t)$ in the invariant way in terms of z_j:

$$-e^{-i\pi b_0} = \frac{\prod_{j=1}^{N+1} z_{2j-1}}{\prod_{j=1}^{N} z_{2j}};$$

$$\frac{-n\pi i}{2} c_n + 1 = \sum_{j=1}^{N+1} z_{2j-1}^n - \sum_{j=1}^{N} z_{2j}^n$$

(for $n > 0$).

Finally we can define the main objects associated with the OPWM problem. Theses are the polynomials $Q_{N+1}(z), P_N(z)$, whose roots are positive (respectively, negative) edges of the pulses in $f(t)$:

$$Q_{N+1}(z) = \prod_{j=1}^{N+1}(z - z_{2j-1});$$

$$P_N(z) = \prod_{j=1}^{N}(z - z_{2j}).$$

From these polynomials a rational 2-point approximation $\frac{Q_{N+1}(z)}{P_N(z)}$ (at $z = 0$ and at $z = \infty$) is formed. The following expansions of this rational function is crucial for spectral and orthogonal polynomial interpretation:

$$\frac{Q_{N+1}(z)}{P_N(z)} = z \cdot e^{-\sum_{M=1}^{\infty} \frac{s_M}{M z^M}} \text{ at } z = \infty.$$

Here we define the sequence of complex numbers $\{s_M\}$ in terms of the Fourier expansion $\{c_M\}$:

$$s_M = \frac{-M\pi i}{2}c_M + 1 \text{ for } M = 1, \ldots.$$

If we denote by $P-(z)$ the reciprocal of a polynomial $P(z)$ (with a leading coefficient one), then we have the second expansion formula of the rational approximation:

$$\frac{Q_{N+1}^-(z)}{P_N^-(z)} = z \cdot e^{-\sum_{M=1}^{\infty} \frac{s_M^*}{M z^M}} \text{ at } z = \infty.$$

Here s_M^* is the complex conjugate to s_M. These expansions follow from Newton's original formula of the sums of powers symmetric functions, and represent the expansions at $z = 0$ and at $z = \infty$ of the rational function $\frac{Q_{N+1}(z)}{P_N(z)}$. One has only to add to it the definition of b_0 in terms of constant coefficients of the polynomials $Q_{N+1}(z)$, $P_N(z)$ (given above):

$$\frac{q_0}{p_0} = e^{-i\pi b_0},$$

where q_0, p_0 are, respectively, constant coefficients of $Q_{N+1}(z)$, $P_N(z)$.

This set of (relatively) classical identities allow us to represent the general OPWM problem of representation of an arbitrary harmonic waveform $\psi(t)$ in terms of the (periodic) ± 1 pulse train as the problem of 2-point Padé approximation (or polynomials, orthogonal on the unit circle). Namely, for harmonic polynomial $\psi(t)$ with $2N + 1$ arbitrary coefficients $B_j, j = 0, \ldots, N$; $A_n, j = 1, \ldots, N$:

$$\psi(t) = B_0 + \sum_{n=1}^{N} A_n \sin nt + B_n \cos nt.$$

we have the theorem characterizing the ± 1 pulse trains with edges α_j that have the same leading $2N + 1$ coefficients in the Fourier expansion:

Theorem. In order for the periodic pulse $f(t)$, as above, to be a harmonic approximation to the waveform $\psi(t)$ to within the first $2N+1$ harmonic terms, it is necessary and sufficient that the rational function

$$\frac{Q_{N+1}(z)}{P_N(z)}$$

with $Q_{N+1}(z) = \prod_{j=1}^{N+1}(z - z_{2j-1})$; $P_N(z) = \prod_{j=1}^{N}(z - z_{2j})$, be a solution to the following 2-point Padé approximation problem:

$$\frac{Q_{N+1}(z)}{P_N(z)} = z \cdot f_\infty(z) + O(z^{-N}) \text{ at } z = \infty$$

$$\frac{Q_{N+1}(z)}{P_N(z)} = f_0(z) + O(z^{N+1}) \text{ at } z = 0$$

where the functions $f_\infty(z), f_0(z)$ are "conjugate" generating functions of Fourier coefficients of $\psi(t)$:

$$f_\infty(z) = e^{-\sum_{M=1}^{\infty} \frac{S_M}{M z^M}};$$

$$S_M = \frac{-M\pi i}{2}(B_M + i \cdot A_M) + 1, \quad M = 1, \ldots;$$

$$f_0(z) = e^{-\sum_{M=0}^{\infty} \frac{S_M^* z^M}{M}};$$

$$S_0^* = -\pi i B_0, \quad S_M^* = \frac{M\pi i}{2}(B_M - i \cdot A_M) + 1, \quad M = 1, \ldots.$$

This particular class of Padé approximations is closely related to the classical class of orthogonal polynomials - polynomials orthogonal on the circle, studied by Szego and others. They posses many important properties, one of which is of particular use in computations is the existence of the three-term recurrence relation, connecting the polynomials $Q_{N+1}(z)$, $P_N(z)$ for consecutive N.

This naturally assumes that we have a "full" harmonic expansion of the basic waveform $\psi(t)$, looking at its first $2N+1$ coefficients, defining the "pulse train approximation of the order N" - $f(t)$ with $N+1$ up-ticks $\alpha_{2j-1}, j = 1, \ldots, N+1$.

The three-term recurrences satisfied by the polynomials $Q_{N+1}(z)$ for $N = 0, \ldots$ and by the polynomials $P_N(z)$ for $N = 0, \ldots$ is the following one familiar from the theory of polynomials orthogonal on the unit circle:

$$X_{N+1} = (z + D_N) \cdot X_N + z \cdot E_N \cdot X_{N-1}$$

(satisfied by $X_N = Q_{N+1}(z)$ and by $X_N = P_N(z)$). An important interpretation of this three-term recurrence is the difference spectral problem of the second order (on X_N), where z plays a role of the spectral parameter (in the complex plane). Such interpretation allows us to use a Riemann boundary value problem and after the reformulation of the Padé approximation problem as the matrix (here, 2×2) factorization problem, to get new numerical methods of solving it. We can determine the polynomials $Q_{N+1}(z)$, $P_N(z)$ (staring from the original Fourier coefficients C_M of $\psi(t)$) in only $O(N \cdot \log N)$ operations. The roots z_j (and thus α_j, whenever they are real) can be determined in $O(N \cdot \log^2 N)$ operations.

We present several simple examples of OPWM approximations to trignometric polynomials. Graphs show the trigonometric polynomial $\psi(t)$ and the corresponding pulse train $f(t)$ on $[0, 2\pi]$.

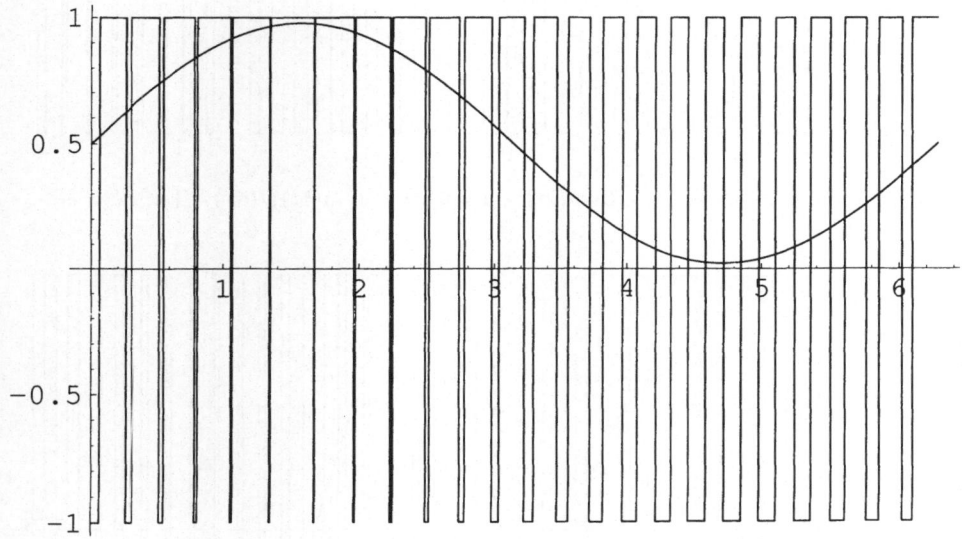

Fig. 1. $\psi(t) = 1/2 + (3\sin(t))/(2\pi)$, $f(t)$, $N = 24$

14.2 Soliton solution of the OPWM Problem

The most interesting, at least in the mathematical sense, thing about the OPWM solution is its deep relationship to completely integrable Hamiltonian (classical and quantum) systems of the isospectral origin. This relationship provides a different characterization of pulse edges in terms of the discrete spectrum information in the scattering matrix, associated with soliton type solutions. The appearance of an infinite family of commuting Hamiltonian flows (representing first integrals of the completely integrable Hamiltonian) is natural, because the OPWM problem depends on Fourier coefficients a_i, b_i as independent variables (representing commuting time flows).

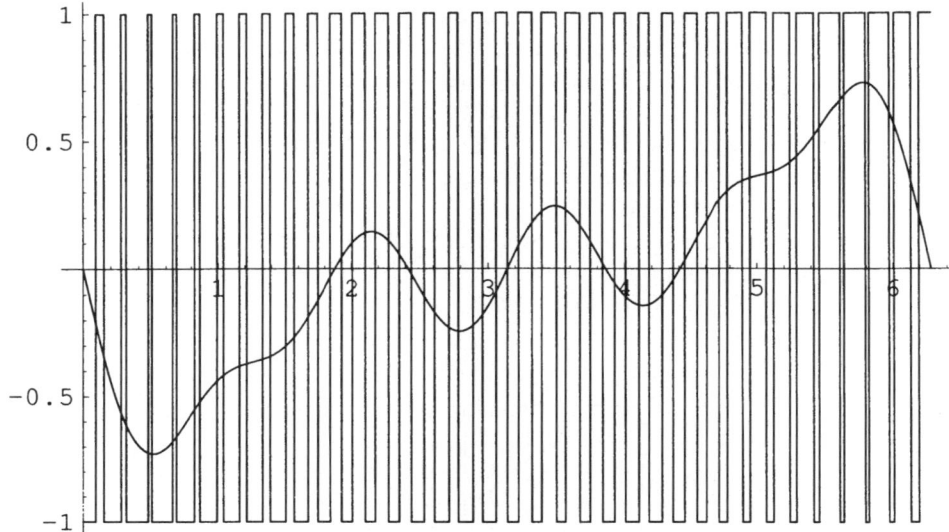

Fig. 2. $\psi(t) = -(6\sin(t) + 6\sin(2t) + 4\sin(3t) + 3\sin(5t))/(6\pi)$, $f(t)$, $N = 36$

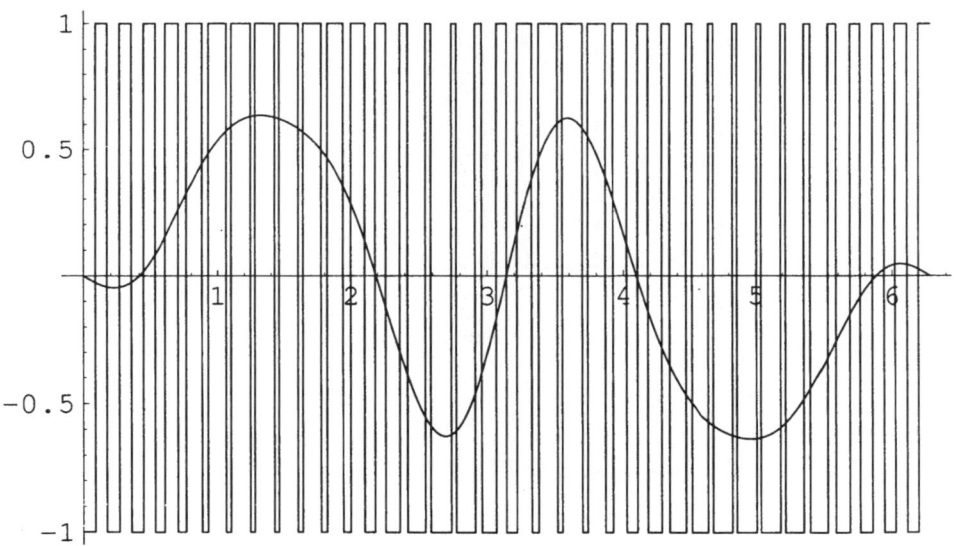

Fig. 3. $\psi(t) = (60\sin(t) + 60\sin(2t) - 70\sin(3t) + 15\sin(4t) - 18\sin(5t))/(60\pi)$, $f(t)$, $N = 36$

The first example of such a "soliton" representation of OPWM solution that we studied arose from the hierarchy of Korteweg- de Vries equations. The KdV case arises from the quarter period waveforms that have odd sine terms only:

$$\psi(t) = \sum_{n=1,n-\text{odd}}^{2N-1} a_n \sin(nt)$$

The corresponding moment problem in the algebraic form is a problem of "the sum of odd powers"

$$\sum_{i=1}^{2m-1} x_i^{2m-1} = s_{2m-1} : m = 1, \ldots, N.$$

The identification of time variables t_m of m-th order KdV Hamiltonians H_m with free parameters is relatively simple

$$t_m = \frac{s_{2m-1}}{(2m-1)2^{m-1}} : m = 1, 2, \ldots$$

where $t_1 = x$ is the spatial variable of the KdV equation.

The fact that the variation of the PWM problem in this case is subject to KdV equations, allows us to derive various approximations to specific classes of PWM expansions. For example, in the most important PWM case of a pure sine wave of an arbitrary amplitude a_1, one gets a difference differential equation in n and a_1, that can be used for an efficient and rapid numerical integration. This and other consequences of solition studies allow for fast and stable numerical methods of computation of pulse edges α_i in the PWM problem even for very large values of N.

Without this additional information the case of large values of N had been considered impractical to study using conventional techniques of numerical solutions of a system of N simultaneous transcendental equations in N unknowns α_i. The reasons for the difficulties include: ill conditioning of the Jacobian matrices in the neighborhood of a generic point in the phase space (these matrices are of Hankel type and resemble famous Hilbert matrices). Straightforward applications of conventional computations of Padé approximants is also unfeasible for large N, due to the presence of a very large number of singular values in the corresponding Hankel matrices. Only because of the control of commuting difference and differential operators, whose eigenfunctions are the polynomials $P_n(x)$ in PWM problem, a stable determination of polynomials $P_n(x)$ and their zeroes $x_i = \cos \alpha_i$ is possible for moderate and large N.

15 The Boundary Value Riemann Problem

The key for the solution of the general Padé approximation problem, as well as to the problem of its isospectral deformation is provided by boundary value problems, and by integral equations associated with them.

A classical complex variables boundary value problem is called the Riemann problem. It can be formulated in the following way. Let Γ be a closed contour (possibly, containing ∞) in the complex plane λ and let $G(\lambda)$ be a matrix function (of order m) defined over Γ. We need to find a matrix function $\psi_1(\lambda)$ analytic inside the contour, and a matrix function $\psi_2(\lambda)$ analytic outside the contour, and ψ_1 and ψ_2 on the contour satisfy the following condition

$$\psi_1 \psi_2 = G(\lambda).$$

Solution of a regular Riemann can be reduced to solution of a system of singular integral equations on the contour Γ. Let the Riemann problem be normalized at a point λ_0 e.g. outside the contour Γ and let

$$\psi_2(\lambda_0) = g.$$

We are looking for the solution of the Riemann problem in the form

$$\psi_1^{-1} = h + \int_\Gamma \frac{\varphi(\xi)}{\xi - \lambda} d\xi$$

inside the contour, and

$$\psi_2 = h + \int_\Gamma \frac{\varphi(\xi)}{\xi - \lambda} d\xi$$

outside the contour. Then the normalization condition yields

$$h = g - \int_\Gamma \frac{\varphi(\xi)}{\xi - \lambda_0} d\xi.$$

Thus we arrive at a singular integral equation on φ for $\lambda \in \Gamma$

$$g - \int_\Gamma \frac{\varphi(\xi)}{\xi - \lambda_0} d\xi + \pi i \varphi(\lambda) T(\lambda) + \int_\Gamma \frac{\varphi}{\xi - \lambda} d\xi = 0,$$

where $T = \frac{G+1}{G-1}$. The solution of this integral equation solves the (regular) Riemann problem.

16 Isospectral Deformation Equations

The appropriate isospectral deformation equations arising from the general OPWM problem is the same one that appears in the chiral σ - model, and, in particular, in Sin - Gordon equation. These equations like other isospectral deformation equations arise from the consistency conditions between (matrix) differential operators, depending on a parameter λ (typically called a spectral parameter). The general form of these consistency type differential spectral problems is

$$(\frac{\delta}{\delta\xi} - U(\lambda))\varphi = 0,$$

$$(\frac{\delta}{\delta\eta} - V(\lambda))\varphi = 0,$$

where $u = U(\lambda), V = V(\lambda)$ are $m \times m$ matrices, rationally depending on the (spectral) parameter λ. The consistency condition for these equations is known as a "Lax" pair and is

$$\frac{\delta}{\delta\eta}U - \frac{\delta}{\delta\xi}V + [U, V] = 0.$$

This is actually a system of nonlinear partial differential equations on residues of $U(\lambda), V(\lambda)$ at their poles independent of η, ξ.

References

[AMM] Airault, H., McKean, H.P., Moser J.: Rational and elliptic solutions of the Korteweg-de Vries and a related many-body problem. Comm. Pure Appl. Math, **30**, 95-148, (1977)

[BGY] Brent, R. P., Gustavson, F. G., Yun, D. Y. Y.: Fast solution of Toeplitz systems of equations and computation of Padé approximants, J. Algorithms, **1**, 259-295, (1980)

[CA74] Candy, J.: A use of limit cycle oscillations to obtain robust analog-to-digital converters. IEEE Trans. Commun., COM-**22**, 298-305, (1974)

[CA97] Candy, J.: An overview of basic concepts. In: Delta-Sigma Converters. IEEE Press, 1-43, (1997)

[CCCS] Chudnovsky, D.V., Chudnovsky, G.V., Czarkowski, D., Selesnik, I.: Solving the Optimal PWM Problem for Single-Phase Inverters, IEEE Trans. Circuits and Systems - I, **49**, 465-475, (2002)

[Ch1] Chudnovsky, D.V., Chudnovsky, G.V.,: Transcendental methods and theta-functions. Proc. Symp. Pure Mathematics, AMS, Providence, RI, v. **49**, part 2, 167-232, (1989)

[Ch2] Chudnovsky, D.V., Chudnovsky, G.V.,: Explicit Continued Fractions and Quantum Gravity, Acta Applic. Math, Kluwer, Netherlands, **36**, 167-185, (1994)

[Ch3] Chudnovsky, D.V., Chudnovsky, G.V.,: Laws of composition of Bäcklund transformations and the universal form of completely integrable systems in dimensions two and three. Proc. Natl. Acad. Sci. USA, **80**, 1774-1777, (1983)

[Ch4] Chudnovsky, D.V., Chudnovsky, G.V.,: Pole expansions of nonlinear partial differential equations. Nuovo Cimento, **40B**, 339-353, (1977)

[Ch5] Chudnovsky, D.V., Chudnovsky, G.V.,: On expansion of algebraic functions in power and Puiseux series, I, II. J. Complexity, **2**, 271-294, (1986); **3**, 1-25, (1987)

[CRGE] Craven, P., Gerzon, M.: Lossless coding for audio discs. J. Audio Eng. Soc., 706-719 (1996)

[CU] Cutler, C.: Transmission systems employing quantization. US Patent 2,927,962, (1960)

[GR87] Gray, R.: Oversampled sigma-delta modulation. IEEE Trans. Commun., COM-**35**, 481-489, (1987)

[GR89] Gray, R: Spectral analysis of quantization noise in single-loop sigma- delta modulation with dc inputs. IEEE Trans. Commun., COM-**36**, 588-599, (1989)

[GR97] Gray, R.: Quantization noise in Delta-Sigma A/D converters. In: Delta-Sigma Converters, IEEE Press, 44-74, (1997)

[GRCW] Gray, R., Chou, W., Wong, P.: Quantization noise in single-loop sigma-delta modulation with sinusoidal inputs. IEEE Trans. Inform. Theory, IT-**35**, 956-968, (1989)

[GRND] Gray, R., Neuhoff, D.: Quantization. IEEE Trans. Inform. Theory, IT-**44**, 2325-2375, (1998)

[GC] Gunturk, C.: Improved error estimates for first order sigma-delta systems. In: International Workshop on Sampling Theory and Applications (Samp TA'99), Norway, (1999)

[IY] Inose, H., Yasuda, Y.: A unit bit coding method by negative feedback. Proc. IEEE, **51**, 1524-1535, (1963)

[K1] Knuth, D.: The Art of Computer Programming, v. **2**, Addison-Wesly, (1981)

[Li] Littlewood, D. E.: The Theory of Group Characters, Oxford University Press, (1958)

[M] MacMahon, P.A.: Combinatory Analysis, v.I,II, Cambridge University Press, (1915)

[NST] S. Norsworthy, R. Schreier, G.Temes, (eds.), Delta-Sigma Data Converters. Theory, Design and Simulation, IEEE Press, (1997)

[PH] Patel, H. S., Hoft, R. G.: Generalized technique of harmonic elimination and voltage control in thyristor inverters: Part I harmonic elimination. IEEE Trans. Ind. Applicat., 310-317, (1973)

[S] Szego, G.: Orthogonal Polynomials, AMS, Providence, RI, (1978)

[Z] Falqui, G, Magri, F., Padroni, M., Zubelli, J.P.: An elementary approach to the polynomial τ-functions of the KP hierarchy. Theor. Math. Physics, **122**, 17-28, (2000)

Use of Padé Approximations in Spline Construction

David V. Chudnovsky[1] and Gregory V. Chudnovsky[2]

[1] IMAS, Polytechnic University, Brooklyn, 6 MetroTech Center, NY 11201
david@imas.poly.edu
[2] IMAS, Polytechnic University, Brooklyn, 6 MetroTech Center, NY 11201
gregory@imas.poly.edu

1 Introduction.

This paper deals with a very important practical problem of constructing the "best" spline approximation to a curve or a surface. Needless to say, this problem has many practical application, especially in cases of data modeling and data fitting. In this paper we are dealing with curves, in one- and multi- dimensional cases, and, in particular, with closed curves. The main methods of this paper are methods of (generalized) Padé approximations and orthogonal polynomials on the interval and the unit circle, and related methods of mechanical quadrature formulas. For references to the approaches to the spline reconstruction of a data curve, where a variety of "fitting" figures of merit is used, see [KH], [MW], [MP], [W], [M95], [C95], [M97], [MM], [LM94], [Z].

The basic approach we use in this paper is that of proper "spectral" match of the spline approximation to the original data curve. This corresponds to the assumption of perceived similarity between objects based on matching according to a certain band-limited frequency characteristic curve. This assumption is used to justify the use of band-limited information extracted from the original data curve to match the (polynomial) spline approximation to such a curve. Even in the simplest (but the most important) case of matching the data curve with piecewise-constant approximation the concept of band-limited approximation leads to interesting generic observations about the number of independent degrees of freedom of the original data that can be captured in the approximations. The simplest data fit is that of fixed locations of N knots of the spline (sample points) where the values can be arbitrary numbers matching N (real) spectral coefficients of the original data. This subject is well covered by Nyquist theorem and related studies of band-limited functions sampled, typically, at uniformly distributed N points. On the other hand, we have recently studied so-called PWM approximations, where the sample points are arbitrary (in a given interval of observations) but the values are limited only to 2 - we approximate data with pulse trains of N

arbitrary width but equal height pulses whose spectral characteristics match N (real) spectral characteristics of the original data (see, e.g. [Ch99]). In this paper we consider the case of variable locations of sample points (spline knots) and values of approximating function. This gives the total of $2N$ degrees of freedom for N knots (sample points) - "twice the Nyquist limit".

The proposed method uses an analytic construction of the spline curve best approximating the data curve by matching the maximum number of Fourier coefficients (trigonometric moments). It is based on technique of generalized (two-point) Padé approximations and orthogonal polynomials on the unit circle, and is very similar to the PWM technique from [Ch99].

We also describe and generalize a somewhat related approach of constructing splines matching the monomial moments of the approximated function on the interval studied in a series of papers starting from [G84]. In the case of an approximation of a single function $f(t)$ on intervals $[0, \infty)$ and $[a, b]$ by splines that match the maximum number of polynomial moments on intervals, the reduction to the classical (Gauss) problem of mechanical quadratures was obtained by Gautschi [G84], Gautschi, Milovanovic, and Frontini [F87], Gautschi, Milovanovic [G86] (see also Micchelli [M88] and related work in [MK88], [M00]). Precise formulations of these results are presented below. An important and related problem of periodic spline approximations was left open in publications quoted above, and is solved.

A spline function of degree $m > 0$, with n distinct knots $\tau_1, \tau_2, \ldots, \tau_n$ in the interior of $[0, 1]$, can be written in terms of truncated powers in the form

$$s_{n,m} = p_m + \sum_{\nu=1}^{n} a_\nu (\tau_\nu - t)_+^m, \ 0 \le t \le 1,$$

where a_ν are real numbers and p_m is a polynomial of degree $\le m$.

Problem. Determine spline $s_{n,m}$ such that

$$\int_0^1 t^j s_{n,m}(t) dt = \int_0^1 t^j f(t) dt,$$

holds for $j = 0, 1, \ldots, 2n - 1$ and

$$s_{n,m}^{(k)}(1) = f^{(k)}(1), \ k = 0, 1, \ldots, m,$$

Here it is assumed that f has m derivatives at $t = 1$, all being known.

Define

$$\mu_j = \frac{(m+j+1)!}{m!j!} \int_0^1 t^j [f(t) - \sum_{k=0}^{m} \frac{f^{(k)}(1)}{k!} (t-1)^k] dt,$$

for $j = 0, 1, \ldots, 2n - 1$. This gives rise to linear functional $\mathcal{L}$ on polynomials of the form $t^{m+1}p(t), p \in \mathbf{P}_{2n-1}$ ($\mathbf{P}_d$ is the set of polynomials in t of degree at most d), defined by its values on monomials:

$$\mathcal{L}(t^{m+1} \cdot t^j) \; = \; \mu_j, \; j = 0, 1, ..., 2n - 1,$$

and the inner product

$$< p, q > = \; \mathcal{L}(t^{m+1} p \cdot q), \; p \cdot q \in \mathbf{P}_{2n-1}.$$

The orthogonal polynomials π_n is defined by

$$\deg \pi_n = n, \; \pi_n = t^n + \dots,$$

$$< \pi_n, q > = \; 0 \text{ for all } q \in \mathbf{P}_{n-1}.$$

Theorem. There exists a unique spline function on $[0.1]$,

$$s_{n,m} \; = \; p_m(t) + \sum_{\nu=1}^{n} a_\nu (\tau_\nu - t)_+^m, \, 0 < \tau_\nu < 1, \, \tau_\nu \neq \tau_\mu \text{ for } \nu \neq \mu,$$

satisfying moment equations of Problem above, if and only if the orthogonal polynomials π_n above exist, are unique, and have n distinct real zeroes τ_ν^n, $\nu = 1, 2, \dots, n$, all contained in an open interval $(0, 1)$. The knots τ_ν of $s_{n,m}$ are then precisely these zeroes,

$$\tau_\nu = \tau_\nu^n, \; \nu = 1, 2, \dots, n;$$

the polynomial p_m is given by

$$p_m(t) \; = \; \sum_{k=0}^{m} \frac{f^{(k)}(1)}{k!} (t - 1)^k,$$

and the coefficients a_ν are obtained uniquely from the linear system

$$\mathcal{L}_0(t^{m+1} p) \; = \; \mathcal{L}(t^{m+!} p) \text{ for all } p \in \mathbf{P}_{n-1},$$

where

$$\mathcal{L}_0(g) \; = \; \sum_{\nu=1}^{n} a_\nu g(\tau_\nu), \tau_\nu = \tau_\nu^{(n)}.$$

More complicated spline approximations where degrees of piecewise approximations vary among the knots also can be described by generalized orthogonal polynomials and Turan's quadrature formulas (see [M00]).

We generalize these and other similar results below for the case of approximating spatial d-dimensional curves $\mathbf{g}(t)$ by spatial splines defined on the interval, that match the maximum (allowed) number of moments on that interval. For this we use the technique of Padé approximations of the Second Kind and related (multi-index) orthogonal polynomials of the Second Kind.

The main problem with the moment-matching (moment-preserving) spline approximations lies with the instabilities in moment problem calculations in cases of moments on infinite and finite intervals. Some of these problems are related to ill-conditioning of related Hankel matrices, and some problems are related to singularities arising from non-normals cases, and cases when roots are not real or lie off the interval that supports the measure.

The trigonometric moments, and related polynomials with zeroes on the unit circle do not have such instabilities; they are more advantageous for construction of spline approximations. They also provide a unique framework for the construction of periodic splines approximating periodic functions (e.g., closed planar or spatial curves). Our results solve the problem of approximating periodic functions (curves) by periodic splines, left open in previous studies. The technique here differs considerable from the traditional approach of Gauss's mechanical quadratures (and their generalizations), and involves two-point Padé approximations (at $z = 0$ and $z = \infty$) and quadratures at the unit circle.

Unlike the corresponding figure of merit based on matching of moments on the interval, figure of merits based on matching of trigonometric moments always has "the best" solution (with the minimal number of knots). This solution is quite practical, fast, and also matches correctly piecewise-polynomial input data. The solution is easily generalized to the case of (closed) multi-dimensional curves.

There is no obvious generalization of this method to the surfaces and other multi-dimensional manifolds due to the absence of any useful general multi-dimensional cubature formulas and related multi-dimensional orthogonal polynomials. Though there are some specific cubature formulas (like Radon's two-dimensional theorem), they seem to work for very specific numbers of knots, and cannot be used with, say, adding one extra knot at a time. Nevertheless, one can use our approach with the curves on multi- dimensional manifolds to find the best knots on the "scan-curves" of the manifolds, constructing this way multi-dimensional spline patches.

2 Multi-dimensional Spline Curves and Padé Approximations.

2.1 Simultaneous Padé Approximations of the Second Kind.

To deal with the problem of analytic construction of d-dimensional spline curves of arbitrary degree m, we apply the methods of simultaneous Padé approximations of the Second Kind, see [Ch85]. Just as the ordinary Padé approximants are expressed in terms of (general) orthogonal polynomials, simultaneous Padé approximants of the Second Kind are also expressed in terms of simultaneous orthogonal polynomials (or a solution to a multiple moments problem). Here are the alternative definitions.

Definition of One-Point Padé approximations of the Second Kind. Let $f_1(z) \ldots f_d(z)$ be a system of functions defined by their Taylor expansions at $z = \infty$. Let $\mathbf{n} = (n_1, \ldots, n_d)$ be a multi-index with non-negative n_i, $i = 1, \ldots, d$. Then the polynomial $P_{\mathbf{n}}(z)$ in z of degree (at most) $|\mathbf{n}| = n_1 + \ldots + n_d$ is called a Padé approximant of the Second Kind to $f_1(z), \ldots, f_d(z)$ at $z = \infty$ of the weight $\mathbf{n}$ if there are polynomials $Q_i(z)\, i = 1, \ldots, d$ in z of degree at most $|\mathbf{n}|$ such that the following approximation holds.

The Laurent expansion at $z = \infty$ of $Q_i(z)/P(z)$ coincides with the Taylor expansion of $f_i(z)$ up to (and including) the term z^{-n_i}, for all $i = 1, \ldots, d$.

An alternative definition of these approximants involve orthogonal polynomials of the Second Kind. Measures defining these polynomials can have support on unions of arbitrary arcs in the complex plane. However, for a typical application that we study here, the support of all d measures is within the same interval $[a, b]$, that can be chosen, without any loss of generality, to be $[0, 1]$. We consider d measures $d\mu_i(t)$ on $[0, 1]$, where often these measures are defined by means of weight functions $w_i(t)dt$ (for $i = 1, \ldots, d$). These measures produce the simultaneous problem of moments and the class of orthogonal polynomials of the Second Kind on the interval, as in the following definition.

Definition of Orthogonal Polynomials of the Second Kind. Let $d\mu_i(t)$ be d measures on the interval $[0, 1]$ $(i = 1, \ldots, d)$. Let $\mathbf{n} = (n_1, \ldots, n_d)$ be a multi-index with non-negative n_i, $i = 1, \ldots, d$. Then the polynomial $P_{\mathbf{n}}(t)$ in t of degree (at most) $|\mathbf{n}| = n_1 + \ldots + n_d$ is called an orthogonal polynomial of the Second Kind with respect to measures $d\mu_1(t)$, $\ldots$, $d\mu_d(t)$ on $[0, 1]$ of the weight $\mathbf{n}$ if the following orthogonality relations hold.

$$\int_0^1 t^j \, P_{\mathbf{n}}(t) \, d\mu_i(t) \; = \; 0, \; j = 0, \ldots, n_i$$

for $i = 1, \ldots, d$.

The relationship between the (one-point) Padé approximants and orthogonal polynomials of the Second Kind is fairly straightforward, just as in the classical case of $d = 1$ of the classical moment problem. For this one converts the (measure) moments into their generating function, expanded at $z = \infty$, as follows.

For $i = 1, \ldots, d$, we define the generating function $f_i(z)$ in terms of the moments of the measure $d\mu_i(t)$ on $[0, 1]$:

$$f_i(z) \; = \; \sum_{k=0}^{\infty} \frac{m_{k,i}}{z^k},$$

where

$$m_{k,i} \; = \; \int_0^1 t^k \, d\mu_i(t), \; k = 0, \ldots.$$

In these notations the problem of constructing simultaneous orthogonal polynomials and (one-point) Padé approximants of the Second Kind for the multi-index $\mathbf{n}$ are equivalent.

The construction of the polynomial $P_\mathbf{n}$ is reduced to the problem of linear algebra via a standard Gramm-Schmidt reduction, and an inversion of the (generalized) Hankel matrix. There are two important conditions on the measures that are often imposed. The first one is the "normality condition" of the Padé approximations. This condition means that the polynomial $P_\mathbf{n}$ is unique (up to a multiplicative constant factor), and has the degree exactly $|\mathbf{n}|$ for all multi-indices $\mathbf{n}$. This condition is equivalent to the condition of non-singularity of all corresponding (generalized) Hankel matrices. The "normality condition" holds generically, and its violation means that there are non-trivial relations between the approximating functions. The simplest necessary condition for "normality" is the linear independence of functions $f_i(z)$, $i = 1, \ldots, d$ over $\mathbf{C}(z)$ (or $\mathbf{R}(z)$).

The second condition concerns the distribution of zeros of orthogonal polynomials. Under very mild conditions on the behavior of functions $f_i(z)$ (related to their singularities), the distribution of zeros of polynomials $P_\mathbf{n}(t)$ approaches the union of supports of measures $d\mu_i(t)$. Stronger conditions ensure that all zeros of polynomials $P_\mathbf{n}(t)$ are simple and are located in the interval $[0, 1]$. For $d = 1$ one such condition is the condition that the weight $w_1(t)$ does not change sign on the interval $[0, 1]$. For $d > 1$ the conditions on weights $w_i(t)$ are stronger, of the form of the Chebyshev system.

Since the construction of orthogonal polynomials implies the solution of linear equations, these polynomials can be expressed in terms of determinants of generalized Hankel matrices. Assuming normality, the monic orthogonal polynomial $P_\mathbf{n}(t)$ corresponding to a multi-index $\mathbf{n} = (n_1, \ldots, n_d)$ of degree $|\mathbf{n}|$ can be expressed as a ratio of two determinants:

$$
P_\mathbf{n}(t) = \frac{1}{D_\mathbf{n}}
\begin{pmatrix}
m_{0,1} & m_{1,1} & \cdots & m_{|\mathbf{n}|,1} \\
\vdots & \vdots & \ddots & \vdots \\
m_{n_1-1,1} & m_{n_1,1} & \cdots & m_{|\mathbf{n}|+n_1-1,1} \\
\vdots & \vdots & \ddots & \vdots \\
m_{0,d} & m_{1,d} & \cdots & m_{|\mathbf{n}|,d} \\
\vdots & \vdots & \ddots & \vdots \\
m_{n_d-1,d} & m_{n_d,d} & \cdots & m_{|\mathbf{n}|+n_d-1,d} \\
1 & t & \cdots & t^{|\mathbf{n}|}
\end{pmatrix},
$$

where

$$D_{\mathbf{n}} = \begin{pmatrix} m_{0,1} & m_{1,1} & \cdots & m_{|\mathbf{n}|-1,1} \\ \vdots & \vdots & \ddots & \vdots \\ m_{n_1-1,1} & m_{n_1,1} & \cdots & m_{|\mathbf{n}|+n_1-2,1} \\ \vdots & \vdots & \ddots & \vdots \\ m_{0,d} & m_{1,d} & \cdots & m_{|\mathbf{n}|-1,d} \\ \vdots & \vdots & \ddots & \vdots \\ m_{n_d-1,d} & m_{n_d,d} & \cdots & m_{|\mathbf{n}|+n_d-2,d} \end{pmatrix},$$

This representation, if used directly, often leeds to ill-conditioned calculations even for moderate $\mathbf{n}$. Faster, and more stable, algorithms are sometimes based on recurrence relations expressing $P_{\mathbf{n}}(t)$. In the case $d = 1$ the relation is a simple three-term linear recurrence, connecting 3 nearby monic orthogonal polynomials:

$$P_n(t) = (t + b_n) \cdot P_{n-1}(t) - c_n \cdot P_{n-2}(t),$$

$n = 2, \ldots$.

This three-term recurrence also allows to reduce the problem of finding zeros of ordinary orthogonal polynomials to the problem of determining eigenvalues of the three-diagonal matrices (formed from the three-term recurrence relation).

In the case of $d > 1$, we have linear recurrences of the order $d+1$ (of length $d+2$) connecting orthogonal polynomials of the Second Kind. The easiest way to express these relations is to look only at diagonal or near-diagonal cases of simultaneous approximations. Namely for every N, we create the multi-index $\mathbf{n}(N) = (q+1, \ldots, q+1, q, \ldots, q)$ ($q+1$ is repeated r times, and q is repeated $d - r$ times) for $N = q \cdot d + r$. Then orthogonal polynomials of the Second Kind corresponding to multi-indices $\mathbf{n}(N)$ satisfy the recurrence of order $d+1$:

$$P_{\mathbf{n}(N)} = t \cdot P_{\mathbf{n}(N-1)} + \sum_{k=1}^{d+1} b_k \cdot P_{\mathbf{n}(N-k)},$$

$N = d+1, \ldots$.

2.2 Multi-dimensional Spline Curves and Moment Problems.

In order to find an analytic solution to the problem of construction of multi-dimensional spline curves using moment approximation we use the method of simultaneous Padé approximations from the previous subsection.

Here we use as a figure of merit of the approximation of the multi-dimensional curve $\mathbf{g}(t) = (g_1(t), \ldots, g_d(t))$ for t in $[0, 1]$ by a spline of degree m a coincidence of maximum allowed number of moments in t. Here d is the dimension of the Euclidian space $\mathbf{R}^d$, where the approximated curve $\mathbf{g}(t)$ lies. In order to avoid trivial cases, that will result in non-"normality" of Padé approximations one should assume that the curve $\mathbf{g}(t)$ is essentially d-dimensional, e.g. functions $g_1(t), \ldots, g_d(t))$ are linearly dependent over $\mathbf{R}$. The minimal assumption on $\mathbf{g}(t)$ is the integrability of $g_i(t)$ on $[0, 1]$.

We are looking at a d-dimensional spline of degree m $\mathbf{sp}(t)$ defined on $[0, 1]$ that approximates $\mathbf{g}(t)$ in the sense of matching the highest (expected) number of moments on $[0, 1]$. Specifically, for a multi-index $\mathbf{n} = (n_1, \ldots, n_d)$, we are looking at the d-dimensional spline of degree m $\mathbf{sp}(t) = (sp_1(t), \ldots, sp_d(t))$ defined on $[0, 1]$ with $N = |\mathbf{n}|$ knots t_i, $i = 1, \ldots, N$ in $[0, 1]$ ($0 \leq t_1 < \ldots < t_N \leq 1$), matching the moments with $\mathbf{g}(t)$ on $[0, 1]$. This matching can be formulated in one of the two following statements (as in the one-dimensional case):

Problem M1. Determine the N-knot spline $\mathbf{sp}(t) = (sp_1(t), \ldots, sp_d(t))$ of degree m, such that

$$\int_0^1 t^k \cdot sp_i(t)\, dt = \int_0^1 t^k \cdot g_i(t)\, dt, \; k = 0, \ldots, N + n_i + m,$$

for $i = 1, \ldots d$.

The second problem assumes that $\mathbf{g}(t)$ has the first m known derivatives at $t = 1$, and we are trying to match these derivatives (at $t = 1$), in addition to the maximal number of moments:

Problem M2. Determine the N-knot spline $\mathbf{sp}(t) = (sp_1(t), \ldots, sp_d(t))$ of degree m, such that

$$\int_0^1 t^k \cdot sp_i(t)\, dt = \int_0^1 t^k \cdot g_i(t)\, dt, \; k = 0, \ldots, N + n_i - 1,$$

and

$$sp_i^{(k)}(t) = g_i^{(k)}(t), \; k = 0, \ldots, m,$$

for $i = 1, \ldots d$.

Any d-dimensional spline of degree m $\mathbf{sp}(t) = (sp_1(t), \ldots, sp_d(t))$ defined on $[0, 1]$ with $N = |\mathbf{n}|$ knots t_j, $j = 1, \ldots, N$ can be represented in the following form:

$$sp_i(t) = p_i(t) + \sum_{j=1}^N a_{j,i} (t_j - t)_+^m,$$

where $p_i(t)$ are polynomials of degree at most m, for $i = 1, \ldots, d$.

Problems M1-M2 of optimal construction of spatial splines $\mathbf{sp}(t)$, approximating original spatial curve $\mathbf{g}(t)$ can be reduced to the problem of finding zeros of orthogonal polynomials of the Second Kind and related (spatial) mechanical quadrature. The corresponding moment problems and functions that are Padé approximated arise from $(m + 1)$-st differentiation, when $\mathbf{g}(t)$ is, essentially, approximated by a linear combination of delta-functions $\delta(t - t_j)$, centered at the splines's knots (zeros of orthogonal polynomials of the Second Kind). This reduction is very similar to the one-dimensional case, and involves mainly integration by parts.

We'll present the d-dimensional moment problems equivalent to Problems M1-M2. These moment problems can be actually reduced to d-dimensional moment problems with d weights on $[0,1]$, provided that $\mathbf{g}(t)$ if $(m+1)$-differentiable: $g_i(t) \in C^{m+1}[0,1]$, $i = 1, \ldots, d$.

First, the case of the Problem M1. The corresponding d sequences of moments $\{m_{k,i}^1\}$, $k = 0, \ldots$ are:

$$m_{k,i}^1 = \frac{(m+k+1)!}{m!k!} \cdot \int_0^1 t^k g_i(t)\, dt, \ k = 0, \ldots,$$

for $i = 1, \ldots, d$. As usual, these moments extend by linearity to linear functionals $\mathcal{L}_i^1$ defined on the set $\mathbf{P}$ of polynomials $q(t)$ in t using the following definition of $\mathcal{L}_i^1$ on monomials t^k:

$$\mathcal{L}_i^1(t^k) = m_{k,i}^1, \ k = 0, \ldots,$$

for $i = 1, \ldots, d$. These linear functionals are extended to the scalar product, defined for all polynomials $p(t)$ and $q(t)$ from $\mathbf{P}$:

$$< p, q >_i^1 = \mathcal{L}_i^1((1-t)^{m+1} \cdot p(t) \cdot q(t)),$$

$i = 1, \ldots, d$.

The orthogonal polynomials of the Second Kind of the multi-index $\mathbf{n}$ corresponding to scalar products $< ., . >_i^1$, $i = 1, \ldots, d$ have zeros that correspond to knots $t_1, \ldots, t_N$ for $N = |\mathbf{n}|$. To determine the parameters $a_{j,i}$, $j = 1, \ldots, N$ and coefficients of polynomials $p_i(t)$ one needs to introduce other linear functionals, arising from $m + 1$-st differentiation of $g_i(t)$:

$$\mathcal{L}_{0,i}^1(h) = \sum_{k=0}^m b_{k,i} h^{(m-k)}(1) + \sum_{j=1}^N a_{j,i} h(t_j),$$

where

$$b_{k,i} = \frac{(-1)^k}{m!} p_i^{(k)}(1), \ k = 0, \ldots, m,$$

for $i = 1, \ldots, d$.

Theorem There exists a unique d-dimensional spline approximation $\mathbf{sp}(t)$, solving the problem M1 for a given multi-index $\mathbf{n} = (n_1, \ldots, n_d)$ iff there exists a unique orthogonal polynomial of the Second Kind $P_\mathbf{n}(t)$ such that

$$< P_\mathbf{n}, q >_i^1 = 0,$$

for all polynomials $q(t)$ of degree $< n_i$ for all $i = 1, \ldots, d$ having $N = |\mathbf{n}|$ real zeros t_j, $j = 1, \ldots, N$ in the interval $(0, 1)$. The parameters $a_{j,i}$ and polynomials $p_i(t)$ (defined by their expansion coefficients $b_{k,i}$) are determined from the block-Vandermonde equations

$$\mathcal{L}^1_{0,i}(t^{m+1} \cdot q) = \mathcal{L}^1_i(q)$$

for all polynomials $q(t)$ of degree $\leq N + n_i$ and all $i = 1, \ldots, d$.

The proof follows from integration by parts that gives:

$$\int_0^{t_j} t^k (t_i - t)^m \, dt = \frac{k! m!}{(m + k + 1)!} t_i^{m+k+1},$$

$$\int_0^1 t^k p_i(t) \, dt = \frac{k! m!}{(m + k + 1)!} \sum_{l=0}^m b_{l,i} \frac{d^{m-l}}{dt^{m-l}} t^{m+1+k} \big|_{t=1},$$

and thus:

$$\mathcal{L}^1_{0,i}(t^{m+1} \cdot t^k) = m^1_{k,i}, \; k = 0, \ldots,$$

for $i = 1, \ldots, d$.

Provided that $\mathbf{g}(t)$ is $(m + 1)$-differentiable: $g_i(t) \in C^{m+1}[0, 1]$, $i = 1, \ldots, d$, one can replace the general moment problem for $m^1_{k,i}$ by a problem, where the measure $d\mu_i(t)$ is a weight:

$$d\mu_i(t) = t^{m+1}(1 - t)^{m+1} \cdot \frac{(-1)^{m+1}}{m!} g_i^{(m+1)}(t) \, dt,$$

for $i = 1, \ldots, d$. This means that the inner product $< .,. >^1_i$ can be represented as:

$$< p, q >^1_i = \int_0^1 p(t) \cdot q(t) \, d\mu_i(t),$$

$i = 1, \ldots, d$.

Then the unique solution to the d-dimensional spline problem M1 exists for a given multi-index $\mathbf{n}$ if and only if the corresponding orthogonal polynomial of the Second Kind $P_{\mathbf{n}}(t)$ is unique of degree $N = |\mathbf{n}|$ (this is true under the normality assumption), and its N roots are all real and lie in the interval $(0, 1)$. Assumptions of normality of the function system $\mathbf{g}(t)$ are necessary for this, and in the case of $d = 1$ the positivity of $g_1^{(m+1)}(t)$ on $[0, 1]$ is a sufficient condition. It is satisfied when, for example, $g_1(t)$ is a monotonic function on $[0, 1]$.

The solution to the problem M2 is very similar, but one introduces slightly different moments and related linear functionals. Also the definition of the polynomials $p_i(t)$, denoted as $p_i^2(t)$ is very simple in terms of values of derivatives of $g_i(t)$ at $t = 1$:

$$p_i^2(t) = \sum_{j=0}^m (t - 1)^j \frac{g_i^{(j)}(1)}{j!}$$

for $i = 1, \ldots, d$. The moments $\{m^2_{k,i}\}$ in the Problem M2 are:

$$m_{k,i}^2 = \frac{(m+k+1)!}{m!k!} \int_0^1 t^k \left(g_i(t) - \sum_{j=0}^m (t-1)^j \frac{g_i^{(j)}(1)}{j!}\right) dt, \; k = 0, \ldots,$$

and these moments extend by linearity to linear functionals $\mathcal{L}_i^2$ defined on the set $\mathbf{P}$ of polynomials $q(t)$ in t using the following definition of $\mathcal{L}_i^2$ on monomials t^k:

$$\mathcal{L}_i^2(t^k) = m_{k,i}^2, \; k = 0, \ldots,$$

for $i = 1, \ldots, d$. These linear functionals are extended to the scalar product, defined for all polynomials $p(t)$ and $q(t)$ from $\mathbf{P}$:

$$< p, q >_i^2 = \mathcal{L}_i^2(p(t) \cdot q(t)).$$

$i = 1, \ldots, d$. To determine the parameters $a_{j,i}$, $j = 1, \ldots, N$ one needs to introduce other linear functionals, arising from $m + 1$-st differentiation of $g_i(t)$:

$$\mathcal{L}_{0,i}^2(h) = \sum_{j=1}^N a_{j,i} h(t_j),$$

$i = 1, \ldots, d$.

Theorem There exists a unique d-dimensional spline approximation $\mathbf{sp}(t)$, solving the problem M2 for a given multi-index $\mathbf{n} = (n_1, \ldots, n_d)$ iff there exists a unique orthogonal polynomial of the Second Kind $P_{\mathbf{n}}(t)$ such that

$$< P_{\mathbf{n}}, q >_i^2 = 0,$$

for all polynomials $q(t)$ of degree $< n_i$ for all $i = 1, \ldots, d$ having $N = |\mathbf{n}|$ real zeros t_j, $j = 1, \ldots, N$ in the interval $(0, 1)$. The parameters $a_{j,i}$ are determined from the block-Vandermonde equations

$$\mathcal{L}_{0,i}^2(t^{m+1} \cdot q) = \mathcal{L}_i^2(q)$$

for all polynomials $q(t)$ of degree $\leq N + n_i$ and all $i = 1, \ldots, d$.

Provided that $\mathbf{g}(t)$ is $(m + 1)$-differentiable: $g_i(t) \in C^{m+1}[0, 1]$, $i = 1, \ldots, d$, one can replace the general moment problem for $m_{k,i}^2$ by a problem, where the measure $d\mu_i(t)$, denoted as $d\mu_i^2(t)$ is a weight:

$$d\mu_i^2(t) = t^{m+1} \cdot \frac{(-1)^{m+1}}{m!} g_i^{(m+1)}(t) \, dt,$$

for $i = 1, \ldots, d$.

The main problem with construction of moment-preserving spline approximations on the interval (infinite or finite) to generic functions is its inherent instability. Corresponding linear problem are notoriously ill-conditioned are require nearly quadratic increase with N in precision of computations, making it prohibitively difficult even for moderate N. These difficulties are significantly less pronounced in the case of periodic splines, or for other similar problem with functions defined on the (part of) the unit circle, when the knots (zeroes) are lying on the unit circle.

3 Trigonometric Moments Problem and Periodic Splines.

3.1 Carathéodory Theorem.

A crucial result used in the solution of the trigonometric moment problem is known as the Carathéodory theorem:

Theorem (Carathéodory). For any N complex numbers $c_1, \ldots, c_N$ there exists a unique $M \leq N$, positive real numbers $\rho_1, \ldots, \rho_M$ and distinct real numbers $\theta_1, \ldots, \theta_M$ such that $-\pi < \theta_j < \pi$, $j = 1, \ldots, M$ and

$$c_k = \sum_{j=1}^{M} \rho_j \cdot e^{ik\theta_j} \ k = 1, \ldots, N.$$

The proof of this Carathéodory theorem is important because the same method is used in some of our proofs as well. We use the classical method described in [GS]. First, one extends the sequence $\{c_k\}_{k=1,\ldots}$ no negative indices as follows: $c_{-k} = c_k^*$ for $k = 1, \ldots$, where c^* is the complex conjugate of c. Then we construct a $(N+1) \times (N+1)$ "zero-diagonal" Toeplitz matrix $\mathbf{T}$ such that $\mathbf{T}_{i,j} = c_{j-i}$ for $i \neq j$ and $i, j = 0, \ldots, N$ and $\mathbf{T}_{i,i} = 0$ for $i = 0, \ldots, N$. Let us denote by λ_0 the smallest eigenvalue of $\mathbf{T}$, and define (for the first time in the context of the Carathéodory theorem) c_0 as $c_0 = -\lambda_0$. Then one can defined a"full" $(N+1) \times (N+1)$ Toeplitz matrix $\mathbf{T}_N$ with $(\mathbf{T}_N)_{i,j} = c_{j-i}$ for $i, j = 0, \ldots, N$.

We can define $M \leq N$ such that $N - M$ is the multiplicity of the eigenvalue λ_0. Then M is the rank of $\mathbf{T}_N$. One looks then at the leading $(M+1) \times (M+1)$ sub-matrix $\mathbf{T}_M$ of $\mathbf{T}_N$ with $(\mathbf{T}_M)_{i,j} = c_{j-i}$ for $i, j = 0, \ldots, M$. Then $\mathbf{T}_M$, as well as $\mathbf{T}_N$ has a zero eigenvalue. Let $\mathbf{e}$ be the zero eigenvalue of $\mathbf{T}_M$, where $\mathbf{e} = (e_0, \ldots, e_M)$.

From $\mathbf{e}$ one can construct the polynomial $\mathbf{Q_e}(z)$, defined via elements of $\mathbf{e}$, $\mathbf{Q_e}(z) = \sum_{i=0}^{M} e_i z^i$. The polynomial $\mathbf{Q_e}(z)$ of degree M has only simple complex roots, and all these roots lie on the unit circle. We denote them as $e^{i\theta_j}$ for $j = 1, \ldots, M$. The weights ρ_j $(j = 1, \ldots, M)$ are defined from the Vandermonde system of equations:

$$c_k = \sum_{j=1}^{M} \rho_j \cdot e^{ik\theta_j} \ k = 1, \ldots, N.$$

The weights ρ_j are positive, and $\sum_{j=1}^{M} \rho_j = c_0$. This proves Carathéodory theorem.

3.2 Approximations of Periodic Function by Periodic Splines.

We start with the one-dimensional case of a real periodic function $f(t)$, defined on $[0, 2\pi]$ (periodic with the period 2π) and having well- defined leading $N+1$

(complex) Fourier coefficients $c_0, \ldots, c_N$. Because $f(t)$ is real, one extends c_k to negative k by defining: $c_{-k} = c_k^*$ for positive k.

Since we consider a problem of approximating a periodic function by periodic splines, we can use instead of $f(t)$ its truncated expansion in the Fourier series:

$$f_N(t) = \sum_{k=-N}^{N} c_k \, e^{ikt}$$

where

$$c_k = \frac{1}{2\pi} \int_0^{2\pi} e^{-ikt} \cdot f(t) \, dt$$

The periodic spline $sp(t)$ on $[0, 2\pi]$ can be defined in two different ways, depending on whether the end points of the period are included in the list of knots. In the simpler definition, the periodic spline $sp(t)$ of degree m on $[0, 2\pi]$ has M knots, starting with $t_0 = 0$. These are:

$$0 = t_0 < \ldots < t_{M-1} < 2\pi,$$

and we can add by periodicity $t_M = 2\pi$. The spline has the generic representation:

$$sp(t) = p_m(t) + \sum_{i=1}^{M-1} a_i (t_i - t)_+^m,$$

with additional periodicity conditions satisfied:

$$sp^{(k)}(2\pi) = sp^{(i)}(0), \quad k = 0, \ldots, m-1.$$

This leaves a total of $2M - 1$ real variables in $sp(t)$ ($M - 1$ for variable knots t_i, $i = 1, \ldots, M-1$ and total M for free coefficients: a_i at $(t - t_i)^m$, $i = 1, \ldots, M - 1$ and for a constant term of $p_m(t)$). This allows us to match a total of $2M - 1$ real Fourier coefficients of $sp(t)$ with Fourier coefficients of $f_{M-1}(t)$ (or $f(t)$), or c_j for $j = 0, \ldots, M-1$. The periodicity conditions uniquely define all but the constant coefficients of $p_m(t)$ in terms of t_i, a_i, $i = 1, \ldots, M - 1$.

Let us write $p_m(t)$ as $\sum_{k=0}^{m} p_{m,k} \cdot t^k$, so that the leading coefficient of $p_m(t)$ (at t^m) is $p_{m,m}$. Then the periodicity conditions uniquely determine all nonconstant coefficients $p_{m,k}$ for $k = 1, \ldots, m$ in terms of t_i, a_i, $i = 1, \ldots, M-1$, from differences of derivatives at $t = 2\pi$ and $t = 0$:

$$\frac{1}{k!} \frac{d^k}{dt^k} p_m(2\pi) = \frac{1}{k!} \frac{d^k}{dt^k} p_m(0) + (-1)^k \binom{m}{k} \cdot \sum_{i=1}^{M-1} a_i t_i^{m-k}, \quad k = 0, \ldots, m-1.$$

The constant coefficient $p_{m,0}$ of $p_m(t)$ is not determined from the periodicity condition, and is determined by equating the constant Fourier coefficients of $sp(t)$ and $f_{M-1}(t)$ (i.e., c_0).

After m-th order differentiation we can reduce the problem of fitting m-th degree spline to the problem of fitting piecewise-constant function (spline of degree 0) with M jump points to the Fourier expansion of $f_{M-1}^{(m)}(t)$. In the notations above, $(t_i - t)_+^0$ is a unit step-function with a jump at $t = t_i$.

Using integration in parts we get for any 2π-periodic function $h(t)$,

$$\int_0^{2\pi} h'(t)e^{int}\, dt = -(in) \int_0^{2\pi} h(t)e^{int}\, dt.$$

This means that in order to determine m-th degree spline $sp(t)$ of the form above (or more specifically, its free parameters: $t_i, i = 1, \ldots, M-1$, coefficients a_i at $(t_i - t)_+^m$ for $i = 1, \ldots, M-1$ and a constant term $p_{m,0}$), we need to solve the problem of matching the $M - 1$ Fourier coefficients c'_k, $k = 1, \ldots, M-1$ of $f_{M-1}^{(m)}(t)$:

$$f_{M-1}^{(m)}(t) = \sum_{k=-(M-1)}^{M-1} (ik)^m c_k e^{ikt},$$

$$c'_k = (ik)^m \cdot c_k$$

with $M - 1$ Fourier coefficients of the piecewise-constant function $s(t)$ on $[0, 2\pi]$ with M jumps at t_i. Assuming, as above, that $t_0 = 0$ and $t_M = 2\pi$, we define $s(t)$ on $[0, 2\pi)$ as follows:

$$s(t) = \beta_i \text{ for } t \in [t_i, t_{i+1}), \ i = 0, \ldots, M-1,$$

and extend $s(t)$ everywhere on the real axis by 2π-periodicity. The relationship between β_i in the definition of $s(t)$ and coefficients a_i $(i = 1, \ldots, M - 1)$ and the leading coefficient $p_{m,m}$ (of $p_m(t)$) in the definition of $sp(t)$ is the following one:

$$\beta_i = m!(p_{m,m} + (-1)^m \sum_{k=i+1}^{M-1} a_k) \text{ for } i = 0, \ldots, M-1.$$

We also define (by periodicity) $\beta_M = \beta_0$.

It is straightforward to compute the Fourier coefficients of the (periodic) piecewise-constant function $s(t)$ on $[0, 2\pi]$. For this let us denote $z_j = e^{-it_j}$, $j = 0, \ldots, M$, so that $z_0 = z_M = 1$ as assumed above. Then

$$\frac{1}{2\pi}\int_0^{2\pi} s(t)e^{-int}\, dt = \frac{1}{2\pi}\frac{1}{in}\sum_{i=0}^{M-1}\beta_i(z_i^n - z_{i+1}^n),$$

for $n \neq 0$. Changing the summation order, remembering that $z_0 = z_M$, $\beta_0 = \beta_M$, and adding a new notation $\beta_{-1} = \beta_{M-1}$, we get:

$$\frac{1}{2\pi}\int_0^{2\pi} s(t)e^{-int}\, dt = \frac{1}{2\pi}\frac{1}{in}\sum_{i=0}^{M-1}\nu_i \cdot z_i^n, \ n \neq 0$$

$$\nu_i = \beta_i - \beta_{i-1}, \ i = 0, \ldots, M-1.$$

In terms of a_i, the definition of ν_i is straightforward:

$$\nu_i = (-1)^{m+1}\, m!\, a_i \text{ for } i = 1, \ldots, M-1,$$

$$\nu_0 = (-1)^m\, m! \sum_{k=1}^{M-1} a_k.$$

This leads to the complete system of equations on t_i, a_i, $i = 1, \ldots, M-1$ in terms of the Fourier coefficients c_k of $f(t)$:

$$2\pi \cdot (in)^{m+1} \cdot c_n = \sum_{j=0}^{M-1} \nu_j \cdot z_j^n, \quad n = 1, \ldots, M-1,$$

$$\sum_{j=0}^{M-1} \nu_j = 0$$

$$2\pi \cdot c_0 = \frac{1}{m+1} \sum_{j=1}^{M-1} a_j t_j^{m+1} + \sum_{k=0}^{m} \frac{1}{k+1} p_{m,k} (2\pi)^{k+1}$$

These equations are supplemented by:

$$\nu_i = (-1)^{m+1}\, m!\, a_i \text{ for } i = 1, \ldots, M-1,$$

$$\frac{1}{k!}\left(\frac{d^k}{dt^k} p_m(2\pi) - \frac{d^k}{dt^k} p_m(0)\right) = (-1)^k \binom{m}{k} \cdot \sum_{i=1}^{M-1} a_i t_i^{m-k}, \quad k = 0, \ldots, m-1.$$

3.3 Solution of Periodic Spline Problem.

The system of equations above provides the solution to the problem of spline approximation using matching of trigonometric moments. As the equations show, the problem is entirely reduced to the solution of the following main trigonometric problem:

$$C_n = \sum_{j=0}^{M-1} \nu_j \cdot z_j^n, \quad n = 1, \ldots, M-1,$$

$$\sum_{j=0}^{M-1} \nu_j = 0$$

Here C_n are properly modified Fourier coefficients:

$$C_n = 2\pi \cdot (in)^{m+1} \cdot c_n \ n \neq 0,$$

and z_j are lying on the unit circle: $z_j = e^{-it_j}$, $j = 0, \ldots, M-1$, so that $z_0 = 1$.

There are two slightly different solutions to this problem. One is based on the standard approach using Padé approximation; this time it is a two- point Padé approximation at $z = 0$ and $z = \infty$. While straightforward, it leaves open the usual questions of normality and rigorous proof of the location of zeroes of Padé approximants on the unit circle. An alternative approach is based on the Carathéodory theorem, and provides the necessary simple criteria of normality and the proof of zeroes' location on the unit circle. We present both solutions that compliment each other.

To arrive at the Padé approximation representation, we introduce the polynomial $Q(z)$ of degree M:

$$Q(z) = \prod_{j=0}^{M-1} (z - z_j),$$

and a rational function $R(z) = \sum_{j=0}^{M-1} \frac{\nu_j}{z - z_j}$, which can be written as a ratio of two polynomials:

$$R(z) = \frac{P(z)}{Q(z)}.$$

Because of the equations on ν_j, $\deg_z(P) \le M - 2$ (or $R(z) \to O(z^{-2})$ as $z \to \infty$). Expanding $R(z)$ at $z = 0$ and $z = \infty$, and using the fact that z_j lie on the unit circle ($z_j^* = z_j^{-1}$) and ν_j are real, we can see that $R(z)$ approximates two series expansions associated with Fourier coefficients $C_{\pm j}$, $j = 1, \ldots, M - 1$. The two expansions are:

$$f^+(z) = \sum_{k=1}^{M-1} C_k z^{-k-1}, \quad f^-(z) = -\sum_{k=1}^{M-1} C_{-k} z^{k-1},$$

at $z = \infty$ and $z = 0$, respectively. Everywhere above and below we set $C_0 = 0$ and $C_{-k} = C_k^*$ for positive k.

The expansions of $R(z)$ at $z = 0$ and $z = \infty$ are

$$R(z) = \sum_{k=0}^{\infty} z^{-k-1} \cdot \sum_{j=0}^{M-1} \nu_j z_j^m \quad \text{at } z \to \infty,$$

$$R(z) = -\sum_{k=0}^{\infty} z^k \cdot \sum_{j=0}^{M-1} \frac{\nu_j}{z_j^{m+1}} \quad \text{at } z \to 0.$$

Thus the approximations properties of $R(z)$, equivalent to the original equations, are the following:

$$R(z) = f^+(z) + O(z^{-M-1}) \quad \text{at } z \to \infty,$$

$$R(z) = f^-(z) + O(z^{M-1}) \quad \text{at } z \to 0.$$

These approximation properties can be replaced by

$$P(z) = Q(z) \cdot f^+(z) + O(z^{-1}) \text{ at } z \to \infty,$$

$$P(z) = Q(z) \cdot f^-(z) + O(z^{M-1}) \text{ at } z \to 0.$$

The two approximations above can be written as a system of linear equations on coefficients of the polynomial $Q(z)$. If $Q(z)$ is expanded in powers of z as:

$$Q(z) = \sum_{j=0}^{M} Q_j \cdot z^j,$$

then we get the system of linear equations on Q_j, $j = 0, \ldots, M$:

$$\sum_{j=0}^{M} Q_j \cdot C_{j-k-1} = 0, \ k = 0, \ldots, M - 2.$$

In the equations above we set $C_0 = 0$ and $C_{-k} = C_k^*$ for positive k.

The system of equations on Q_j above ($M - 1$ of them) is supplemented by the following one:

$$\sum_{j=0}^{M} Q_j = 0,$$

equivalent to the statement that $Q(1) = 0$ or $z_0 = 1$. In the case of normality of two-point Padé approximation, the system of M equations on Q_j uniquely determines the polynomial $Q(z)$ up to the (normalization of) the leading coefficient Q_M; so one can add $M + 1$-st equation

$$Q_M = 1.$$

In the generic case the two-point Padé approximations are normal and polynomials $Q(z)$ have all roots distinct and lying on the unit circle. However, this approach based on Padé approximations alone (or solution of the system of Toeplitz equations) does not always guarantee the solution of the problem of finding the periodic spline with proper matching trigonometric moments. Here the situation is similar to that of splines matching moments on the interval, where sometimes the orthogonal polynomials have zeroes outside of the interval that supports the measure. The problem of matching the trigonometric moments is different - it **always** has solution. Moreover, this solution (as a function of coefficients c_k) is stable.

The explicit solution to the problem of finding the periodic spline with proper matching trigonometric moments can be derived from the Carathéodory theorem (or, more precisely, from the method used in the proof of the Carathéodory theorem). In this approach we start with the system of equations, as in the proof of Carathéodory theorem:

$$C_n^1 = \sum_{j=1}^{M-1} \nu_j \cdot z_j^n, \ n = 0, \ldots, M - 1,$$

Here C_n^1, $n = 1, \ldots, M - 1$ are variables, while C_0^1 is dependent on C_n^1, $n = 1, \ldots, M - 1$ and is defined from the eigenvalue equation:

$$\left| \mathbf{T}^1 + C_0^1 \cdot \mathbf{I}_M \right| = 0,$$

for

$$(\mathbf{T}^1)_{i,j} = C_{j-i}^1, \quad i, j = 0, \ldots, M - 1, \quad i \neq j,$$

$$(\mathbf{T}^1)_{i,i} = 0.$$

(More precisely, C_0^1 is the largest real number satisfying this eigenvalue equation.)

We add $z_0 = 1$ (for $t_0 = 0$) to the list of roots $\{z_k\}_{k=1}^{M-1}$ on the unit circle with a variable ν_0 as its weight. Thus we set $C_n^1 + n u_0 = C_n$ for all $n = 1, \ldots, M - 1$. In these notations the system of nonlinear equations

$$C_n = \sum_{j=0}^{M-1} \nu_j \cdot z_j^n, \quad n = 1, \ldots, M - 1,$$

$$\sum_{j=0}^{M-1} \nu_j = 0$$

can be represented, according to the (proof of) Carathéodory theorem as a determinantal identity:

$$\begin{vmatrix} -\nu_0 & C_1 - \nu_0 \ldots C_{M-1} - \nu_0 \\ C_{-1} - \nu_0 & -\nu_0 \ldots C_{M-2} - \nu_0 \\ \vdots & \vdots \ddots \quad \vdots \\ C_{-M+1} - \nu_0 \ C_{-M+2} - \nu_0 \ldots & -\nu_0 \end{vmatrix} = 0.$$

This equation can be written in the form

$$\left| \mathbf{T} - \nu_0 \cdot \mathbf{U}_M \right| = 0,$$

where, as above, $(\mathbf{T})_{i,j} = C_{j-i}$, $i, j = 0, \ldots, M-1$, for $i \neq j$ and $(\mathbf{T})_{i,i} = 0$, and $\mathbf{U}_M$ is "all ones" matrix: $(\mathbf{U}_M)_{i,j} = 1$, $i, j = 0, \ldots, M-1$. Consequently, we get a simple linear equation, defining ν_0:

$$\left| \mathbf{T} \right| - \nu_0 \cdot \left| \mathbf{T}' \right| = 0,$$

where $\mathbf{T}'$ is an algebraic compliment of (elements of) $\mathbf{T}$.

Once ν_0 is determined from this equation, all other ν_j and $z_j = e^{i\theta_j}$ for $j = 1, \ldots, M - 1$ are determined from the Carathéodory theorem (as above) with $C_j^1 = C_j - \nu_0$ for $j = 1, \ldots, M - 1$.

This method is stable and also has the advantage that it determines the **minimal** number M of knots t_i with $t_0 = 0$ of the periodic spline $sp(t)$ of degree m that matches N Fourier coefficients c_i of $f(t)$ (for $M \leq N$).

There are, however, some cases when $|\mathbf{T}'| = 0$ and in those cases one cannot get any periodic spline $sp(t)$ of degree m that matches M Fourier coefficients c_i of $f(t)$ and has M knots on $[0, 2\pi]$ with $t_0 = 0$. It turns out that there is still a stable solution in this case as well; it just requires M knots to be in a general position in $[0, 2\pi]$.

Here we will use a second definition of a periodic spline $sp_2(t)$ on $[0, 2\pi]$ that does not fix a knot at $t_0 = 0$. In the second, a more complex, definition, the periodic spline $sp_2(t)$ of degree m on $[0, 2\pi]$ has M knots:

$$0 \leq t_0 < \ldots < t_{M-1} \leq 2\pi,$$

and we can add by periodicity $t_M = t_0$. The spline has the form:

$$sp_2(t) = p_m(t) + \sum_{i=0}^{M-1} a_i(t_i - t)_+^m,$$

satisfying very strong periodicity conditions (of C^m-smoothness):

$$sp_2^{(k)}(2\pi) = sp_2^{(i)}(0), \quad k = 0, \ldots, m.$$

This leaves total of $2M$ real variables in $sp_2(t)$: M for variable knots t_i, $i = 0, \ldots, M-1$, $M-1$ for coefficients: a_i at $(t - t_i)^m$, $i = 0, \ldots, M-1$ and for a constant term of $p_m(t)$. The single relation between a_i is the following one:

$$a_0 + \ldots + a_{M-1} = 0.$$

This allows us to match $2M - 1$ real Fourier coefficients of $sp(t)$ with Fourier coefficients of $f_{M-1}(t)$ (or $f(t)$), or c_j for $j = 0, \ldots, M-1$, and still leaves one extra degree of freedom.

This extra degree of freedom is very easy to observe. Since $f(t)$ is 2π - periodic, so is $f(t - t_0)$ for any (real) t_0, and Fourier coefficients of $f(t - t_0)$ are related to Fourier coefficients of $f(t)$ by multiplicative factors of e^{int_0}. Then we apply essentially the "simple periodic splines" methods, described above, to $f(t - t_0)$ but now with M knots $t_i \leftarrow t_i - t_0$, $i = 0, \ldots, M-1$ (so one of the knots is a fixed one at $t = 0$). As long as the corresponding $\mathbf{T}' = \mathbf{T}'(t_0)$ is non-singular (and that is the generic case), one has an extra parameter t_0 to fit. It can be used, for example, to get a better L_2-approximation by a periodic spline $sp_2(t)$ of degree m with M knots of $f(t)$ that matches M Fourier coefficients of $f(t)$. It also can be used to match one extra real or imaginary part of the Fourier coefficient c_M.

There are a variety of algorithms that can be used for the computations of the periodic spline approximations $sp(t)$ and $sp_2(t)$. For a fixed degree m the basic algorithm is reduced to operations with Toeplitz matrices (or the construction of two-point Padé approximations). The standard (slow) method of computation are of complexity $O(M^2)$. These methods are either based on direct solution of Toeplitz equations or the use of the recurrence relations connecting Padé approximants $Q(z)$ for increasing M. The recurrence relations

follow from the classical three-term recurrence relations connecting polynomials orthogonal on the unit circle, see [S]. Those polynomials have the form:

$$\rho_n(z) = \begin{vmatrix} C_0 & C_{-1} & \dots & C_{-n} \\ C_1 & C_0 & \dots & C_{-n+1} \\ \vdots & \vdots & \ddots & \vdots \\ C_{n-1} & C_{n-2} & \dots & C_{-1} \\ 1 & z & \dots & z^n \end{vmatrix}.$$

While all zeroes of $\rho_n(z)$ lie within the unit circle $|z| < 1$, one constructs the
"para-orthogonal" polynomials [J89] that have their zeroes on the unit circle
$|z| = 1$:

$$\rho_n(z) + w \cdot \hat{\rho}_n(z) \text{ for } |w| = 1,$$

where $\hat{\rho}_n(z) = z^n \cdot (\rho_n)^*(1/z)$. Two-point Padé approximants $Q(z)$ and polynomials $\mathbf{Q_e}(z)$ from the proof of Carathéodory theorem are examples of para-
orthogonal polynomials, see [J89].

The recurrence relations arising from polynomials orthogonal on the unit
circle provide a recursive approach to the construction of optimal periodic
splines $sp(t)$ and $sp_2(t)$. In this approach one can increase the number M
of knots in a way such that previous knots's locations interleave the new
ones. The recursive approach based on Padé approximants also provides a
fast method of computation of splines $sp(t)$ for a fixed $f(t)$ and m but increasing M. This recursive approach based on a conventional "divide and
conquer" polynomial interpolation with fast polynomial multiplication and
division results in $O\left(M \log^2(M)\right)$ complexity.

3.4 A Very Simple Example.

We present a very simple example of a single harmonics case:

$$f(t) = \cos(t),$$

or $c_k = 0$ unless $k = \pm 1$, where $c_{\pm 1} = \frac{1}{2}$. In this case one can very easily
determine analytically the knots t_j and weights a_j, and the quality of the
spline approximation $sp(t)$, as described in the previous section, to $f(t)$ in
various norms. There are 3 cases of different parity of m and M.

The first, typical, case is of even m and odd M. We put $M = 2n - 1$, and
the set of M knots t_j, $j = 0, \dots, M - 1$ which is:

$$\{0\} \cup \{\frac{j\pi}{n} : j = 1, \dots, n - 1\} \cup \{\frac{j\pi}{n} : j = n + 1, \dots, 2n - 1\}$$

(with $t_0 = 0$). The corresponding ν_j are:

$$\nu_j = (-1)^{m/2+1} \cdot \frac{\pi}{n} \cdot \sin(t_j) \text{ for } j = 1, \dots, M$$

(m is even).

The second case is that of odd m and even M. We put $M = 2n - 2$ for even n. (The same solution also covers the case $M = 2n - 1$ for even n.) Here the set of M knots t_j, $j = 0, \ldots, M - 1$:

$$\{0, \pi\} \cup \{\frac{j\pi}{n} : j = 1, \ldots, 2n - 1 \text{ and } j \text{ not divisible by } \frac{n}{2}\},$$

($\frac{n}{2}$ is an integer). The corresponding ν_j are:

$$\nu_j = (-1)^{(m+1)/2} \cdot \frac{\pi}{n} \cdot \cos(t_j) \text{ for } j = 0, \ldots, M$$

(m is odd).

The third (and last) case is that of odd m and odd M (mod 4) $= 1$. We put $M = 2n - 1$ for odd n. Here the set of M knots t_j, $j = 0, \ldots, M - 1$:

$$\{0, \pi\} \cup \{\frac{j\pi}{2n} : j = 1, \ldots, 2n - 1 \text{ for odd } j \text{ and } j \neq n, 3n\},$$

The corresponding ν_j are:

$$\nu_j = (-1)^{(m+1)/2} \cdot \frac{\pi}{n} \cdot \cos(t_j) \text{ for } j = 1, \ldots, M$$

(m is odd) and $\nu_0 = 0$ corresponding to $t_0 = 0$.

In all these cases, t_j and ν_j uniquely determine the spline approximation $sp(t)$ (from formulas presented above for a_j, $j = 1, \ldots, M - 1$ and $p_m(t)$).

The standard measure of the average quality approximation of $sp(t)$ to $f(t)$ for an arbitrary $f(t)$ is based on L_2-norm:

$$\epsilon = \left(\frac{1}{2\pi} \int_0^{2\pi} ((f(t) - sp(t))^2 \, dt \right)^{\frac{1}{2}}.$$

This quantity can be determined analytically using the Parceval identity and the Fourier expansion of $f(t) - sp(t)$ that is easily computable in terms of trigonometric sums:

$$\sum_{j=0}^{M-1} \nu_j e^{-ikt_j}, \ k = M, \ldots$$

and Fourier coefficients of $f(t)$. In turn, trigonometric sums are computed using $P(z), Q(z)$ polynomials from two-point Padé approximation to $f^{(\pm)}(z)$.

We define

$$G(m, n) = \frac{\psi^{(2m+1)}(1 - \frac{1}{2n}) + \psi^{(2m+1)}(1 + \frac{1}{2n})}{(2m+1)! \cdot (2n)^{2m+2}},$$

for a polygamma function $\psi(z)$. This can be simplified to:

82

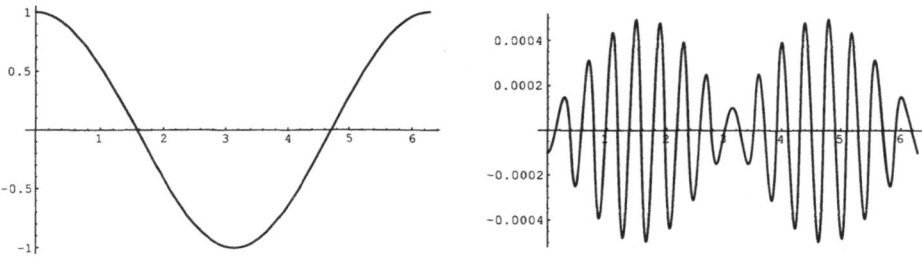

Fig. 1. $m = 2$, $n = 8$, $M = 15$. Plots of $\cos(t)$, $sp(t)$ and $\cos(t) - sp(t)$.

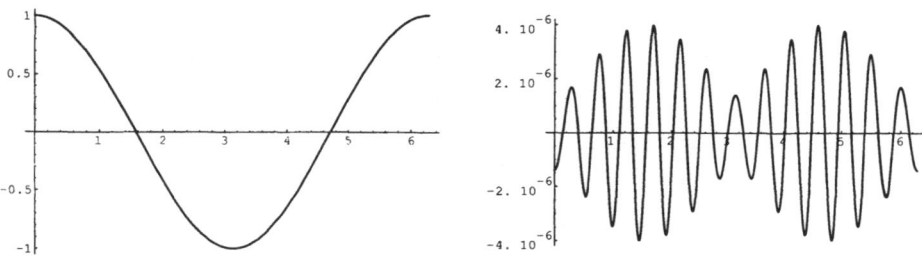

Fig. 2. $m = 4$, $n = 7$, $M = 13$. Plots of $\cos(t)$, $sp(t)$ and $\cos(t) - sp(t)$.

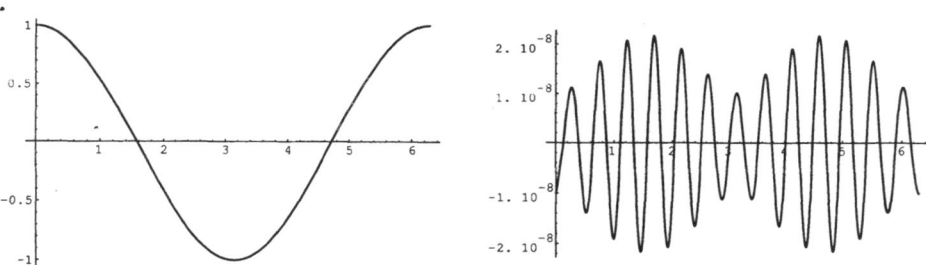

Fig. 3. $m = 6$, $n = 7$, $M = 13$. Plots of $\cos(t)$, $sp(t)$ and $\cos(t) - sp(t)$.

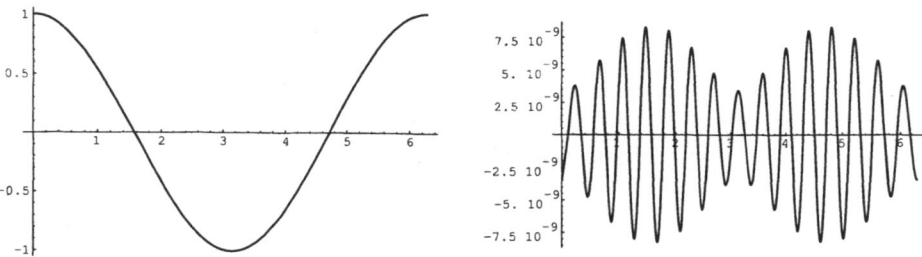

Fig. 4. $m = 6$, $n = 8$, $M = 15$. Plots of $\cos(t)$, $sp(t)$ and $\cos(t) - sp(t)$.

$$G(m,n) = -1 - \left[\pi \frac{d^{2m+1}}{dz^{2m+1}}(\cot(\pi z))\right]_{z=\frac{1}{2n}} \cdot \frac{1}{(2m+1)!(2n)^{2m+2}}.$$

When $f(t) = \cos(t)$ in all three cases above we get the following expression for ϵ as a function of m and n:

$$\epsilon = \sqrt{\frac{1}{2}G(m,n)}.$$

As above, in the case 1 we have m even and $M = 2n - 1$; in the case 2 we have m odd and $M = 2n - 2$ for even n; in the case 3 we have m odd and $M = 2n - 1$ for odd n.

Looking at $\log_2 \epsilon$, we can see that $\log_2 \epsilon \leq -24$ for $m = 3, n \geq 33; m = 4, n \geq 15; m = 5, n \geq 9; m = 6, n \geq 6; m = 7, n \geq 5; m \geq 8, n \geq 4; m \geq 10, n \geq 3; m \geq 14; n \geq 2$.

We include several plots of $f(t) = \cos(t)$ and of the corresponding spline approximation $sp(t)$ in Figs. $1-4$. Fig. 5 shows the plot of the figure of merit $\log_2 \epsilon$ as a function of m, n.

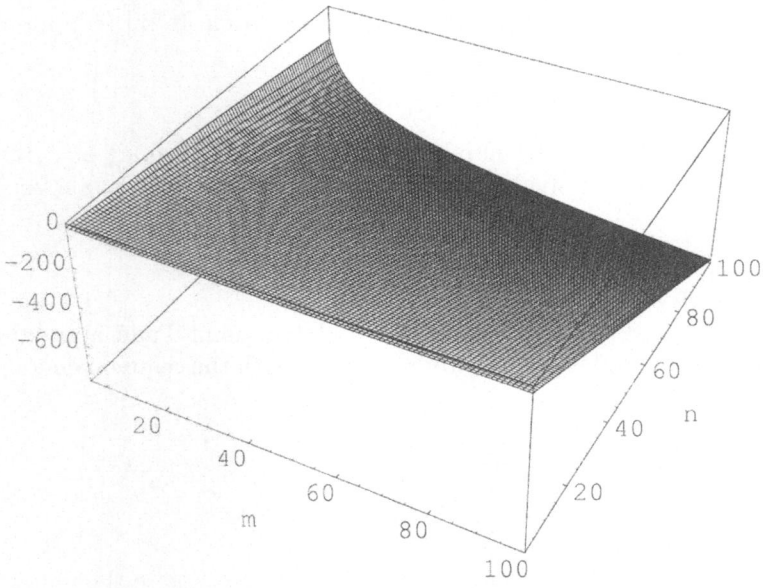

Fig. 5. $\log_2 \epsilon(m, n)$ as a function of m, n.

3.5 Multidimensional Closed Curves and Periodic Splines.

The case of d-dimensional closed curves $\mathbf{f}(t) = (f_1(t), \ldots, f_d(t))$ approximated by a periodic spline curve $\mathbf{sp}(t) = (sp_1(t), \ldots, sp_d(t))$ is similar to the case of d-dimensional curves approximated by d-dimensional splines on the interval that matches the maximal number of moments. However, here we are using the two-point generalized Padé approximants (at $z = 0$ and $z = \infty$) of the Second Kind. This problem also leads to an interesting generalization of the Carathéodory theorem where one expresses d sequences $\{c_n^\alpha\}$ of Fourier coefficients by trigonometric exponentials

$$\sum_{j=1}^{M} \nu_j^\alpha \cdot e^{in\theta_j} \text{ for } \alpha = 1, \ldots, d$$

with M harmonics $e^{i\theta_j}$, $j = 1, \ldots, M$ common to all d sequences (but different weights ν_j^α). Just as Carathéodory theorem is used for a spectral estimation of a time series, the multidimensional problem is used for a joint spectral estimation of multiple time series (e.g., multi-channel data collected from the same source at the same time).

In this problem, as above, we look at the d-dimensional curves $\mathbf{f}(t) = (f_1(t), \ldots, f_d(t))$ that is 2π-periodic and is defined by its Fourier coefficients c_n^α with the usual reality condition $c_{-k}^\alpha = \overline{c_k^\alpha}$ for $\alpha = 1, \ldots, d$. We are trying to find a d-dimensional spline of degree m: $\mathbf{sp}(t) = (sp_1(t), \ldots, sp_d(t))$, defined on $[0, 2\pi]$ and 2π-periodic that has M knots on $[0, 2\pi]$ (common to all individual spline functions $sp_\alpha(t)$):

$$0 \leq t_0 < \ldots < t_{M-1} \leq 2\pi,$$

and matching the maximum number of Fourier coefficients c_n^α of $f_\alpha(t)$, $\alpha = 1, \ldots, d$. The form of individual spline functions is the same as above:

$$sp_\alpha(t) = p_{\alpha,m}(t) + \sum_{i=0}^{M-1} a_{\alpha,i}(t_i - t)_+^m.$$

The problem is reduced to the generalized two-point Padé approximation problem of the Second Kind (i.e., approximants with the common denominator $Q(z)$) for functions

$$f_\alpha^+(z) = \sum_{k=1}^{M-1} C_k^\alpha z^{-k-1}, \quad f_\alpha^-(z) = -\sum_{k=1}^{M-1} C_{-k}^\alpha z^{k-1},$$

at $z = \infty$ and $z = 0$, respectively.

The solution is reduced to linear algebra problems and to the following set of Toeplitz-like equations on the Fourier coefficients C_k^α:

$$\sum_{i=0}^{M} q_i \cdot C_{i+l}^\alpha = 0, \, l = 0, \ldots, N^\alpha$$

for $\alpha = 1, \ldots, d$ and $\sum_{\alpha=0}^{d}(N^\alpha + 1) \geq M$. The knots t_j are determined from roots $z = z_j$ of the associated polynomial $Q(z) = \sum_{i=0}^{M} q_i \cdot z^i$ that lie on the unit circle by identification $z_j = e^{-it_j}$, $j = 1, \ldots, M$.

References

[Ch85] Chudnovsky, D.V., and Chudnovsky, G.V.: Applications of Padé approximations to diophantine inequalities in values of G-functions. Lecture Notes Math., Springer-Verlag, v. **1135**, 9-51 (1985)

[Ch99] Chudnovsky, D.V., and Chudnovsky, G.V.: Solution of the pulse width modulation problem using orthogonal polynomials and Korteweg-de Vries equations. Proc. Nat'l Acad. Sci. USA, **98**, no. 22, 12263-12268 (1999)

[C95] Cohen, F., Huang, Z., Yang, Z.: Invariant matching and identification of curves using B-splines curve representation. IEEE Trans. Image Processing. **4(1)**, (1995).

[F87] Frontini, M., Gautschi, W., Milovanovic, G.M.,: Moment preserving spline approximation on finite intervals. Numer. Math., **50**, 503-518 (1987)

[G84] Gautschi, W.,: Discrete approximation to spherically symmetric distributions. Numer. Math., **44**, 53-60 (1984)

[G86] Gautschi, W., Milovanovic, G.M.,: Spline approximations to spherically symmetric distributions. Numer. Math., **49**, 111-121 (1986)

[GS] Grenander, U., Szego, G.: Toeplitz Forms and their Applications. U. of California Press, Berkeley, (1958)

[J89] Jones, W.B., Njasstad, O., Thorn, W.J.,: Moment theory, orthogonal, polynomials, quadrature, and continued fractions associated with the unit circle. Bull. London Math. Soc., **21**, 113-152 (1989)

[KH] Kreylos, O., Hamann, B.,: On Simulated Annealing and the Construction of Linear Spline Approximations for Scattered Data. IEEE Trans. on Visualization and Computer Graphics, v. **7**, no. 1, 17 (2001)

[LM94] Lu, F., Milios, E.E.,: Optimal spline fitting to planar shape, Signal Processing. **37**, 129-140 (1994)

[M88] Micchelli, C.A. : Monosplines and moment preserving spline approximation. In: Brass, H., Hammerlin, G. (eds) Numerical Integration III, Birkhauser, Basel, 130-139 (1988)

[MK88] Milovanovic, G.V., Kovacevic, M.A.,: Moment-preserving spline approximation and Turan quadratures. In: Agarwal, R.P., Chow, Y.M., Wilson, S.J. (eds.), Numerical Mathematics Singapore, 1988, ISNM v. **86**, Birkhauser, Basel, 357-365 (1988)

[M00] Milovanovic, G.M.,: Quadratures with multiple nodes, power orthogonality, and moment-preserving spline approximation. In: Gautschi, W., Marcellan, F., Reichel, L. (eds.) Numerical Analysis 2000, Vol. V, Quadrature and orthogonal polynomials. J. Comput. Appl. Math. **127**, 267-286 (2001)

[M95] Milanfar, P., Verghese, G., Karl, W., Willsky, A.: Reconstruction of polygons from moments with connection to array processing. IEEE Trans. Signal Processing, **43(2)**, (1995)

[M97] Meier, F.W., Shuster, G.M., Katsaggelos, A.K.: An efficient boundary encoding scheme using B-spline curves which is optimal in the rate distortion sense. In: Proc. of VCIP, (1997)

[MM] Mokhtarian, F., Mackworth, A.K.,: A theory of multiscale, curvature-based shape representation for planar curves. IEEE Trans. PAMI, **14(8)**, 789-805

[MP] Marciniak, K. ,Putz, B.: Approximation of spirals by piecewise curves of fewest circular arc segments. Computer Aided Design, **16(2)**, 87-90 (1984)

86

[MW] Meek, D. S., Walton, D. J.: Approximating smooth planar curves by arc splines. Journal of Computational and Applied Mathematics, **59**, 221-231, (1995)

[N] Nyquist, H.: Certain topics in telegraph transmission theory. AIEE Trans., 617-644 (1928)

[S] Szego, G.: Orthogonal Polynomials. AMS, Providence, RI, (1978)

[W] Wallner, J.: Generalized multiresolution analysis for arc splines. In: Dahlen, M., Lyche, T., Schumaker, L. L. (eds.) Mathematical Methods for Curves and Surfaces II, Vanderbilt University Press, 537-544

[Z] Zaletelj, J., Pecci, R., Spaan, F., Hanjalic, A., Lagendijk, R.I.,: Rate Distortion Optimal Contour Compression Using Cubic B-Splines, European Signal Processing Conference (EUSIPCO '98), Rhodos, GR, (1998)

Interactions between number theory and operator algebras in the study of the Riemann zeta function (d'après Bost–Connes and Connes)

PAULA B. COHEN

August 8, 2000

Abstract

This paper arose as a written version of a lecture given at the City University of New York Number Theory Seminar in April 1997. The lecture was an overview of ideas of Bost-Connes and of Connes on possible ways to approach the study of the Riemann zeta function using ideas inspired by noncommutative geometry. The work of Connes is directly aimed at a solution of the Generalised Riemann Hypothesis and has undergone substantial improvement in the intervening time. In this article we give an overview of the original work and these later developments.

1 Introduction

In 1859 Riemann published an important foundational paper on the Riemann zeta function. Recall that this function is given for $\mathrm{Re}(s) > 1$ by

$$\zeta(s) = \sum_{n=1}^{\infty} \frac{1}{n^s}$$

and that it has a continuation to all the complex plane which is analytic except for a simple pole at $s = 1$. It is straightforward to show that the Riemann zeta function has zeros at the negative even integers and these are called the trivial zeros of the Riemann zeta function. The Riemann Hypothesis predicts that the remaining zeros lie on the line $\mathrm{Re}(s) = 1/2$. One knows that the non-trivial zeros of $\zeta(s)$ lie in the band $\mathrm{Re}(s) \in]0, 1[$. The generalisation of this function for a number field is known as the Dedekind zeta function. It encodes much important arithmetical information about the field. One of the major motivations of number theory is to understand more fully the Dedekind zeta function, the most famous challenge being to understand the locus of the zeros of this function and in particular to settle the validity of the Generalised Riemann Hypothesis for these functions. Much powerful work has been done in analytic number theory in the attempt to solve the Generalised Riemann Hypothesis directly.

As the study of the zeros of the Riemann zeta function and its generalisations is so difficult, one may ask how it is possible to recast the problem. For example, Polya and

Hilbert proposed that if one can construct a Hilbert space $\mathcal{H}$ and an operator D in $\mathcal{H}$ whose spectrum comprises the zeros of the Riemann zeta function in the band $\mathrm{Re}(s) \in]0,1[$, then possibly one can settle whether or not $\sqrt{-1}(D - 1/2)$ is self-adjoint or whether $D(1 - D)$ is positive, which would imply the Riemann hypothesis. The point here is that the properties of self-adjointness or of positivity are hopefully easier to check. It is important that the construction of the Hilbert space and the operator should not depend *a priori* on the zeta function, to avoid tautologies. An as yet non-rigorous approach to the Riemann hypothesis initiated by Connes in [5] and further developed in [6],[7], includes an interesting and rigorous interpretation of the zeros of the L-functions with Grossencharacter of a global field in terms of the action of the idele class group on the coset space of the adeles modulo the principal ideles.

This work of Connes derives in particular from his work with Bost [2] which was in turn inspired by ideas of Julia [11] and others. The aim is to enrich our knowledge of the Riemann zeta function by creating a dictionary between its properties and phenomena in statistical mechanics. The starting point of these approaches is the observation that, just as the zeta functions encode arithmetic information, the partition functions of quantum statistical mechanical systems encode their large-scale thermodynamical properties. The first step therefore is to construct a quantum dynamical system with partition function the Riemann zeta function, or the Dedekind zeta function in the general number field case. In order for the quantum dynamical system to reflect the arithmetic of the primes, it must capture also some sort of interaction between them. This last feature translates in the statistical mechanical language into the phenomenon of spontaneous symmetry breaking at a critical temperature with respect to a natural symmetry group. In the region of high temperature, there is a unique equilibrium state as the system is in disorder and symmetric with respect to the action of the symmetry group. In the region of low temperature, a phase transition occurs and the symmetry is broken. This symmetry group acts transitively on a family of possible extremal equilibrium states. The construction of a quantum dynamical system with partition function the Riemann zeta function $\zeta(\beta)$ and spontaneous symmetry breaking or phase transition at its pole $\beta = 1$ with respect to a natural symmetry group was achieved by Bost and Connes in [2]. A different construction of the basic algebra using crossed products was proposed by Laca and Raeburn and extended to the number field case by them with Arledge in [1]. An extension of the work of Bost and Connes to general global fields was done by Harari and Leichtnam in [10]. The generalisation proposed by Harari and Leichtnam in [10] fails to capture the Dedekind zeta function as partition function in the case of a number field with class number greater than 1. Their partition function in that case is the Dedekind zeta function with a finite number of non-canonically chosen Euler factors removed. This prompted the author's paper [3] where the full Dedekind zeta function is recovered as partition function. This is achieved by recasting the original construction of Bost and Connes more completely in terms of adeles and ideles. The symmetry group of the system constructed by Bost and Connes is a Galois group, in fact the Galois group over the rational number field of its maximal abelian extension. Using the Artin isomorphism, which says that this symmetry group is also the unit group of the finite ideles, Bost and Connes recover the actual Galois action on the elements of the maximal abelian extension via its action on the equilibrium states of the system. In the general number field case, the symmetry group is again the unit group of the finite ideles, but this group does not in general have a Galois interpretation. See [10] for a discussion of this point.

In [7], Connes outlines a tantilising analogy between the Galois correspondence in number theory and the classification of factors in the theory of von-Neumann algebras. This is in the spirit of the idea of André Weil that the key to understanding the Riemann hypothesis may well lie in a decent Galois interpretation of the idèle class group of a number field. The difficulty here lies in the intervention of the archimedean valuations and lies beyond, for example, what is known about the case of function fields of varieties over finite fields. Connes maintains that aspects of the theory of the classification of factors represents a type of Galois theory adapted to the real and complex numbers.

In this paper we provide an introduction both to the work of Bost–Connes and to the work of Connes on the Riemann hypothesis. Our purpose is to assist the interested reader to understand these works. We restrict our attention to the rational numbers and to the Riemann zeta function. It should be noted that a solution of the Riemann hypothesis may well necessarily involve studying the zeta function as part of a family of zeta functions in the spirit of [12].

Acknowledgements: The author thanks the Ellentuck Fund and the School of Mathematics of the Institute for Advanced Study, Princeton, for their support during the preparation of this paper.

2   The problem studied by Bost–Connes

Before stating the problem solved by Bost and Connes in [2] and its analogue for number fields, we recall a few basic notions from the C^*-algebraic formulation of quantum statistical mechanics. For the background, see [4]. Recall that a C^*-algebra B is an algebra over the complex numbers $\mathbb{C}$ with an adjoint $x \mapsto x^*$, $x \in B$, that is, an anti-linear map with $x^{**} = x$, $(xy)^* = y^*x^*$, $x, y \in B$, and a norm $\|.\|$ with respect to which B is complete and addition and multiplication are continuous operations. One requires in addition that $\|xx^*\| = \|x\|^2$ for all $x \in B$. All our C^*-algebras will be assumed unital. The most basic example of a noncommutative C^*-algebra is $B = M_N(\mathbb{C})$ for $N \geq 2$ an integer. The C^*-algebra plays the role of the "space" on which the system evolves, the evolution itself being described by a 1-parameter group of C^*-automorphisms $\sigma : \mathbb{R} \mapsto \operatorname{Aut}(B)$. The quantum dynamical system is therefore the pair (B, σ_t). It is customary to use the inverse temperature $\beta = 1/kT$ rather than the temperature T, where k is Boltzmann's constant. Then, one has the definition of Kubo-Martin-Schwinger (KMS) of an equilibrium state at inverse temperature β. Recall that a state φ on a C^*-algebra B is a positive linear functional on B satisfying $\varphi(1) = 1$. It is the generalisation of a probability distribution.

Definition 1 *Let (B, σ_t) be a dynamical system, and φ a state on B. Then φ is an equilibrium state at inverse temperature β, or KMS_β-state, if for each $x, y \in B$ there is a function $F_{x,y}(z)$, bounded and holomorphic in the band $0 < \operatorname{Im}(z) < \beta$ and continuous on its closure, such that for all $t \in \mathbb{R}$,*

$$F_{x,y}(t) = \varphi(x\sigma_t(y)), \qquad F_{x,y}(t + \sqrt{-1}\beta) = \varphi(\sigma_t(y)x). \tag{1}$$

In the case where $B = M_N(\mathbb{C})$, every 1-parameter group σ_t of automorphisms of B can be written in the form,

$$\sigma_t(x) = e^{itH} x e^{-itH}, \qquad x \in B, \qquad t \in \mathbb{R},$$

for a self-adjoint matrix $H = H^*$. Then for $H \geq 0$ and for all $\beta > 0$, there is a unique KMS_β equilibrium state for (B, σ_t) given by

$$\phi_\beta(x) = \operatorname{Trace}(xe^{-\beta H})/\operatorname{Trace}(e^{-\beta H}), \qquad x \in M_N(\mathbb{C}). \tag{2}$$

This has the familiar form of a Gibbs state and is easily seen to satisfy the KMS_β condition of 1. The KMS_β states can therefore be seen as generalisations of Gibbs states. The normalisation constant $\text{Trace}(e^{-\beta H})$ is known as the partition function of the system. A symmetry group G of the dynamical system (B, σ_t) is a subgroup of $\text{Aut}(B)$ commuting with σ:

$$g \circ \sigma_t = \sigma_t \circ g, \qquad g \in G, t \in \mathbb{R}.$$

Consider now a system (B, σ_t) with interaction. Then, guided by quantum statistical mechanics, we expect to see the following features. When the temperature is high, so that β is small, the system is in disorder, there is no interaction between its constituents and the state of the system does not see the action of the symmetry group G: the KMS_β-state is unique. As the temperature is lowered, the constituents of the system begin to interact. At a critical temperature β_0 a phase transition occurs and the symmetry is broken. The symmetry group G then permutes transitively a family of extremal KMS_β- states generating the possible states of the system after phase transition: the KMS_β-state is no longer unique. This phase transition phenomenon is known as spontaneous symmetry breaking at the critical inverse temperature β_0. The partition function should have a pole at β_0. For a fuller explanation, see [2]. The problem solved by Bost and Connes was the following.

Problem 1: *Construct a dynamical system (B, σ_t) with partition function the zeta function $\zeta(\beta)$ of Riemann, where $\beta > 0$ is the inverse temperature, having spontaneous symmetry breaking at the pole $\beta = 1$ of the zeta function with respect to a natural symmetry group.*

As mentioned in the introduction, the symmetry group is the unit group of the ideles, given by $W = \prod_p \mathbb{Z}_p^*$ where the product is over the primes p and $\mathbb{Z}_p^* = \{u_p \in \mathbb{Q}_p : |u_p|_p = 1\}$. We use here the normalisation $|p|_p = p^{-1}$. This is the same as the Galois group $\text{Gal}(\mathbb{Q}^{ab}/\mathbb{Q})$. Here $\mathbb{Q}^{ab}$ is the maximal abelian extension of the rational number field $\mathbb{Q}$, which in turn is isomorphic to its maximal cyclotomic extension, that is the extension obtained by adjoining to $\mathbb{Q}$ all the roots of unity. The interaction detected in the phase transition comes about from the interaction between the primes coming from considering at once all the embeddings of the non-zero rational numbers $\mathbb{Q}^*$ into the completions $\mathbb{Q}_p$ of $\mathbb{Q}$ with respect to the prime valuations $|.|_p$. The natural generalisation of this problem to the number field case was solved in [3] and is the following.

Problem 2: *Given a number field K, construct a dynamical system (B, σ_t) with partition function the Dedekind zeta function $\zeta_K(\beta)$, where $\beta > 0$ is the inverse temperature, having spontaneous symmetry breaking at the pole $\beta = 1$ of the Dedekind function with respect to a natural symmetry group.*

Recall that the Dedekind zeta function is given by

$$\zeta_K(s) = \sum_{C \subset \mathcal{O}} \frac{1}{N(C)^s}, \qquad \text{Re}(s) > 1. \tag{3}$$

Here $\mathcal{O}$ is the ring of integers of K and the summation is over the ideals C of K contained in $\mathcal{O}$. The symmetry group is the unit group of the finite ideles of K. The main new difficulty that arose in the number field case was to generalise to the case of class number greater than 1. One has to use the fact that the principal prime ideals already provide enough interaction to engender a phenomenon of spontaneous symmetry breaking.

For the natural generalisation to the function field case see [10]. For the sake of exposition, we restrict ourselves in the sequel to the case of the rational numbers, that is to a discussion of Problem 1.

3 Construction of the C^*-algebra

We give a different construction of the C^*-algebra of [2] to that found in their original paper. It is essentially equivalent to the construction of [1], except that we work with adeles and ideles. In the generalisation to the number field case, this makes quite a difference. Let $\mathcal{A}$ denote the finite adeles of $\mathbb{Q}$, that is the restricted product of $\mathbb{Q}_p$ with respect to $\mathbb{Z}_p$. Recall that this restricted product consists of the infinite vectors $(a_p)_p$, indexed by the primes p, such that $a_p \in \mathbb{Q}_p$ with $a_p \in \mathbb{Z}_p$ for almost all primes p. The (finite) adeles form a ring under componentwise addition and multiplication. The (finite) ideles $\mathcal{J}$ are the invertible elements of the adeles. They form a group under componentwise multiplication. Let $\mathbb{Z}_p^*$ be those elements of $u_p \in \mathbb{Z}_p$ with $|u_p|_p = 1$. Notice that an idele $(u_p)_p$ has $u_p \in \mathbb{Q}_p^*$ with $u_p \in \mathbb{Z}_p^*$ for almost all primes p. Let

$$\mathcal{R} = \prod_p \mathbb{Z}_p, \qquad I = \mathcal{J} \cap \mathcal{R}, \qquad W = \prod_p \mathbb{Z}_p^*.$$

Further, let I denoted the semigroup of integral ideals of $\mathbb{Z}$. It is the semigroup of $\mathbb{Z}$-modules of the form $m\mathbb{Z}$ where $m \in \mathbb{Z}$. Notice that I as above is also a semigroup. We have a natural short exact sequence,

$$1 \to W \to I \to \mathrm{I} \to 1. \tag{4}$$

The map $I \to \mathrm{I}$ in this short exact sequence is given as follows. To $(u_p)_p \in I$ associate the ideal $\prod_p p^{\mathrm{ord}_p(u_p)}$ where $\mathrm{ord}_p(u_p)$ is determined by the formula $|u_p|_p = p^{-\mathrm{ord}_p(u_p)}$. It is clear that this map is surjective with kernel W, that is that the above sequence is indeed short exact. By the Strong Approximation Theorem we have

$$\mathbb{Q}/\mathbb{Z} \simeq \mathcal{A}/\mathcal{R} \simeq \oplus_p \mathbb{Q}_p/\mathbb{Z}_p \tag{5}$$

and we have therefore a natural action of I on $\mathbb{Q}/\mathbb{Z}$ by multiplication in $\mathcal{A}/\mathcal{R}$ and transport of structure. We use here that $I\mathcal{R} \subset \mathcal{R}$. Mostly we shall work in $\mathcal{A}/\mathcal{R}$ rather than $\mathbb{Q}/\mathbb{Z}$. We have the following straightforward Lemma (see [3]).

Lemma 1 For $a = (a_p)_p \in I$ and $y \in \mathcal{A}/\mathcal{R}$, the equation

$$ax = y$$

has $n(a) =: \prod_p p^{\mathrm{ord}_p(a_p)}$ solutions in $x \in \mathcal{A}/\mathcal{R}$. Denote these solutions by $[x : ax = y]$.

In the above lemma it is important to bear in mind that we are computing modulo $\mathcal{R}$. Now, let $\mathbb{C}[\mathcal{A}/\mathcal{R}] =: \mathrm{span}\{\delta_x : x \in \mathcal{A}/\mathcal{R}\}$ be the group algebra of $\mathcal{A}/\mathcal{R}$ over $\mathbb{C}$, so that $\delta_x \delta_{x'} = \delta_{x+x'}$ for $x, x' \in \mathcal{A}/\mathcal{R}$. We have (see for comparison [1]),

Lemma 2 The formula

$$\alpha_a(\delta_y) = \frac{1}{n(a)} \sum_{[x:ax=y]} \delta_x$$

for $a \in I$ defines an action of I by endomorphisms of $C^*(\mathcal{A}/\mathcal{R})$.

The endomorphism α_a for $a \in I$ is a one-sided inverse of the map $\delta_x \mapsto \delta_{ax}$ for $x \in \mathcal{A}/\mathcal{R}$, so it is like a semigroup "division". The C^*-algebra can be thought of as the operator norm closure of $\mathbb{C}[\mathcal{A}/\mathcal{R}]$ in its natural left regular representation in $l^2(\mathcal{A}/\mathcal{R})$. We now appeal to the notion of semigroup crossed product developed by Laca and Raeburn and used in [1], applying it to our situation. A covariant representation of $(C^*(\mathcal{A}/\mathcal{R}), I, \alpha)$ is a pair (π, V) where

$$\pi : C^*(\mathcal{A}/\mathcal{R}) \to B(\mathcal{H})$$

is a unital representation and

$$V : I \to B(\mathcal{H})$$

is an isometric representation in the bounded operators in a Hilbert space $\mathcal{H}$. The pair (π, V) is required to satisfy,

$$\pi(\alpha_a(f)) = V_a \pi(f) V_a^*, \qquad a \in I, \quad f \in C^*(\mathcal{A}/\mathcal{R}).$$

Notice that the V_a are not in general unitary. Such a representation is given by (λ, L) on $l^2(\mathcal{A}/\mathcal{R})$ with orthonormal basis $\{e_x : x \in \mathcal{A}/\mathcal{R}\}$ where λ is the left regular representation of $C^*(\mathcal{A}/\mathcal{R})$ on $l^2(\mathcal{A}/\mathcal{R})$ and

$$L_a e_y = \frac{1}{\sqrt{n(a)}} \sum_{[x:ax=y]} e_x.$$

The universal covariant representation, through which all other covariant representations factor, is called the (semigroup) crossed product $C^*(\mathcal{A}/\mathcal{R}) \times_\alpha I$. This algebra is the universal C^*-algebra generated by the symbols $\{e(x) : x \in \mathcal{A}/\mathcal{R}\}$ and $\{\mu_a : a \in I\}$ subject to the relations

$$\mu_a^* \mu_a = 1, \quad \mu_a \mu_b = \mu_{ab}, \qquad a, b \in I, \tag{6}$$

$$e(0) = 1, \quad e(x)^* = e(-x), \quad e(x)e(y) = e(x+y), \qquad x, y \in \mathcal{A}/\mathcal{R}, \tag{7}$$

$$\frac{1}{n(a)} \sum_{[x:ax=y]} e(x) = \mu_a e(y) \mu_a^*, \qquad a \in I, y \in \mathcal{A}/\mathcal{R}. \tag{8}$$

The relations in (6) reflect a multiplicative structure, those in (7) an additive structure and those in (8) how these multiplicative and additive structures are related via the crossed product action. Julia [11] observed that by using only the multiplicative structure of the integers one cannot hope to capture an interaction between the different primes. When $u \in W$ then μ_u is a unitary, so that $\mu_u^* \mu_u = \mu_u \mu_u^* = 1$ and we have for all $x \in \mathcal{A}/\mathcal{R}$,

$$\mu_u e(x) \mu_u^* = e(u^{-1} x), \qquad \mu_u^* e(x) \mu_u = e(ux). \tag{9}$$

Therefore we have a natural action of W as inner automorphisms of $C^*(\mathcal{A}/\mathcal{R}) \times_\alpha I$ using (9).

To recover the C^*-algebra of [2] we must split the short exact sequence (4). The ideals in I are all of the form $m\mathbb{Z}$ for some $m \in \mathbb{Z}$. This generator m is determined up to sign. Consider the image of $|m|$ in I under the diagonal embedding $q \mapsto (q)_p$ of $\mathbb{Q}^*$ into I, where the p-th component of $(q)_p$ is the image of q in $\mathbb{Q}_p^*$ under the natural embedding of $\mathbb{Q}^*$ in $\mathbb{Q}_p^*$. The map

$$+ : m\mathbb{Z} \mapsto (|m|)_p \tag{10}$$

defines a splitting of (4). Let I_+ denote the image and define B to be the semigroup crossed product $C^*(\mathcal{A}/\mathcal{R}) \times_\alpha I_+$ with the restricted action α from I to I_+. By transport of structure using (5), this algebra is easily seen to be isomorphic to a semigroup crossed product of $C^*(\mathbb{Q}/\mathbb{Z})$ by N_+, where N_+ denotes the positive natural numbers. This is the algebra constructed in [2] (see also [1]). ¿From now on, we use the symbols $\{e(x) : x \in \mathbb{Q}/\mathbb{Z}\}$ and $\{\mu_a : a \in \mathrm{N}_+\}$. It is essential to split the short exact sequence in this way in order to obtain the symmetry breaking phenomenon. In particular, this replacement of I by I_+ now means that the group W acts by outer automorphisms. For $x \in B$, one has that $\mu_u^* x \mu_u$ is still in B (computing in the larger algebra $C^*(\mathcal{A}/\mathcal{R}) \times_\alpha I$), but now this defines an outer action of W. This coincides with the definition of W as the symmetry group as in [2].

4   The Theorem of Bost–Connes

Using the abstract description of the C^*-algebra B of §3, to define the time evolution σ of our dynamical system (B, σ) it suffices to define it on the symbols $\{e(x) : x \in \mathbb{Q}/\mathbb{Z}\}$ and $\{\mu_a : a \in \mathrm{N}_+\}$. For $t \in \mathbb{R}$, let σ_t be the automorphism of B defined by

$$\sigma_t(\mu_m) = m^{it}, \quad m \in \mathrm{N}_+, \qquad \sigma_t(e(x)) = e(x), \quad x \in \mathbb{Q}/\mathbb{Z}. \tag{11}$$

By (6) and (9) we clearly have that the action of W commutes with this 1-parameter group σ_t. Hence W will permute the extremal KMS$_\beta$-states of (B, σ_t). To describe the KMS$_\beta$-states for $\beta > 1$, we shall represent (B, σ_t) on a Hilbert space. Namely, following [2], let $\mathcal{H}$ be the Hilbert space $l^2(\mathrm{N}_+)$ with canonical orthonormal basis $\{\varepsilon_m, m \in \mathrm{N}_+\}$. For each $u \in W$, one has a representation π_u of B in $B(\mathcal{H})$ given by,

$$\pi_u(\mu_m)\varepsilon_n = \varepsilon_{mn}, \quad m, n \in \mathrm{N}_+$$

$$\pi_u(e(x))\varepsilon_n = \exp(2i\pi n u \circ x)\varepsilon_n, \quad n \in \mathrm{N}_+, x \in \mathbb{Q}/\mathbb{Z}. \tag{12}$$

Here $u \circ x$ for $u \in W$ and $x \in \mathbb{Q}/\mathbb{Z}$ is the multiplication induced by transport of structure using (5). One verifies easily that (12) does indeed give a C^*-algebra representation of B. Let H be the unbounded operator in $\mathcal{H}$ whose action on the canonical basis is given by

$$H\varepsilon_n = (\log n)\varepsilon_n, \quad n \in \mathrm{N}_+. \tag{13}$$

Then clearly, for each $u \in W$, we have

$$\pi_u(\sigma_t(x)) = e^{itH}\pi_u(x)e^{-itH}, \qquad t \in \mathbb{R}, x \in B.$$

Notice that, for $\beta > 1$,

$$\mathrm{Trace}(e^{-\beta H}) = \sum_{n=1}^{\infty}\langle e^{-\beta H}\varepsilon_n, \varepsilon_n\rangle = \sum_{n=1}^{\infty} n^{-\beta}\langle\varepsilon_n, \varepsilon_n\rangle = \sum_{n=1}^{\infty} n^{-\beta},$$

so that the Riemann zeta function appears as a partition function of Gibbs state type. We can now state the main result of [2].

Theorem 1 *(Bost-Connes) The dynamical system (B, σ_t) has symmetry group W. The action of $u \in W$ is given by $[u] \in \mathrm{Aut}(B)$ where*

$$[u] : e(y) \mapsto e(u \circ y), \quad y \in \mathbb{Q}/\mathbb{Z}, \qquad [u] : \mu_a \mapsto \mu_a, \quad a \in \mathrm{N}.$$

This action commutes with σ,

$$[u] \circ \sigma_t = \sigma_t \circ [u], \qquad u \in W, \quad t \in \mathbb{R}.$$

Moreover, (1) for $0 < \beta \leq 1$, *there is a unique* KMS_β *state. (It is a factor state of Type* III_1 *with associated factor the Araki-Woods factor* R_∞.) *(2) for* $\beta > 1$ *and* $u \in W$, *the state*

$$\phi_{\beta,u}(x) = \zeta(\beta)^{-1}\mathrm{Trace}(\pi_u(x)e^{-\beta H}), \qquad x \in B$$

is a KMS_β *state for* (B, σ_t). *(It is a factor state of Type* I_∞). *The action of* W *on* B *induces an action on these* KMS_β *states which permutes them transitively and the map* $u \mapsto \phi_{\beta,u}$ *is a homomorphism of the compact group* W *onto the space* $\mathcal{E}_\beta$ *of extremal points of the simplex of* KMS_β *states for* (B, σ_t). *(3) the* ζ *function of Riemann is the partition function of* (B, σ_t).

Part (1) of the above theorem is difficult and the reader is referred to [2] for complete details, as for a full proof of (2). That for $\beta > 1$ the KMS_β-states given in part (2) fulfil 1 of §2 is a straightforward exercise. Notice that they have the form of Gibbs equilibrium states.

5 The space and group action in Connes' approach to the Riemann hypothesis

Theorem 1 solves Problem 1 of §2. More information is contained in its proof however. As mentioned in the §1, given the existence of the Artin isomorphism in class field theory for the rationals, one can recover the Galois action of W explicitly. It is still an open problem to exhibit this Galois action in terms of an analogue of (B, σ_t) in a satisfactory way for general number fields. Indeed, in [3] the analogue of W is again an infinite product over unit groups in the integers of local fields, completions of the number field at the prime ideals. But this is no longer in general isomorphic to the Galois group of the maximal abelian extension of the number field, which is the symmetry group one wants to recover. It appears that one must generalise the approach of Bost–Connes so that the extremal KMS states, when restricted to the basic symbols generating the C^*-algebra, take values in the maximal abelian extension of the number field. The action of the symmetry group on these extremal states should then induce the action of the Galois group of the maximal abelian extension on these values.

Another feature occurs in the analysis of the proof of part (1) of Theorem 1. One can treat the infinite places in a similar way to that already described for the finite places, so working with the (full) adeles A and (full) ideles J. The ring of adeles A of $\mathbb{Q}$ consists of the infinite vectors $(a_\infty, a_p)_p$ indexed by the archimedean place and the primes p of $\mathbb{Q}$ with $a_p \in \mathbb{Z}_p$ for all but finitely many p. The group J of ideles consists of the infinite vectors $(u_\infty, u_p)_p$ with $u_\infty \in \mathbb{R}, u_\infty \neq 0$ and $u_p \in \mathbb{Q}_p, u_p \neq 0$ and $|u_p|_p = 1$ for all but finitely many primes p. There is a norm $|\cdot|$ defined on J given by $|u| = |u_\infty| \prod_p |u_p|_p$. We have natural diagonal embeddings of $\mathbb{Q}$ in A and $\mathbb{Q}^* = \mathbb{Q} \setminus \{0\}$ in J induced by the embeddings of $\mathbb{Q}$ into its completions. Notice that by the product formula $\mathbb{Q}^* \subset \mathrm{Ker}|\cdot|$. We define an equivalence relation on A by $a \equiv b$ if and only if there exists a $q \in \mathbb{Q}^*$ with $a = qb$. With respect to this equivalence, we form the coset space $X = A/\mathbb{Q}^*$. The ideles J act on A by componentwise multiplication, which induces an action of $C = J/\mathbb{Q}^*$ on X. Notice that this action has fixed points. For example, whenever an adele a has $a_p = 0$ it is a fixed point of the embedding of $\mathbb{Q}_p^*$ into J (to $q_p \in \mathbb{Q}_p^*$ one assigns the idele with 1 in every place except the pth place.) On the other hand, every Type III_1 factor has a continuous decomposition, that is it can be written as a cross-product of $\mathbb{R}$ with a Type II_∞ factor. Connes has observed that the von-Neumann algebra of Type III_1 in the region $0 < \beta \leq 1$ of Theorem 1 has in its continuous decomposition the Type II_∞ factor given by the crossed product of $L^\infty(A)$ by the action of $\mathbb{Q}^*$ by multiplication. The associated von Neumann algebra has orbit space $X = A/\mathbb{Q}^*$.

The pair (X, C) plays a fundamental role in Connes's proposed approach to the Riemann hypothesis in [5]. We remark that Connes's approach does not as yet reprove the established analogue of the Riemann hypothesis in this case: if it did so this would provide an alternative proof. Indeed, even for the case of the projective line, where the Riemann hypothesis is trivial, the approach of Connes demands an analysis not only of the counting of points over finite fields on the projective line, but also an understanding of the moduli space of vector bundles over the projective line, due to the fact that all the characters of the idèle class group also intervene. In order for his approach to work, Connes would need to be able to prove an asymptotic formula for the trace of the action of C on X. A "spectral" analysis of this action, related to older ideas of Weil [17] and Tate, shows it to be related to the non-trivial zeros of the L-functions with Grossencharacter, and a heurisitic "geometric" analysis relates this action to the Weil distribution. The "geometric" heuristics aim at suggesting a proof for the positivity of the Weil distribution which is known to imply the Riemann hypothesis. In the subsequent sections we give a leisurely introduction to these ideas.

6 A trace formula over the reals

There is a rigorous local version of the geometric side of the asymptotic trace formula of [6]. For simplicity we work at the infinite place, the discussion at the prime places being analogous, and justify the formula in terms of the simple computation of the distributional trace. Consider the action of $\mathbb{R}^* = \mathbb{R} \setminus \{0\}$ on $\mathbb{R}$ by multiplication. This induces the action on smooth functions on $\mathbb{R}$,

$$U(u)g = g \circ u^{-1}, \qquad u \in \mathbb{R}^*, \quad g \in C^\infty(\mathbb{R}). \tag{14}$$

We average this action over a function of rapid decay. Namely, for $h \in \mathcal{S}(\mathbb{R}^*)$, we let $U(h)$ be the operator in $L^2(\mathbb{R})$ given by,

$$U(h) = \int_{\mathbb{R}^*} h(u)U(u)d^*u, \tag{15}$$

where the multiplicative Haar measure d^*u is normalised by,

$$\int_{|u|\in[1,\Lambda]} d^*u \simeq \log \Lambda, \qquad \Lambda \to \infty. \tag{16}$$

The associated kernel of $U(u)$ is

$$k(x, u^{-1}, y) = \delta(y - u^{-1}x), \tag{17}$$

and for $u \neq 1$,

$$\int_{\mathbb{R}} k(x, u^{-1}, x)dx = |1 - u^{-1}|^{-1}. \tag{18}$$

If $h \in \mathcal{S}(\mathbb{R}^*)$ and $h(1) = 0$, we define the distributional trace of $U(h)$ to be,

$$\text{"Trace"}(U(h)) = \int_{\mathbb{R}^*} \frac{h(u^{-1})}{|1 - u|}d^*u. \tag{19}$$

To state an ordinary trace formula in the general case $h(1) \neq 0$, we introduce the principal part of the integral in (19), defined by

$$\int' \frac{h(u^{-1})}{|1-u|} d^*u = <L, \frac{h(u^{-1})}{|u|}>, \tag{20}$$

where L is the unique distribution on $\mathbb{R}$ which agrees with $\frac{du}{|1-u|}$ for $u \neq 1$ and whose Fourier transform vanishes at 1. If $g(u) = h((u+1)^{-1})/|u+1|^{-1}$, then from [6], §V, we have the formula

$$\int' g(a) \frac{da}{|a|} = -\int_{\mathbb{R}} \hat{g}(u) \log |u| du. \tag{21}$$

Let $\Lambda > 0$ and let P_Λ be the projection given by multiplication by the characteristic function of the set

$$\{\xi \in L^2(\mathbb{R}) : \xi(x) = 0, \text{ for } |x| > \Lambda\}. \tag{22}$$

Let F denote the Fourier transform in $L^2(\mathbb{R})$ and $\hat{P}_\Lambda = FP_\Lambda F^{-1}$. Let $R_\Lambda = \hat{P}_\Lambda P_\Lambda$. In [6] the following is proved.

Proposition 1 *If $h \in S_c(\mathbb{R}^*)$, then $R_\Lambda U(h)$ is a trace class operator in $L^2(\mathbb{R})$ and as $\Lambda \to \infty$ we have the asymptotic formula*

$$\text{Trace}(R_\Lambda U(h)) = 2h(1) \log \Lambda + \int' \frac{h(u^{-1})}{|1-u|} d^*u + o(1). \tag{23}$$

An analogous result holds for the local fields $\mathbb{Q}_p$ and also if one considers a finite number of places at a time. The problem is that the error term then implied by the "$o(1)$" is not uniform in the set of primes chosen so that one cannot pass to the limit over all primes. This seems to represent a deep difficulty of an analytic number theoretic nature.

7 The Polya-Hilbert space

Connes proposes a Polya-Hilbert space for the problem in the following way. Let $S(A)_0$ denote the subspace of $S(A)$ given by,

$$S(A)_0 = \{f \in S(A) : f(0) = \int f dx = 0\}. \tag{24}$$

Let E be the averaging over $\mathbb{Q}^*$ operator which to $f \in S(A)_0$ associates the element of $S(C)$ given by

$$E(f)(u) = |u|^{1/2} \sum_{q \in \mathbb{Q}^*} f(qu). \tag{25}$$

For $\delta \geq 0$, let $L^2(X)_{0,\delta}$ be the completion of $S(A)_0$ with respect to the norm given by,

$$\|f\|_\delta^2 = \int_C |E(f)(u)|^2 (1 + \log^2 |u|)^{\delta/2} d^*u, \quad \text{for} \quad \int_{|u| \in [1,\Lambda]} d^*u \simeq \log \Lambda, \quad \Lambda \to \infty. \tag{26}$$

If $g(x) = f(qx)$ for some fixed $q \in \mathbb{Q}^*$, then $\|g\|_\delta = \|f\|_\delta$ and so one sees that this norm respects, in this sense, the passage to the quotient $A/\mathbb{Q}^*$. We define $L^2(X)_\delta$ by the short exact sequence

$$0 \to L^2(X)_{0,\delta} \to L^2(X)_\delta \to \mathbb{C} \oplus \mathbb{C}(1) \to 0. \tag{27}$$

When $\delta = 0$ we write $L^2(X)_0$ and $L^2(X)$ for the first two terms. Here $\mathbb{C}$ is the trivial C-module and $\mathbb{C}(1)$ is the C-module for which $u \in C$ acts by $|u|$, where $|\cdot|$ is the norm on C. Multiplication of C on A induces a representation of C on $L^2(X)$ given by,

$$(U(\lambda)\xi)(x) = \xi(\lambda^{-1}x). \tag{28}$$

We introduce a Hilbert space H_δ via another short exact sequence,

$$0 \to L^2(X)_{0,\delta} \to L^2(C)_\delta \to H_\delta \to 0. \tag{29}$$

Here $L^2(C)_\delta$ is the completion with respect to the weighted Haar measure as in (26), where we write $L^2(C)$ when $\delta = 0$. The spectral interpretation on H_δ of the critical zeros of the L-functions in [5] relies on taking $\delta > 0$. Indeed, this is needed to control the growth of the functions on the non-compact quotient X: ultimately this parameter is eliminated from conjectural trace formula by using cut-offs. It is important here to use the measure $|u|d^*u$ (implicit in (26)) instead of the additive Haar measure dx, this difference being a veritable one for global fields, where one has $dx = \lim_{\epsilon \to 0} \epsilon |x|^{1+\epsilon} d^*x$.

The regular representation descends to H_δ (it commutes with E up to a phase as an easy calculation shows) and we denote it by W. Connes describes (H_δ, W) as the Polya-Hilbert space with group action for his approach to the Riemann hypothesis. He proves in [5] and [6] the following remarkable result relating the trace of this action to the zeros on the critical line of the L-functions with Grossencharacter.

Theorem 2 *For any Schwartz function $h \in \mathcal{S}(C)$ the operator*

$$W(h) = \int_C W(u)h(u)d^*u$$

in H is trace class, and its trace is given by

$$\text{Trace}(W(h)) = \sum_{L(\chi,\frac{1}{2}+is)=0, s \in \mathbb{R}} \widehat{h}(\chi, is)$$

where the sum is over the characters χ of C with the multiplicity of the zero being counted as the largest integer $n < \frac{1}{2}(1 + \delta)$ with n at most the multiplicity of $\frac{1}{2} + is$ as a zero of $L(\chi, z)$. Moreover, we define,

$$h(\chi, z) := \int_C h(u)\chi(u)|u|^z d^*u.$$

Now, the action of C is free on $L^2(C)_\delta$ so that the short exact sequence (29) tells us that the trace of the action of C on H_δ should be, up to a correction due to a regularisation, the negative of the trace of the action of C on $L^2(X)_{0,\delta}$. From (27), we see that the regularised trace of the action of C on $L^2(X)_\delta$ should involve the sum of the corresponding trace on $L^2(X)_{0,\delta}$ and the trace on $\mathbb{C} \oplus \mathbb{C}(1)$. Therefore the regularised trace of the action of C on $L^2(X)_\delta$ should involve the trace of the action on $\mathbb{C} \oplus \mathbb{C}(1)$ *minus* the trace of this action on H_δ. This *minus* sign is crucial for the comparison with the Weil distribution.

The following result going back to Weil [17] shows how the operation E in (25) brings in the zeros of the L functions in the critical strip and provides the key to proof of Theorem 2 and the appearance of the non-trivial zeros of the $L(\chi, s)$ in Proposition 3. It indicates that the non-trivial zeros of the L-functions should "span" H_δ as they are "orthogonal" to the image of E

Proposition 2 *Let χ be a character on C. For any $\rho \in \mathbb{C}$ with $\mathrm{Re}(\rho) \in] - \frac{1}{2}, \frac{1}{2}[$, we have*

$$\int_C E(\xi)(u)\chi(u)|u|^\rho d^*u = 0, \qquad \text{for all } \xi \in \mathcal{S}(A)_0,$$

precisely when $L(\chi, \frac{1}{2} + \rho) = 0$.

8 Relation to the Weil distribution

In [6] Connes works with a cut-off in order to avoid the parameter δ of §7. He introduces a family of subspaces $B_{\Lambda,0}$ of $L^2(X)_0$, depending on a real parameter $\Lambda > 0$, such that $E(B_{\Lambda,0}) \subset \mathbf{S}_\Lambda$ where,

$$\mathbf{S}_\Lambda = \{\xi \in L^2(C) : \xi(u) = 0, \text{ for } |u| \notin [\Lambda^{-1}, \Lambda]\}. \tag{30}$$

In the case of global fields of positive characteristic it makes sense to introduce $B_{\Lambda,0}$ as the space spanned by the $f \in \mathcal{S}(A)_0$ that together with their Fourier transform vanish for $|x| > \Lambda$. In the number field case, all such functions (on the reals) are trivial, and $B_{\Lambda,0}$ is described in terms of prolate spheroidal functions. This aspect of [6] is very technical and we do not attempt to go into it here as it offers little additional insight into the underlying ideas. As the asymptotic trace formula involving them is conjectural, we *assume* for our present purpose the existence of an appropriate family of $B_{\Lambda,0}$. Let $Q_{\Lambda,0}$ be the orthogonal projection onto $B_{\Lambda,0}$ and $Q'_{\Lambda,0} = E Q_{\Lambda,0} E^{-1}$. By assumption,

$$Q'_{\Lambda,0} \leq S_\Lambda, \tag{31}$$

where S_Λ denotes the projection onto the set $\mathbf{S}_\Lambda$ in (30). Therefore, for all $\Lambda > 0$, the following distribution is positive,

$$\Delta_\Lambda(f) = \mathrm{Trace}((S_\Lambda - Q'_{\Lambda,0})V(f)), \qquad f \in \mathcal{S}(C), \tag{32}$$

where V is the regular representation of C on $L^2(C)$ and $V(f) = \int_C f(h)V(u)d^*u$. The positivity of Δ_Λ signifies that for $f \in \mathcal{S}(C)$ we have

$$\Delta_\Lambda(f * f^*) \geq 0, \tag{33}$$

where $f^*(u) = \overline{f(u^{-1})}$. Therefore the limiting distribution is also positive,

$$\Delta_\infty = \lim_{\Lambda \to \infty} \Delta_\Lambda \geq 0. \tag{34}$$

In [6] the following conjecture is formulated, together with a proposed construction of the family $B_{\Lambda,0}$.

Conjecture 1 *One can find subspaces $B_{\Lambda,0}$ of $L^2(X)_0$ such that the distributions Δ_Λ are of positive type and converge to the Weil distribution, so that for $h \in \mathcal{S}(C)$ of compact support,*

$$\Delta_\infty(h) = \int_C h(u)(|u|^{1/2} + |u|^{-1/2})d^*u - \sum_v \int'_{\mathbb{Q}_v^*} \frac{h(u)}{|1 - u|}|u|^{1/2}d^*u, \tag{35}$$

where v runs over the places of $\mathbb{Q}$.

It is known that the positivity of the Weil distribution, which would follow from Conjecture 1 using (34), implies the Riemann hypothesis (see for example [14]). One can view Δ_Λ, via (32), as a distribution associated to the regularised trace of the action of C on the Hilbert space H_δ of §7 with the cut-off at $|u| = \Lambda$ enabling one to take $\delta = 0$. The factor $|u|^{1/2}$ in the integrands of (35) as opposed to those of (23) is due to the passage to the quotient by the image of E.

We can reformulate Conjecture 1 in terms of the sequence of projections Q_Λ where Q_Λ is the projection in $L^2(X)$ onto $B_{\Lambda,0} \oplus \mathbb{C} \oplus \mathbb{C}(1)$ as follows.

Conjecture 2 *There is a sequence of closed projections Q_Λ, $\Lambda > 0$, in $L^2(X)$ extending the $Q_{\Lambda,0}$ above such that for $h \in S(C)$ of compact support we have, as $\Lambda \to \infty$,*

$$\operatorname{Trace}(Q_\Lambda U(h)) = 2h(1) \log' \Lambda + \sum_v \int_{\mathbb{Q}_v^*}' \frac{h(u^{-1})}{|1-u|} d^*u + o(1), \tag{36}$$

where the sum is over the places v of $\mathbb{Q}$. Here $2\log' \Lambda = \int_{|\lambda| \in [\Lambda^{-1}, \Lambda]} d^*u$ which is asymptotic to $2\log \Lambda$ for the correct choice of Haar measure d^*u on C.

9   Relation of the limit distribution to the zeros

The distributions Δ_Λ can be regarded as a sequence of distributions attached to the Polya-Hilbert space and action (H_δ, W) of §7. We have the following interpretation of Δ_∞ in terms of the zeros of the L-functions with Grossencharacter: compare with Theorem 2 and notice how the use of the asymptotics of the cut-off has eliminated the parameter δ.

Proposition 3 *Let h be a function of compact support on $S(C)$. Then,*

$$\Delta_\infty(h) = \sum_{\chi,\rho} N(\chi, \tfrac{1}{2} + \rho) \int_{z \in i\mathbb{R}} \widehat{h}(\chi, z) d\mu_\rho(z) \tag{37}$$

where the sum is over the pairs (χ, ρ) of characters of C and zeros ρ of $L(\chi, \tfrac{1}{2} + \rho)$ with $\operatorname{Re}(\rho) \in\,]-\tfrac{1}{2}, \tfrac{1}{2}[$. The number $N(\chi, \tfrac{1}{2} + \rho)$ is the multiplicity of the zero, the measure $d\mu_\rho(z)$ is the harmonic measure of ρ with respect to the line $i\mathbb{R} \subset \mathbb{C}$ and the Fourier transform $\widehat{h}$ of h is defined by

$$\widehat{h}(\chi, \rho) = \int_C h(u)\chi(u)|u|^\rho d^*u. \tag{38}$$

In [6] Proposition 3 is proven in the function field case, and the proof for number fields is outlined. It allows one to derive under Conjecture 1, the following conjectural explicit formula.

Conjecture 3 *Let $h \in S(C)$ be of compact support, then*

$$\sum_v \int_{\mathbb{Q}_v^*}' \frac{h(u)}{|1-u|} |u|^{1/2} d^*u$$

$$= \int_C h(u)(|u|^{1/2} + |u|^{-1/2})d^*u - \sum_{\chi,\rho} N(\chi, \tfrac{1}{2} + \rho) \int_{z \in i\mathbb{R}} \widehat{h}(\chi, z) d\mu_\rho(z),$$

where v runs over the places of $\mathbb{Q}$.

Proposition 3 gives an interpretation of the non-trivial zeros of the Riemann zeta function with the non-critical zeros appearing as resonances for the harmonic measure with respect to the real line. Conjecture 3 is the global analogue of Proposition 1 and can be thought of as a conjectural trace formula for the action of C on $L^2(X)$. In accordance with the comments preceding Proposition 2, one sees that for the action of C on $L^2(X)$, the critical zeros would enter the trace formula with a *minus* sign, which Connes interprets as their being an absorption spectrum for the action of C on $A/\mathbb{Q}^*$. This represents an important *new* feature over other proposed spectral models to date coming from physics. It is in complete agreement with the intuitions gleaned form the proofs in the function field case that yield a spectral interpretation of the zeros as the eigenvalues of the action of Frobenius on l-adic cohomology. The analogue of the Polya-Hilbert space is given, in the curve case (if one replaces $\mathbb{C}$ by $\mathbb{Q}_l$, with $l \neq p$ where p is the base field characteristic) by the 1st degree étale cohomology group on the curve with coefficients in $\mathbb{Q}_l$. This group appears with an overall *minus* sign in the Lefchetz formula. In other words, the spectral interpretation of the zeros of the Riemann zeta function should be as an absorption spectrum rather than as an emission spectrum, if one uses the language of spectroscopy.

10   A global geometric trace formula

One of the most interesting heuristic aspects of Connes's ideas, which appears already in [5], is the interpretation of the global analogue of Proposition 1 as a non-commutative trace formula for a flow given in Conjecture 2. In [5], the validity of Conjecture 2 was interpreted as the existence of a noncommutative generalisation of a distributional trace formula for flows on compact manifolds (see [9]). If M is a smooth compact manifold with an everywhere non-vanishing vector field ξ, then the associated flow is given by the 1-parameter group $\{F_t = \exp t\xi\}_{t \in \mathbb{R}}$ which induces an action of $\mathbb{R}$ on smooth functions on M by,

$$U(t)f = f \circ F_t, \qquad f \in C^\infty(M). \tag{39}$$

For $h \in \mathcal{S}(M)$, consider the operator,

$$U(h) = \int_{\mathbb{R}} U(t)h(t)dt, \tag{40}$$

where we suppose for simplicity that $h(0) = 0$. Then we know from [9] that the distributional trace is given by,

$$\text{``Trace''}(U(h)) = \sum_{\pi} \int_{H_\pi} h(t)\frac{1}{|1 - P_t|}d_\pi t, \tag{41}$$

where the sum is over the primitive periodic orbits π and where H_π is the isotropy group of any $x \in \pi$ with measure $d_\pi t$ normalised so that H_π has covolume 1. We have denoted by P_t the restriction of the differential $d(\exp(t\xi))_x$ to the space transverse to the orbits (the Poincaré map) and by $|1 - P_t|$ the absolute value of the determinant of $1 - P_t$ which is assumed non-degenerate. Now, observe the following fact about the pair (X, C) proved in [5].

Lemma 3 *For $x \in X, x \neq 0$, the isotropy group of x in C is compact if and only if there exists exactly one place v of $\mathbb{Q}$ with $x_v = 0$ where $(x_\infty, x_p)_p$ is any lift of x to A.*

If $x \in A$ with $x_v = 0$ at only one place v of $\mathbb{Q}$, then the isotropy group of x in C is isomorphic to $\mathbb{Q}_v^*$ and the transverse space is isomorphic to $\mathbb{Q}_v$. Hence the analogue of the

Poincaré return map for this fixed point is multiplication of $\mathbb{Q}_v^*$ on $\mathbb{Q}_v$ whose trace formula we discussed in §6 for $v = \infty$. We therefore write suggestively $|1 - P_v| = |1 - u|$, $u \in \mathbb{Q}_v^*$. Let $h \in S(C)$ with compact support and $h(1) = 0$. The analogue of the trace formula (41) becomes, once we agree to single out the fixed points with compact isotropy group determined by Lemma 3,

$$\text{``Trace''}(U(h)) = \sum_v \int_{\mathbb{Q}_v^*} \frac{h(u^{-1})}{|1 - u|} d^* u. \tag{42}$$

This is consistent with the local computation of §6 when $h(1) = 0$ and introduces the sum over local terms in (35) as coming from an analogue for $A/\mathbb{Q}^*$ of a geometric trace formula for flows.

These heuristics show that the validity of the Riemann hypothesis would imply that the above analogue of the distributional trace formula for flows on manifolds makes sense for the pair (X, C). This is remarkable in light of the fact that the space X displays none of the regularity properties needed in the proof of the manifold case.

11 Concluding remarks

The work of Bost–Connes and Connes leads naturally to investigating $(A/\mathbb{Q}^*, C)$ as the geometric site and (H_δ, W) as the spectral site for the study of the generalised Riemann Hypothesis. Although much of Connes's suggested approach to the Riemann hypothesis is conjectural, it seems likely that a deeper understanding of the space $A/\mathbb{Q}^*$ should have interesting consequences. The analogy between the trace formula for flows on smooth compact manifolds of [9] and the conjectured trace formula in (42), suggests the interpretation of Weil's explicit formula as a trace formula on $A/\mathbb{Q}^*$. It should be noted that Paul Cohen proposed in unpublished remarks in the 1970's that in order to understand the Riemann Hypothesis one should understand the nature of measurable $\mathbb{Q}^*$-invariant functions on A. The fact that the von-Neumann algebra associated to $A/\mathbb{Q}^*$ is of Type II_∞, as shown by the work of Bost–Connes, indicates that it is not possible to do classical measure (Type I) theory on $A/\mathbb{Q}^*$. Non-classical ideas are therefore needed and the work of Connes is a bold step in that direction.

Some very recent ideas of Connes expressed at the AMS/Clay Institute conference on Noncommutative Geometry in Mount Holyoke in June 2000 indicate that he may be able to reexpress the conjectured trace formulae in terms of indices associated to Fredholm modules in the sense of Connes's noncommutative geometry. This would lay the investigation open to the techniques of that theory.

To date, no spectral interpretation of the zeros, even in the function field case, has yielded in itself a proof of the Riemann hypothesis: something else is always needed. In the function field case this extra ingredient is Castelnuovo positivity. In one of Weil's proofs of the curve case over finite fields, he uses the explicit formula (which Connes conjectures can be described as an appropriate trace formula) together with the proof that the Weil distribution is positive. By contrast, Connes in [6] worked by construction with the Hilbert space H_δ where positivity was ensured by definition and conjectured an appropriate explicit formula, thereby reversing in some sense the logic of the proofs in the curve case. In the more recent approach [7], positivity is part of the problem. One constructs a sequence of projections which are positive and the difficult part is to prove that their limit gives the Weil distribution. The spectral interpretation is there, but once more it is not enough.

The conjectures and analogies in Connes's work seem to offer much more than an approach to the Riemann hypothesis. They suggest a fruitful interaction between the theories of operator algebras and C^*-algebras and the analytic and algebraic theories of numbers. A likely source for the immediate future of a new input into number theory is Connes's outline of a noncommutative Brauer theory given in [7]. Even if the Riemann hypothesis is proven by techniques independent of those proposed by Connes, the set-up he is proposing could then step in armed with the validity of the Riemann hypothesis to view arithmetic in a new and exciting way. An immediate challenge is therefore to understand and exploit Connes's approach in the function field case where the Riemann hypothesis is known. This has not even been carried out for the case of the projective line where the analysis should be most tractable.

References

[1] J. ARLEDGE, M. LACA, I. RAEBURN, *Semigroup crossed products and Hecke algebras arising from number fields*, Doc. Mathematica **2** (1997) 115–138.

[2] J-B. BOST, A. CONNES, *Hecke Algebras, Type III factors and phase transitions with spontaneous symmetry breaking in number theory*, Selecta Math. (New Series), 1, 411–457 (1995).

[3] P.B. COHEN, *A C^*-dynamical system with Dedekind zeta partition function and spontaneous symmetry breaking*, J. Théorie des Nombres de Bordeaux, **11** (1999), 15–30

[4] A. CONNES, Noncommutative Geometry, Academic Press, 1994. (Version française: Géométrie non commutative, InterEditions, Paris, 1990.)

[5] A. CONNES, *Formule de trace en géométrie non commutative et hypothese de Riemann*, C.R. Acad. Sci. Paris, **t.323**, Série 1 (Analyse),1231–1236 (1996).

[6] A. CONNES, *Trace formula in Noncommutative Geometry and the zeros of the Riemann zeta function*, Selecta Mathematica, New Series **5**, n. 1 (1999), 29–106.

[7] A. CONNES, *Noncommutative geometry and the Riemann zeta function*, in Mathematics: frontiers and perspectives, eds, Vladimir Arnold...[et al], IMU, AMS 1999, 35–54.

[8] J. DIXMIER, Les C^*-algebres et leurs représentations, Gauthier-Villars, Paris, 1964.

[9] V. GUILLEMIN, *Lectures on spectral theory of elliptic operators*, Duke Math. J., **44** no. 3, 485–517 (1977).

[10] D. HARARI, E. LEICHTNAM, *Extension du phénomene de brisure spontanée de symétrie de Bost-Connes au cas des corps globaux quelconques*, Selecta Mathematica, New Series 3 (1997), 205-243.

103

[11] B. JULIA, *Statistical Theory of Numbers*, in Number Theory and Physics, Springer Proceedings in Physics, Vol. 47, 1990.

[12] N. KATZ, P. SARNAK, Random matrices, Frobenius eigenvalues and Monodromy, *AMS Colloquium Publications* **45** 1999.

[13] K.R. PARTHASARATHY, An Introduction to Quantum Stochastic Calculus, *Monographs in Mathematics* Vol. 85, Birkhaüser, 1992.

[14] S. PATTERSON, An Introduction to the Riemann zeta function, *Cambridge studies in advanced mathematics* Vol 14, Cambridge Uni. Press, 1988.

[15] B. RIEMANN, Ueber die Anzahl der Primzahlen unter einer gegebenen Grösse, *Monat der Königl. Preuss. Akad. der Wissen. zu Berlin aus der Jahre 1859* (1860), 671–680; also, *Gesammelte math. Werke und wissensch. Nachlass*, 2. Aufl. 1892, 145–155. English translation in H.M. EDWARDS, *Riemann's zeta function*, Academic Press New York–London 1974.

[16] E.C. TITCHMARSH, The Theory of the Riemann Zeta-function, Second Edition revised by D.R. Heath-Brown, Oxford University Press, New York, 1988.

[17] A. WEIL, *Fonctions zeta et distributions*, Séminaire Bourbaki **312** (1966).

Paula B. Cohen
School of Mathematics   UMR AGAT au CNRS
Institute for Advanced Study   Mathématiques, Bât M2
Einstein Drive   UFR de Mathématiques
Princeton, 08540 NJ, USA   Villeneuve d'Ascq, 59655, FRANCE
pcohen@math.ias.edu   pcohen@agat.univ-lille1.fr

A HYPERELLIPTIC CURVE
WITH REAL MULTIPLICATION OF DEGREE TWO

HARVEY COHN

Abstract. The analogue of complex multiplication in an elliptic curve is illustrated for a hyperelliptic curve with real multiplication of degree two (over C). Humbert's equation, which characterizes such a curve, was derived in 1899 by elegant *tour de force* of contemporary analysis and geometry; but this equation can be derived very simply by use of computer algebra on the abelian integrals, leading directly to a sufficient proof of Humbert's equation from his remarkable conic configuration. A set of suitable hyperelliptic coefficient parameters is also introduced. The relation to Hilbert modular functions is outlined in an appendix.

0. Summary. This all began with the AGM (arithmetic-geometric mean) of Gauss (1799), (see [1], [5]), which became classically interpreted as a mapping of one *elliptic* curve in two-to-one fashion onto another. Since elliptic curves have one (coefficient) parameter, this means, loosely speaking, for some $s \to s_0$:

$$\{y^2 = x(x-1)(x-s^2)\} \to \{y_0^2 = x_0(x_0-1)(x_0-s_0^2)\}.$$

Then when $s = s_0$, the curves are the same so these *(discrete) values* of s become "singular values" with "complex multiplication." (See §1-4 for definitions and other details).

If we deal with *hyperelliptic* curves (of genus two), then there was a corresponding AGM found by Richelot (1837), (see [1]), also later interpreted as a correspondence between two such curves (giving rise to an isogeny between jacobians with kernel isomorphic to $\mathbf{Z}/2\mathbf{Z} \times \mathbf{Z}/2\mathbf{Z}$). Since these curves have three (coefficient) parameters (see [3]), this means, more loosely speaking, for some $(s,t,u) \to (s_0,t_0,u_0)$:

$$\{y^2 = x(x-1)(x-s^2)(x-t^2)(x-u^2)\} \to \{y_0^2 = x_0(x_0-1)(x_0-s_0^2)(x_0-t_0^2)(x_0-u_0^2)\}.$$

This can not be a two-to-one mapping as in the case of genus one, but a mapping of abelian integrals in a manner described below. We then have a relationship due to Georges Humbert (1899), (see [10]), reducing the curves to *two parameters,* which he called "singular cases" with "real multiplication." (See §5-7 for definitions and other details).

Either complex or real multiplication is "singular" in the sense that there is a loss of dimension of the parameters.

The real cases were derived by Humbert as a culmination of nineteenth century state of the art, with the use of Kummer surfaces, theta functions, and projective invariants. In this paper, we shall justify Humbert's results by computer on a more elementary and (basically) self-contained level, by working with the abelian integrals using MAPLE.

Presented to the New York Number Theory Seminar under the title "Analogies between real and complex multiplication."

Mathematics Subject Classification. Primary 14K22, 11F41. *Key words and phrases.* Complex multiplication, real multiplication, Jacobi manifold, Humbert's criterion.

Typeset by $\mathcal{A}_{\mathcal{M}}$S-TEX

Actually, the two singular hyperelliptic parameters are algebraically related to the two variables of the field of Hilbert modular functions for $Q(\sqrt{2})$ (as developed by Hecke and Siegel, see [7], [6] [4], [8], [16], [18]). This relation apparently is not known explicitly, but no further use is made of Hilbert modular functions here.

The author had previously worked in this area, mostly on modular equations [4] for the (transcendental) Hilbert modular invariants. He is therefore all the more cognizant of his debt to Armand Brumer, Mark Heiligman, Everett Howe, Bjorn Poonen, and Jerome Solinas for useful conversations which eased the transition to (algebraic) coefficient parameters.

There is obviously a prepossessing need to know that the transcendental and algebraic results are part of the same "language," and the author notes that Armand Brumer informs him of prior (unpublished) results in this area. The author also notes previous papers of J.-F. Mestre ([13], [14], [15]) exploring Humbert's methods and also [3], [17] sources of reference.

COMPLEX MULTIPLICATION OF DEGREE TWO

1. The period structure. To recapitulate the classic result, (see [5]), consider a lattice generated by two complex vectors, 1 and $\sqrt{-2}$,

$$(1.1a) \qquad R = [1, \sqrt{-2}] = \{n + m\sqrt{-2} \mid n, m \in Z\}.$$

Then for this period structure, U an abelian integral (always of the first kind) is defined in the period parallelogram, i.e.,

$$(1.1b) \qquad U \in C/R.$$

Complex multiplication of R by $\sqrt{-2}$ produces a sublattice of R. In matrix form,

$$(1.2a) \qquad R = (1 \ \sqrt{-2}),$$

$$(1.2b) \qquad (\sqrt{-2})\,R = (\sqrt{-2} \ -2) = R \begin{pmatrix} 0 & -2 \\ 1 & 0 \end{pmatrix}.$$

If we were looking for multipliers of norm two, we should find two others:

$$(1.2c) \qquad (1+i)\,(1\ i) = (1\ i) \begin{pmatrix} 1 & -1 \\ 1 & 1 \end{pmatrix},$$

$$(1.2d) \qquad (\tfrac{1}{2}(1+\sqrt{-7}))\,(1\ \tfrac{1}{2}(1+\sqrt{-7})) = (1\ \tfrac{1}{2}(1+\sqrt{-7})) \begin{pmatrix} 0 & -2 \\ 1 & 1 \end{pmatrix}.$$

Of course, we note the above matrices are of determinant 2, and likewise

$$(1.2e) \qquad N(\sqrt{-2}) = N(1+i) = N(\tfrac{1}{2}(1+\sqrt{-7})) = 2.$$

2. The elliptic curve. If an elliptic curve has a two-to-one mapping onto itself (a two-isogeny), then its period lattice must be equivalent to itself by a complex multiplier mapping this lattice into a sublattice of index two. Again classically, this is an isogeny found as follows:

The elliptic curve is written with (cross-ratio) parameter $\lambda (= s^2)$,

$$(2.1a) \qquad y^2 = x(x-1)(x-s^2), \quad (s^2 \neq 0, 1).$$

We write the parameter λ as a square with the hindsight that its square root s will emerge in the computation. There are many equivalent forms of the elliptic curve (2.1a). The j-invariant, defined as

$$(2.1b) \qquad j[\lambda] = 256(1 - \lambda + \lambda^2)^3 / \lambda^2 (\lambda - 1)^2,$$

is *algebraic* in the coefficients of (2.1a) and it classifies elliptic curves up to isomorphism. Note that $j[\lambda]$ is invariant under the transformation group (S_3) permuting the points $\{0, 1, \infty\}$:

$$(2.1c) \qquad \lambda \rightarrow \lambda' \in \{\lambda, 1/\lambda, 1 - \lambda, (\lambda - 1)/\lambda, \lambda/(\lambda - 1), 1/(1 - \lambda)\}$$

If the period parallelogram has ratio τ (with $\Im\tau > 0$), then j has the well-known *transcendental* expansion in $q = \exp 2\pi i \tau$,

$$(2.1d) \qquad j(\tau) = 1/q + 744 + 196884q + 21493760q^2 + \cdots.$$

(The direct connection with the periods like τ and the coefficients like s^2 is not easily available in the hyperelliptic case which follows).

The traditional Gauss-Landen transformation,

$$(2.2a) \qquad x_0 = -\frac{(x-1)(x-s^2)}{x(1+s)^2}$$

$$(2.2b) \qquad y_0 = i\frac{y(x^2 - s^2)}{x^2(s+1)^3},$$

generates the *unique* correspondence of x_0 with the pair $\{x, s^2/x\}$ which extends to

$$(2.2c) \qquad (x_0, y_0) \overset{1 \text{ to } 2}{\longleftrightarrow} (x, y).$$

With the new parameter,

$$(2.2d) \qquad s_0 = \frac{s-1}{s+1},$$

we verify by substitution,

$$(2.2e) \qquad y_0^2 = x_0(x_0 - 1)(x_0 - s_0^2).$$

Note that we replace x_0 by $1 - x_0$ then s_0 would be replaced by

(2.2f)
$$(1 - s_0^2)^{1/2} = \sqrt{s}/(\frac{1+s}{2}),$$

the ratio of geometric-to-arithmetic means of 1 and s.

Since $x = \infty$ corresponds to $x_0 = \infty$, we see from (2.1a) and (2.2de) that

(2.3)
$$i(s+1) \int_{\infty}^{x} \frac{dx}{y} = \int_{\infty}^{x_0} \frac{dx_0}{y_0} \quad \text{(modulo periods)}.$$

In terms of the j-function, we have the two values

(2.4a)
$$X = j[s^2], \quad Y = j[s_0^2],$$

and on eliminating s and s_0 using (2.2d), we obtain the modular equation of order two,

$$(\Phi_2(X, Y) =) \; X^3 + Y^3 - X^2Y^2 + 1488XY(X + Y) - 162000(X^2 + Y^2) + 40773375XY$$
(2.4b)
$$+ 8748000000(X + Y) - 157464000000000 = 0.$$

3. Verification of Gauss-Landen. We return to the substitution (2.2a) to see how it might be used as a model for the hyperelliptic case where the abelian differentials (always of the first kind) are mapped rather than curves.

We try to reconstruct the relation (2.2a), starting with

(3.1a)
$$x_0 = \frac{(x-1)(x-s^2)c}{x}.$$

This is reasonable since the singularities $x \in \{0, 1, \infty, s^2\}$ go into the subset $x_0 \in \{0, \infty\}$. To find the constant c we try a substitution into the abelian differential of the first kind, namely

(3.1b)
$$\frac{dx_0}{\sqrt{x_0(x_0 - 1)(x_0 - s_0^2)}} = \frac{\sqrt{c}\,dx(x - s)(x + s)}{\sqrt{x(x-1)(x-s^2)q_1(x)q_2(x)}},$$

where we designate

(3.1c)
$$q_1(x) = cx^2 - cxs^2 - cx + cs^2 - x,$$
(3.1d)
$$q_2(x) = cx^2 - cxs^2 - cx + cs^2 - s_0^2 x.$$

It is now clear that the quadratics $q_1(x), q_2(x)$ can be conveniently taken as perfect squares to make sure the right-hand radical in (3.1b) remains elliptic. Thus

(3.2a)
$$\text{disc } q_1 = (cs^2 - 2cs + c + 1)(cs^2 + 2cs + c + 1) = 0,$$
(3.2b)
$$\text{disc } q_2 = (cs^2 + 2cs + c + s_0^2)(cs^2 - 2cs + c + s_0^2) = 0.$$

There are four ways the two discriminants can vanish, we choose the second factor from each of (3.2ab) and solve each for c. Thus a good choice is

(3.2c)
$$c = -\frac{1}{(s+1)^2} = -\frac{s_0^2}{(s-1)^2},$$

and the common value of c also yields the value of s_0 in (2.2d).

In essence we derive the mapping (2.2abd) as one of many mappings for producing a relation like (2.3). The parametric relations on the coefficients are only proved sufficient. The fact that they cover all cases of degree two is unimportant here since a corresponding result is not attempted for the hyperelliptic generalization.

4. Singular values. The singular cases of (2.2abd) are those discrete values of the parameter s where

$$(4.1a) \qquad j[s^2] = j[s_0^2],$$

so the elliptic curve is the same in (2.1a) and (2.2e). Now (2.3) leads to

$$(4.1b) \qquad \mu \int_\infty^1 \frac{dx}{y} = \int_\infty^0 \frac{dx}{y} \quad \text{modulo periods},$$

where μ is a complex multiplier (to be determined below). Thus we have two periods of the *same* elliptic curve with ratio equal to the complex multiplier μ.

When (4.1a) leads to $s_0 = \pm s$ then $\mu = i(s + 1)$. When, because of (2.1c), we use $\lambda' = 1 - \lambda$, or $1 - s^2 = s_0^2$, then $\mu = (s + 1)$ (as the transformation $x \to 1 - x$ must be used to cancel $s^2 \to 1 - s^2$, and this removes the i). The information is all contained in this table:

$$(4.2a)$$

equation	root	value of μ	$j[s^2]$
$s_0 = -s$	$s = -1 + \sqrt{2}$	$i(s + 1) = \sqrt{-2}$	8000
$s_0 = s$	$s = i$	$i(s + 1) = -1 + i$	1728
$s_0^2 = 1 - s^2$	$s = (-3 + \sqrt{-7})/2$	$s + 1 = (-1 + \sqrt{-7})/2$	$-3375.$

REAL MULTIPLICATION OF DEGREE TWO

5. Real multiplication. Here we are dealing with a period matrix arising from (hyperelliptic) curves of genus two. There are two independent abelian integrals (U, U') of the first kind with four pairs of periods for four independent paths on the Riemann surface. Analogously with §1 (above), there are four pairs of period vectors forming a lattice

$$(5.1a) \qquad R = [R_{.1}, R_{.2}, R_{.3}, R_{.4}] = \{n_1 R_{.1} + n_2 R_{.2} + n_3 R_{.3} + n_4 R_{.4}\} \quad (n_i \in \mathbf{Z}).$$

Then U, U' are defined modulo periods as follows:

$$(5.1b) \qquad \begin{pmatrix} U \\ U' \end{pmatrix} \in \mathbf{C}^2/R.$$

To simplify the lattice R, we select linear combinations of U, U' using $\mathrm{GL}(2, \mathbf{C})$ to produce suitable independent abelian integrals and from $\mathrm{GL}(4, \mathbf{Z})$ for the periods to reduce (5.1a) to a standard form showing three parameters ρ, ρ', ρ'' (subject to sign conditions deferred to §13),

$$(5.1c) \qquad R = \begin{pmatrix} 1 & 0 & \rho & \rho' \\ 0 & 1 & \rho' & \rho'' \end{pmatrix}.$$

(This is possible since the periods are those of abelian integrals).

The essence of real multiplication is that another application of $S \in \mathrm{GL}(2, \mathbf{R})$ on R can produce a subbasis (as in (1.2b)),

$$(5.2a) \qquad SR = RM, \quad M \in \mathrm{GL}(4, \mathbf{Z}).$$

This can happen only in special (singular) cases: Assume, e.g., $2\rho'' = \rho$. Then

$$(5.2b) \qquad S = \begin{pmatrix} 0 & 2 \\ 1 & 0 \end{pmatrix}: \quad SR = \begin{pmatrix} 0 & 2 & 2\rho' & \rho \\ 1 & 0 & \rho & \rho' \end{pmatrix} = R \begin{pmatrix} S & 0 \\ 0 & S^{tr} \end{pmatrix}.$$

This is called a "real" multiplication because e.,g., the relation

$$(5.2c) \qquad S^2 = \begin{pmatrix} 2 & 0 \\ 0 & 2 \end{pmatrix} = 2E,$$

shows S has *real* eigenvalues $\pm\sqrt{2}$. Also the period vectors of SR form a lattice as in (5.1a) with the property that

$$(5.2d) \qquad R/SR \approx \mathbf{Z}/2\mathbf{Z} \times \mathbf{Z}/2\mathbf{Z}.$$

The (absolute) value of the determinant S leads to the characteristic property (5.2d) of lattices (and jacobians).

6. Hyperelliptic curves. The hyperelliptic curve (always of genus two here), has three parameters (again written as squares),

$$(6.1a) \qquad H: \quad y^2 = x(x-1)(x-s^2)(x-t^2)(x-u^2).$$

We refer frequently to the "quintic" roots $\{\infty, 0, 1, s^2, t^2, u^2\}$, or the *Weierstrass points*, (which are assumed distinct, of course). The abelian differentials form a linear space of dimension two. Generators are chosen for convenience as dU and dU^* with

$$(6.1b) \qquad U = \int_\infty^x \frac{(x-s)dx}{y}, \quad U^* = \int_\infty^x \frac{(x+s)dx}{y}.$$

We are considering a correspondence of H with H_0, where

$$(6.1c) \qquad H_0: \quad y_0^2 = x_0(x_0-1)(x_0-s_0^2)(x_0-t_0^2)(x_0-u_0^2),$$

We try the mapping

$$(6.2a) \qquad \frac{(x-1)(x-s^2)}{x} = c\frac{(x_0-1)(x_0-s_0^2)}{x_0}.$$

It clearly is analogous to (3.1a), but (2.2) fails. For now, we only say

$$(6.2b) \qquad x_0 \overset{2 \text{ to } 2}{\longleftrightarrow} x,$$

from the correspondence of the pairing

$$(6.2c) \qquad \{x_0, s^2/x_0\} \overset{1 \text{ to } 1}{\longleftrightarrow} \{x, s^2/x\}.$$

We may not, however, write an analog for the correspondence (2.2c); the best result is blemished by a ± sign:

(6.2d) $$(x_0, y_0) \overset{1 \text{ to } 2}{\longleftrightarrow} (x, \pm y).$$

6.3 LEMMA. *A curve C of genus g can not have an N-fold cover by a curve C' of genus g' unless $g' > 1 + N(g - 1)$.*

The proof follows from the Riemann-Hurwitz formula

(6.3a) $$2(g' - 1) = 2N(g - 1) + W,$$

where W is the (nonnegative) number of branchings of the covering. We show some simple illustrations when $N = 2$:

g'	g	W	$C' \to C$ (two-fold cover)
0	0	2	$\{y = x^2\} \to \{y = X\}$
1	0	4	$\{y^2 = x^3 - 1\} \to \{Y = x^3 - 1\}$
1	1	0	"Gauss-Landen"
2	0	6	$\{y^2 = x^6 - 1\} \to \{Y = x^6 - 1\}$
2	1	2	$\{y^2 = x^6 - 1\} \to \{y^2 = X^3 - 1\}$
2	2	$W < 0(?)$	impossible!

Yet when we construct the *differentials* (in §10-11 below) we shall see the correspondence of jacobians is in one-to-two fashion (without sign ambiguities).

The correspondence (6.2a) comes indirectly from Richelot who first broached the generalization of the Gauss-Landen transformation. His work is reconstructed in [1] with an ingenious formalism. The explicit form of (6.2a) is deducible from it but comes more directly from Königsberger [12]. In Rohn [19], the two-to-two relation is described, essentially as a four-to-one mapping, not of curves but of the pair $\{U, U^*\}$ into its image, i.e., an isogeny of the two jacobians. We reexamine such matters in the light of Humbert's concepts.

Our objective for now is to specify c and s_0 (6.2a) so as to produce *singular* real multiplication. This will mean a self-isogenous jacobian.

The field determined by the coefficients of H in (6.1a), namely $Q(s^2, t^2, u^2)$ is not large enough for the coefficients of H_0, even the field of the parameters of H,

(6.4a) $$k = Q(s, t, u),$$

requires an extension field, $[K : k] = 2$ (see §7 below), so that

(6.4b) $$c, s_0, t_0^2, u_0^2 \in K.$$

7. The transformation parameters. We note that (6.2a) yields two values of x for each x_0. Rewriting it as

(7.1a) $$c\left(x_0 + \frac{s_0^2}{x_0} - 1 - s_0^2\right) = \left(x + \frac{s^2}{x} - 1 - s^2\right)$$

we find the conjugates x, x' satisfy

(7.1b) $$xx' = s^2.$$

We work directly (as in §3) with the differential from (6.1b),

(7.1c) $$dU = \frac{dx(x-s)}{\sqrt{x(x-1)(x-s^2)(x-t^2)(x-u^2)}}.$$

Our purpose is to transform it into a differential in x_0.

From (6.2a),

(7.2a) $$c\frac{dx_0(x_0^2 - s_0^2)}{x_0^2} = \frac{dx(x^2 - s^2)}{x^2}$$

So, dU may be split into two factors,

(7.2b) $$dU = d\Omega\, A,$$

where

(7.2c) $$d\Omega = \frac{dx_0(x_0^2 - s_0^2)\sqrt{c}}{x_0\sqrt{x_0(x_0-1)(x_0-s_0^2)}}, \qquad A = \frac{x}{(x+s)\sqrt{(x-t^2)(x-u^2)}}.$$

Using the conjugation notation of (7.1b), we see $x \to x'$ leads to $dU \to dU'$, $A \to A'$, while $d\Omega = d\Omega'$. Then invoking Abel's symmetry theorem, the sum of differentials

(7.2d) $$dU + dU' = d\Omega(A + A'),$$

(7.2e) $$A + A' = \frac{x}{(x+s)\sqrt{(x-t^2)(x-u^2)}} + \frac{x'}{(x'+s)\sqrt{(x'-t^2)(x'-u^2)}},$$

should be a differential on H_0 in (6.1c) from the mapping (6.2b).

It is a well-known trick to write the sum of radicals as a radical, e.g.,

(7.3a) $$(A + A') = \sqrt{A^2 + A'^2 + 2AA'}.$$

Consider the internal radical,

(7.3b) $$AA' = \frac{xx'}{(x+s)(x'+s)\sqrt{(x-t^2)(x'-t^2)(x-u^2)(x'-u^2)}}.$$

Use is now made of the symmetric functions

(7.3c) $$xx' = s^2, \quad x+x' = s^2 + 1 + cx_0 - cs_0^2 - c + cs_0^2/x_0 \; (= \sigma_1).$$

As in (3.1cd), we obtain quadratic expressions which we make into perfect squares, such as

$$(7.4a) \qquad (x - t^2)(x' - t^2) = s^2 + t^4 - \sigma_1 t^2 = q_1(x_0)/x_0,$$

where

$$(7.4b) \qquad q_1(x_0) = -t^2 c x_0^2 - x_0(-s^2 - t^4 + t^2 s^2 + t^2 - t^2 c s_0^2 - t^2 c) - t^2 c s_0^2$$

The discriminant of $q_1(x_0)$ vanishes, i.e.,

$$(7.4c) \qquad \Delta(t) = (-t^2 c s_0^2 + 2t^2 c s_0 - t^2 c + t^2 - t^4 - s^2 + t^2 s^2)$$

$$(7.4d) \qquad (-t^2 c s_0^2 - 2t^2 c s_0 - t^2 c + t^2 - t^4 - s^2 + t^2 s^2) = 0.$$

If we worked with the factor $(x - u^2)(x' - u^2)$ instead, we should obtain similarly $\Delta(u) = 0$. In either case there is the symmetry $s_0 \to -s_0$.

If we solve for c in the (first) factor (7.4c) from $\Delta(t)$ (as shown), and if we also solve for c in the (second) factor (7.4d) from $\Delta(u)$, we obtain two equations

$$(7.4e) \qquad c = \frac{(-1+t)(t+1)(s-t)(s+t)}{(s_0-1)^2 t^2} = \frac{(-1+u)(u+1)(s-u)(s+u)}{(-s_0-1)^2 u^2}.$$

The final result is now seen. We need $K = k(\omega)$ where

$$(7.5a) \qquad \omega^2 = (t^2 - 1)(u^2 - 1)(s^2 - t^2)(s^2 - u^2).$$

Then, comparing the two fractions in (7.4e), we find

$$(7.5b) \qquad s_0 = -\frac{ut^2 - ut^4 - us^2 + ut^2 s^2 + \omega t}{ut^2 - ut^4 - us^2 + ut^2 s^2 - \omega t},$$

and by taking the mean of the values in (7.4e), we find

$$(7.5c) \qquad c = \frac{\omega}{tu(s_0^2 - 1)}.$$

Finally, AA' is rational (no square roots), i.e.,

$$(7.6a) \qquad AA' = \frac{x_0^2 s}{q_2(x_0) tuc(x_0 - s_0)(x_0 + s_0)},$$

$$(7.6b) \qquad q_2(x_0) = cx_0^2 + 2sx_0 + s^2 x_0 + x_0 - cs_0^2 x_0 - cx_0 + cs_0^2,$$

It would seem purely "routine" as a problem in computer algebra (e.g., MAPLE V) to find the square root of $A^2 + A'^2 \pm 2AA'$ but unfortunately the field of $K = k(\omega)$ presents too many difficulties because of the four parameters s, t, u, ω. The fact that ω is a dependent parameter makes it even harder! (The persistant error message is "object too long").

We could resort to the parallel formalism of Richelot to find expressions for t_0^2 and u_0^2, using the four parameters, s, t, u, ω (and an additional radical), but there is no exact analogue of the Gauss-Landen transformation (2.2ab).

We do not show this work here, as our purpose is to specialize such results to the singular case only. It will be characterized by $s = s_0$ although there are many equivalent ways, e.g., $s = 1/s_0$, $s = t_0$, etc. (Recall that we only claim a sufficient condition for real multiplication.)

HUMBERT'S CRITERION

8. Humbert's equations. We shall derive Humbert's criterion [10] as a sufficient condition for real multiplication of degree two. Humbert showed it is also necessary, (by stronger techniques using the Kummer surface and the theta-function).

We start with the condition that $s = s_0$. Then eliminating ω in (7.5ab), we obtain the

$$(8.1a) \qquad \text{resultant} = (t^2 - 1)(s^2 - t^2)h_0h_2 \ (= 0).$$

Here $h_0 = 0$ and $h_2 = 0$ are two of these four equations generated under changes $s \to -s, t \to -t, u \to -u$:

$$(8.1b) \qquad h_0 = s^2t - s^2u - st - su + t^2su + t^2u + su^2t - u^2t,$$

$$(8.1c) \qquad h_1 = s^2t - s^2u + st + su - t^2su + t^2u - su^2t - u^2t,$$

$$(8.1d) \qquad h_2 = -s^2t - s^2u + st - su + t^2su + t^2u - su^2t + u^2t,$$

$$(8.1e) \qquad h_3 = s^2t + s^2u + st - su + t^2su - t^2u - su^2t - u^2t.$$

Thus, if we are dealing with the actual coefficients (s^2, t^2, u^2), we should have to multiply all four and get Humbert's equation,

$$
\begin{aligned}
(8.1f) \quad H(s^2,t^2,u^2) = h_0h_1h_2h_3 = \ & 2s^6u^6t^2 + 12s^2u^4t^4 - u^4t^8s^4 - u^4s^4 \\
& - u^8t^4 - t^4s^8 - s^8u^4 + 2u^6t^6 + 2t^4s^6 + 2s^6u^4 - s^4u^8t^4 \\
& - 16u^4t^6s^4 + 2u^4t^8s^2 + 12u^6t^6s^2 + 12u^2t^6s^4 + 2s^4u^2t^2 \\
& + 2s^2u^2t^6 + 2s^2u^6t^2 - 16s^4u^2t^4 + 12s^6u^2t^2 - 16s^4u^4t^2 \\
& + 40u^4t^4s^4 + 2u^6t^6s^4 - 16u^4t^6s^2 + 2u^2t^6s^6 + 12u^4t^4s^6 \\
& - 16s^2u^6t^4 + 2s^2u^8t^4 + 12s^4u^6t^2 - 16t^4s^6u^2 - 16u^6t^4s^4 \\
& - 16s^6u^4t^2 + 2t^2s^8u^2 - t^4s^4 - u^4t^8 \ (= 0).
\end{aligned}
$$

Curiously, Humbert failed to use square parameters, so his equation did not have this factorization. It was written (essentially) as homogeneous of degree eight.

To see the symmetries involved, number the branch points of the hyperelliptic curve (6.1a), and introduce $v(= 1)$ for homogeniety. (Thus $s \to s/v, t \to t/v, u \to u/v$ in (8.1b-f) with the $h_0, \cdots, h_3$ and H replaced by their numerators $v^4h_0, \cdots, v^4h_3$ and $v^{16}H$). Then the (coefficient) parameters are

$$(8.2a)$$

"quintic" roots	0	∞	v^2	t^2	s^2	u^2
vertices	1	2	3	4	5	6

Accordingly, the equations (8.1bcde) and (8.1f) are invariant under the dihedral group (of order eight):

$$(8.2b) \qquad D_4 = \ <(3456), (35)> .$$

Therefore, with $4!/8 = 3$ equations conjugate to (8.1f), and 15 ways of selecting the two points $\{1, 2\}$ for 0 and ∞, there is a total of 45 equations of degree 8 in the quintic coefficients which constitutes Humbert's criterion. The process of defining the manifold of real multiplications of degree two itself is far from easy. (For the total manifold of hyperelliptic curves, see [11]).

A simplified version of the factors of Humbert's equation, can be given with the help of the symbol

$$(8.3a) \qquad [ab] = \frac{a^2 - b^2}{ab}.$$

Then with $v(= 1$ for homogeniety), we can see

$$(8.3b) \qquad h_0 = stuv([ts] + [su] + [uv] - [vt]),$$
$$(8.3c) \qquad = uv(t^2 - s^2) + vt(s^2 - u^2) + ts(u^2 - v^2) - su(v^2 - t^2),$$

and (ignoring sign), h_2, h_3, h_1 all come from h_0 by the cyclic permutation of (v, t, s, u), from (8.2b).

Finally, from $h_0 = 0$, we obtain a rational parametrization, using

$$(8.4a) \qquad r = ut,$$

then with *only two* parameters r and s, we express all coefficients of the quintic, i.e.,

$$(8.4b) \qquad t^2 = r\frac{r - rs + s^2 + s}{r + rs + s^2 - s} \ (= a_1),$$

$$(8.4c) \qquad u^2 = r\frac{r + rs + s^2 - s}{r - rs + s^2 + s} \ (= a_2),$$

and the singular quintic (from (6.1a)) now becomes

$$(8.4d) \qquad y^2 = x(x - 1)(x - s^2)(x - a_1)(x - a_2).$$

9. **Humbert's geometric criterion.** From the singularities of Kummer surfaces, Humbert was able to express his criterion as a property of conic sections.

9.1 HUMBERT'S CONIC CONSTRUCTION. *Take an arbitrary (nondegenerate) conic C in the x, y, z projective plane and parametrize it in any way, e.g.,*

$$(9.1a) \qquad C: \qquad z : x : y = \phi_0(w) : \phi_1(w) : \phi_2(w),$$

where $\phi_j(w)$ represent suitable polynomials of degree at most two. Then let $\{1, 2, 3, 4, 5, 6\}$ represent the points found by substituting for w the values from the table (8.2a), (with $v = 1$). Construct the tangents at all six points. The tangents at $3, 4, 5, 6$ (in that order) form a quadrilateral by the successive intersections, $34, 45, 56, 63$, as in Figure 1.

Then the condition for real multiplication of degree two is that another conic C' exist which has a common tangent with C from points 1 and 2 and which passes through the vertices of the quadrilateral $34, 45, 56, 63$.

The details were omitted by Humbert, but we can make a straightforward computation to verify that the conic construction is equivalent to Humbert's equation (8.1f).

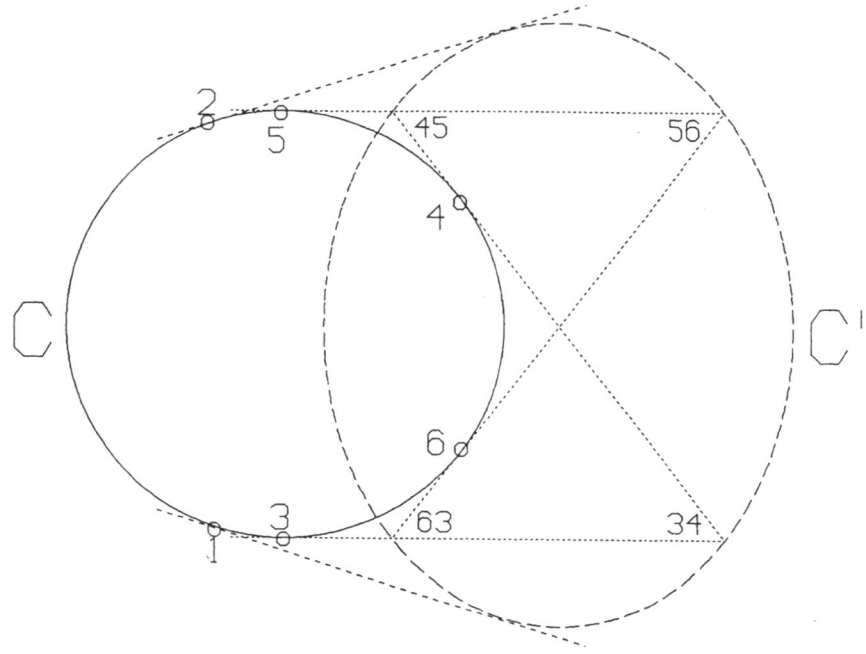

Figure 1. Humbert's Criterion

Actually, all conics have (in C) parametrizations projectively equivalent with each other, hence with the simplest:

(9.1b)
$$C: \quad y = w^2, \; x = w, \; (z = 1),$$

(in affine coordinates). We note that on C,

(9.1c)
$$\{\text{tangent line at } (a, a^2)\} = \{y + a^2 = 2xa\}.$$

The intersection (x, y) of the tangents at (g_1, g_1^2) and (g_2, g_2^2) is

(9.1d)
$$\{y + g_1^2 = 2xg_1\} \cap \{y + g_2^2 = 2xg_2\} = \left(\frac{g_1 + g_2}{2}, g_1 g_2\right).$$

Therefore we look for the second conic C' having the common tangent to C at ∞ and 0. These two conditions leave three parameters in

(9.2a) C' : $(q(x,y) =)$ $x^2 + 2Bxy + B^2y^2 + 2Ax + A^2 + 2Gy = 0.$

The parameters A, B, G result from the common tangencies as follows:

(9.2b) $(x,y) \to \infty \implies q(x,y) \approx (x + By)^2,$ $y \to 0 \implies q(x,y) \approx (x + A)^2.$

We now have four equations for the three unknowns A, B, G:

$$(q_{34} =) \; q(\frac{1+t^2}{2}, t^2) = 0, \; (q_{45} =) \; q(\frac{t^2+s^2}{2}, t^2s^2) = 0$$

(9.2c) $$(q_{56} =) \; q(\frac{s^2+u^2}{2}, s^2u^2) = 0, \; (q_{63} =) \; q(\frac{u^2+1}{2}, u^2) = 0.$$

We first eliminate G, (which is linear), between the *three* successive pairs of the equations $q_{34}, q_{45}, q_{56}, q_{63}$, so q_{34}, q_{45} result in (say) q_{345}, etc.

We thereby obtain a trio of equations, $q_{345}, q_{456}, q_{563}$, nonlinear in A and B. Now we have a choice of which of A or B to eliminate next.

So we eliminate A between the *two* successive pairs and q_{345}, q_{456} result in (say) q_{3456}, etc. We now have the two equations q_{3456}, q_{4563}. The remaining variable, namely B, is then eliminated from this *one* last pair to obtain the (partial) "resultant-GAB."

We next eliminate B first (before A) so we obtain the corresponding (partial) "resultant-GBA."

It turns out that each of the partial resultants has extraneous factors, which are eliminated by comparison. Thus to within a numerical factor,

(9.2d) $$\gcd(\text{resultant-}GAB, \text{resultant-}GBA) = H(s^2, t^2, u^2).$$

INTEGRAL IDENTITIES

10. Transformation of abelian integrals. We complete the identification of the singular case (8.4bcd) by showing the two-to-two mapping of each of the hyperelliptic curves H and H_0 of §6 and the mapping of the abelian integrals.

Now with $s = s_0$, the computer algebra simplifies sufficiently. We state the results and sketch the computations in §11. Starting again with a relation of type (6.2a),

(10.1a) $$\frac{(x-1)(x-s^2)}{x} = \kappa\frac{(x_0-1)(x_0-s^2)}{x_0},$$

and we look for a constant to provide the self-mapping of the abelian integrals. We shall show in the next section

(10.1b) $$\kappa = \frac{(r^2-s^2)(r+s^2)(r-1)}{r(r-rs+s^2+s)(r+rs+s^2-s)}.$$

We shall also see that the elliptic curves are

(10.1c) $\qquad H: \quad y^2 = x(x-1)(x-s^2)(x-a_1)(x-a_2),$

(10.1d) $\qquad H_0: \quad y_0^2 = x_0(x_0-1)(x_0-s^2)(x_0-b_1)(x_0-b_2),$

where a_1 and a_2 are given in (8.4bc) and b_1 and b_2 are derived by the involution

(10.1e) $\qquad r \rightarrow s^2(1-r)/(r+s^2), \quad s \rightarrow s,$

which transforms

(10.1f) $\qquad a_1 \rightarrow \dfrac{(r-1)s^2(r+s)}{(r+s^2)(r-s)} = b_1, \quad a_2 \rightarrow \dfrac{(r-1)s^2(r-s)}{(r+s^2)(r+s)} = b_2.$

We further note from (10.1a), for given x_0 (or x) that two conjugate values of x (resp. x_0) are uniquely determined by

(10.2) $\qquad xx' = s^2, \quad x_0x_0' = s^2,$

while at x' (or x_0') the corresponding values y' (resp. y_0') is determined with ambiguity of sign.

10.3 MAIN IDENTITY. *The abelian differentials at the conjugate points add as follows:*

(10.3a) $\qquad dx_0(x_0-s)/y_0 + dx_0'(x_0'-s)/y_0' = \gamma_1 \, dx(x-a_1)/y,$

(10.3b) $\qquad dx_0(x_0-s)/y_0 - dx_0'(x_0'-s)/y_0' = \gamma_1 \, dx'(x'-a_1)/y',$

(10.3c) $\qquad dx_0(x_0+s)/y_0 + dx_0'(x_0'+s)/y_0' = \gamma_2 \, dx(x-a_2)/y,$

(10.3d) $\qquad dx_0(x_0+s)/y_0 - dx_0'(x_0'+s)/y_0' = \gamma_2 \, dx'(x'-a_2)/y'.$

The constants satisfy

(10.3e) $\qquad \gamma_1^2 = -\dfrac{(r+rs+s^2-s)(r^2-s^2)(r+s^2)}{(r-rs+s^2+s)rs^2(r-1)}, \quad \gamma_1\gamma_2 = \dfrac{(r^2-s^2)(r+s^2)}{rs^2(r-1)}$

Note that the relation between (10.3ab) and (10.3cd) is $s \rightarrow -s$, leading to $(a_1,a_2) \rightarrow (a_2,a_1)$ and $\gamma_1 \rightarrow \gamma_2$.

We must consider the ambiguity of signs. First of all, γ_1 has an arbitrary sign, but it determines γ_2. Furthermore y_0 and y_0' can also be chosen arbitrarily, but then yy' has a sign determined by the product of (10.3a) and (10.3b) (which cancels the ambiguity of y_0 and y_0'). The signs are consistent for (10.3cd).

We define the jacobians of H and H_0 on each Riemann Surface (modulo periods of course) as column vectors with components consisting of abelian integrals with upper limits $P = (x,y)$, $P_0 = (x_0,y_0)$, etc., and some convenient lower limit, say ∞. Thus

(10.4a)

$$H: \quad u(P) = \gamma_1 \int^P dx(x-a_1)/y, \quad v(P) = \gamma_2 \int^P dx(x-a_2)/y,$$

(10.4b)

$$H_0: \quad u_0(P_0) = \int^{P_0} dx_0(x_0 - s)/y_0, \quad v_0(P_0) = \int^{P_0} dx_0(x_0 + s)/y_0.$$

Thus the jacobians are the column vectors

(10.4c) $$H: \quad J(P) = (u(P), v(P))^{tr},$$

(10.4d) $$H_0: \quad J_0(P_0) = (u(P_0), v(P_0))^{tr}.$$

Note that a change of (say) $(x, y) \rightarrow (x, -y)$ will cause $u \rightarrow -u$, $v \rightarrow -v$, $J(P) \rightarrow -J(P)$.

10.5 MAPPING OF JACOBIANS. *Theorem (10.3) may be rewritten with (10.3ac) and (10.3bd) grouped as follows:*

(10.5a) $$J_0(x_0, y_0) + J_0(x_0', y_0') = J(x, y),$$

(10.5b) $$J_0(x_0, y_0) + J_0(x_0', -y_0') = J(x', y').$$

The Jacobi inversion theorem leads to

(10.5c)
$$(u(x, y), v(x, y)) \rightarrow \{(u_0(x_0, y_0), v_0(x_0, y_0)), (u_0(x_0', y_0'), v_0(x_0', y_0'))\},$$

(10.5d)
$$(u(x', y'), v(x', y')) \rightarrow \{(u_0(x_0, y_0), v_0(x_0, y_0)), (u_0(x_0', y_0'), v_0(x_0', y_0'))\}.$$

In jacobian notation, more simply,

(10.5e) $$J(P) \rightarrow \{J_0(P_0), J_0(P_0')\},$$

(10.5f) $$J(P') \rightarrow \{J_0(P_0), -J_0'(P_0')\}.$$

Thus the jacobians of H and H_0 have a one-to-two correspondence that the curves could not have (recall §6).

11. Proof of the main identity. The steps follow the pattern of §7 (with $s = s_0$), so a brief sketch might suffice. From (10.1a), (with κ as yet unknown),

(11.1a) $$\kappa dx_0(x_0^2 - s^2)/x_0^2 = dx(x^2 - s^2)/x^2.$$

Together with the internal substitution of (10.1a) into (10.1c), this gives us

(11.1b) $$\text{left hand (10.3ab)} = \frac{dx(x^2 - s^2)}{\sqrt{\kappa x}\sqrt{x(x-1)(x - s^2)}} (A \pm A'),$$

(11.1c) $$A = \frac{x_0}{(x_0 + s)\sqrt{(x_0 - b_1)(x_0 - b_2)}}, \quad A' = \frac{x_0'}{(x_0' + s)\sqrt{(x_0' - b_1)(x_0' - b_2)}}.$$

What remains is a smaller computer algebra exercise than in §7 (fewer parameters): We compute $(AA')^2$ and $A^2 + A'^2$ as rational functions of x using (10.1a), e.g.,

(11.2a) $\qquad x_0 x_0' = s^2, \quad x_0 + x_0' = 1 + s^2 + (x-1)(x-s^2)/(x\kappa).$

As in §7, AA' is rational when $(x_0 - b_1)(x_0' - b_1)$ is a perfect square. The discriminant condition (as in (7.4cd)) gives the value of κ in (10.1b). We have two values $\pm AA'$, so we are simultaneously computing both of (10.3ab),

(11.2b) $\qquad A \pm A' = \sqrt{A^2 + A'^2 \pm 2AA'}.$

The radicand becomes a perfect square again, and this gives the differential shown in (10.3ab).

Finally, (10.3cd) follows from the substitution $s \to -s$.

12. Self-mapping of jacobians. The mapping of jacobians (10.5) from $H_0(x_0, y_0)$ to $H(x, y)$ can be recast as a mapping from $H(x, y)$ to itself (actually $H(z, w)$) by the biunique change of variables

(12.1a) $\qquad x_0 = (z-1)s^2/(z-s^2),$

(12.1b) $\qquad y_0 = w(s+1)(s-1)^2 s^3/(z-s^2)^3 \gamma_0,$

so the equation (10.1d) for H_0 becomes that of H again, i.e.,

(12.1c) $\qquad H: \quad w^2 = z(z-1)(z-s^2)(z-a_1)(z-a_2).$

The transformation (12.1a) comes from (10.1a) with x and z symmetrically disposed:

(12.1d) $\qquad \dfrac{(x-1)(x-s^2)}{x} \cdot \dfrac{(z-1)(z-s^2)}{z} = \kappa(s^2 - 1)^2.$

Here, κ comes from (10.1b) and

(12.1e) $\qquad \gamma_0^2 = -\dfrac{((r^2 - s^2)(r+s^2)^2(s-1)}{(r+rs+s^2-s)(r-rs+s^2+s)(s+1)s^2}.$

The abelian differentials on the left in (10.3a-d) can be transformed back to $H(z, w)$ by means of

(12.2a) $\qquad dx_0(x_0 - s)/y_0 = \gamma_0 dz(z+s)/w.$

At the same time the differentials on the right can be transformed to the "standard" ones in (10.4a) by

(12.2b) $\qquad (x - a_1) = (x+s)\alpha + (x-s)\beta, \quad (x - a_2) = (x-s)\alpha^* + (x+s)\beta^*,$

(12.2c) $\qquad \alpha = (s - a_1)/2s, \quad \beta = (s + a_1)/2s,$

(with the operator $*$ denoting the involution $s \to -s$).

The transformation (10.5ab) of jacobians (as column vectors) now becomes

(12.3a) $\qquad J(z, w) + J(z', w') = MJ(x, y),$

(12.3b)
$$J(z,w) + J(z',-w') = MJ(x',y').$$

Here M is the matrix

(12.3c)
$$M = \begin{pmatrix} \gamma^*\alpha & \gamma^*\beta \\ \gamma\beta^* & \gamma\alpha^* \end{pmatrix},$$

with

(12.3d)
$$\gamma^2 = \frac{(s+1)(r+rs+s^2-s)^2}{(s-1)r(r-1)(r+s^2)},$$

(12.3e)
$$\gamma\gamma^* = \frac{(r+rs+s^2-s)(r-rs+s^2+s)}{r(r-1)(r+s^2)}.$$

It can then be seen (using the choice of signs from (12.3de)), that M had determinant -1 and trace zero. Hence its eigenvalues are ± 1.

We now choose new jacobian coordinates to follow the eigenvectors, (say)

(12.4a)
$$j(z,w) = (\int dz(\epsilon_{11}z + \epsilon_{12})/w, \int dz(\epsilon_{21}z + \epsilon_{22})/w)^{tr},$$

and likewise (with the same eigenvectors) for $j(x,y)$.

With this choice of jacobian (12.3ab) becomes

(12.4c)
$$j(z,w) + j(z',w') = Tj(x,y),$$

(12.4d)
$$j(z,w) + j(z',-w') = Tj(x',y'),$$

(12.4e)
$$T = \begin{pmatrix} 1 & 0 \\ 0 & -1 \end{pmatrix}.$$

As before, (12.4c) and (12.4d) are equivalent (only one need be used).

The involution T and the correspondence of (z,w) with (z',w') justify the kernel $\mathbf{Z}/2\mathbf{Z} \times \mathbf{Z}/2\mathbf{Z}$ in the isogeny of the jacobian $j(z,w)$.

APPENDIX: THE HILBERT MODULAR INVARIANTS

13. Hecke's parameters. The period matrix R in (5.1c) must satisfy the sign condition

(13.1a)
$$\Im\rho\Im\rho'' - (\Im\rho')^2 > 0,$$

when the periods are those of abelian integrals. A period matrix with such sign conditions is called a *Riemann matrix*. We had the further restriction to produce real mutiplication

(13.1b)
$$\rho = 2\rho''.$$

Hecke [7] introduced new variables

(13.1c)
$$(\rho/2 =) \rho'' = (\tau - \tau')/2\sqrt{2}, \quad \rho' = -(\tau + \tau')/2,$$

for which the sign condition (13.1a) is satisfied by

(13.1d)
$$\Im\tau > 0 > \Im\tau'.$$

In (5.1c) we now replace R by SR, (where the matrix S denotes a linear combination of the abelian integrals),

(13.1e)
$$S = \begin{pmatrix} 1 & -\sqrt{2} \\ 1 & \sqrt{2} \end{pmatrix}.$$

Then the Riemann matrix R is rewritten as

(13.2a)
$$R = \begin{pmatrix} 1 & -\sqrt{2} & \sqrt{2}\tau & -\tau \\ 1 & \sqrt{2} & -\sqrt{2}\tau' & -\tau' \end{pmatrix}.$$

There are (again) four periods shown as column vectors for the two abelian integrals represented by the rows. There is an obvious equivalence under change of variables through multiplication by $\mathbf{GL_2(C)}$ on the left and unimodular change of period basis through multiplication by $\mathbf{GL_2(Z)}$ on the right, i.e.,

(13.2b)
$$R \approx \mathbf{GL_2(C)} \cdot R \cdot \mathbf{SL_4(Z)}.$$

The real multiplier ring isomorphic to $\mathbf{Z}(\sqrt{2})$ is put in evidence (again) by the *real* action on the variables:

(13.2c)
$$\begin{pmatrix} \sqrt{2} & 0 \\ 0 & -\sqrt{2} \end{pmatrix} \cdot R = \begin{pmatrix} \sqrt{2} & -2 & 2\tau & -\sqrt{2}\tau \\ -\sqrt{2} & -2 & 2\tau' & \sqrt{2}\tau' \end{pmatrix}.$$

The right-hand matrix columns clearly show only a subgroup of the periods of (13.2b) of index 4, i.e., of structure $\mathbf{Z}/2\mathbf{Z} \times \mathbf{Z}/2\mathbf{Z}$ as in (5.2d).

14. The Hilbert modular group. Generically, this is the group G acting on the pair (τ, τ') which preserves the equivalence classes of the Riemann matrices with real multiplication.

In our context, first there is a symmetric subgroup G^* of index 2,

(14.1a)
$$G^*: \quad \tau \to \frac{\alpha\tau + \beta}{\gamma\tau + \delta}, \quad \tau' \to \frac{\alpha'\tau' + \beta'}{\gamma'\tau' + \delta'}.$$

Here $\alpha, \beta, \gamma, \delta$ lie in $\mathbf{Z}(\sqrt{2})$, and $\alpha', \beta', \gamma', \delta'$ are the conjugates under $\sqrt{2} \to -\sqrt{2}$. The determinants also must satisfy

(14.1b)
$$\alpha\delta - \beta\gamma = \lambda, \quad \alpha'\delta' - \beta'\gamma' = \lambda',$$

for λ a totally positive unit, actually $\Omega = (1 + \sqrt{2})^{2m}$, $(m \in \mathbf{Z})$. (In this case, it suffices to take $\lambda = \lambda' = 1$ by using a factor of $\Omega^{1/2}$ on $\alpha, \beta, \gamma, \delta$, – likewise for the conjugates.)

The full Hilbert modular group has the alternating operation

(14.1c)
$$A: \quad \tau \to -\tau', \ \tau' \to -\tau.$$

So $G = <G^*, A>$. This group preserves the half-planes $\Im\tau > 0 > \Im\tau'$.

To see the action on the Riemann matrices (as in Hecke [7]), first let the element of G^* act so $R \to G^*(R)$. Then we remove the denominators by row multiplications,

(14.2a)
$$\begin{pmatrix} \gamma\tau + \delta & 0 \\ 0 & \gamma'\tau' + \delta' \end{pmatrix} \cdot G^*(R) = R^*,$$

where,

(14.2b) $\quad R^* = \begin{pmatrix} (\gamma\tau + \delta) & -\sqrt{2}(\gamma\tau + \delta) & \sqrt{2}(\alpha\tau + \beta) & -(\alpha\tau + \beta) \\ (\gamma'\tau' + \delta') & \sqrt{2}(\gamma'\tau' + \delta') & -\sqrt{2}(\alpha'\tau' + \beta') & -(\alpha'\tau' + \beta') \end{pmatrix}.$

Using integral components $\{a_0, \cdots, d_1\}$ such that

(14.2c) $\qquad \alpha = a_0 + a_1\sqrt{2}, \ \beta = b_0 + b_1\sqrt{2}, \ \gamma = c_0 + c_1\sqrt{2}, \delta = d_0 + d_1\sqrt{2},$

we can rewrite R^* in terms of rational integers,

(14.2d) $\qquad R^* = R \cdot M, \text{ where } M = \begin{pmatrix} d_0 & -2d_1 & 2b_1 & -b_0 \\ -d_1 & d_0 & -b_0 & b_1 \\ c_1 & -c_0 & a_0 & -a_1 \\ -c_0 & 2c_1 & -2a_1 & a_0 \end{pmatrix}.$

Since it can be verified that

(14.2e) $\qquad \text{determinant}(M) = (\alpha\delta - \beta\gamma)(\alpha'\delta' - \beta'\gamma') = 1,$

we see the substitutions of G^* preserves the equivalence class of R.

It is a trivial matter to see that the alternating substitution

(14.3) $\qquad A: \ \tau \to -\tau', \ \tau' \to -\tau, \ \sqrt{2} \to -\sqrt{2}$

simply interchanges the rows of R. So the Hilbert modular group $G = \langle G^*, A \rangle$ is a group preserving the equivalence classes of R.

15. The modular function field. The Hilbert modular function field for $Q(\sqrt{2})$ is the analogue of $Q(j(\tau))$ for the Klein modular group acting on τ.

We start with the *symmetric subfield* $Q(x, y)$ (see [6]). There

(15.1a) $\qquad (x =) \ x(\tau, \tau') = H_2^2/H_4, \quad (y =) \ y(\tau, \tau') = H_2 H_4/H_6,$

and H_k denotes special modular forms of (even) weight k, e.g.,

(15.1b) $\qquad G^*: \ H_k((\alpha\tau + \beta)/(\gamma\tau + \delta)) = ((\gamma\tau + \delta)(\gamma'\tau' + \delta'))^k H(\tau, \tau').$

The symmetry is derived from

(15.1c) $\qquad A: \ H_k(\tau, \tau') = H_k(-\tau', -\tau).$

The selection of the forms H_k of weight k is made so as to have *simplest infinite behavior*. In effect, we have the following diagonals:

(15.2a) $\qquad D_1: \tau(1 + \sqrt{2}) = \tau'(1 - \sqrt{2}) = \omega_1,$

(15.2b) $\qquad D_2: \tau = -\tau' = \omega_2.$

Then H_2, H_4, H_6 are chosen uniquely as the modular forms of indicated weight for which the diagonal behavior is like the Klein-Weber modular function $j(\omega_2)$ on $\mathbf{D}_2$ and like the Hecke $\sqrt{2}$-modular function $j_2(\omega_1)$ on $\mathbf{D}_1$. (Recall $j_2(\omega_1) = j_2(-1/\omega_1) = j_2(\omega_1 + \sqrt{2})$.) Thus (see [6] or [4])

(15.2c) $\qquad \mathbf{D}_1 : 1/x = H_4/H_2^2 = j_2(\omega_1), \quad H_6(\omega_1) \equiv 0,$

(15.2d) $\qquad \mathbf{D}_2 : 1/(xy) = H_6/H_2^3 = j(\omega_2), \quad H_4(\omega_2) \equiv 0.$

The field $\mathbf{Q}(x,y)$ is characterized by the symmetry (of $\tau \leftrightarrow -\tau', \tau' \leftrightarrow -\tau$). The *extended* or *alternating* field is $\mathbf{Q}(x,y,a)$ with a determined by choice of sign in

(15.3a) $\qquad (a^2 =) \; a(x,y)^2 = \dfrac{1}{2}(y+4)(x^2 y - 1728x - 288xy - 1024y^2 + 4xy^2),$

according to work of Gundlach [6] and Nagaoka [16]. Here

(15.3b) $\qquad a(x(\tau,\tau'), y(\tau,\tau')) = -a(x(-\tau',-\tau), y(-\tau',-\tau)).$

(The factor of $\frac{1}{2}$ will be needed to avoid irrationalities later on).

16. Explicit two-isogenies.

Three 2-isogenies come from $\tau \to (2+\sqrt{2})\tau, \tau' \to (2-\sqrt{2})\tau'$, (note the multipliers $2 \pm \sqrt{2}$ have norm 2 and preserve the signs). Actually, the three mappings arise concurrently by coset properties in the Hilbert modular group

(16.1) $\qquad (\tau, \tau') \to \begin{cases} ((2+\sqrt{2})\tau, (2-\sqrt{2})\tau')) \\ (\tau/(2+\sqrt{2}), \tau'/(2-\sqrt{2})) \\ ((\tau+1)/(2+\sqrt{2}), (\tau'+1)/(2-\sqrt{2})) \end{cases} \qquad (x,y,a)^1 \to^3 (X,Y,A).$

This is analogous to the triple $\tau \to \{2\tau, \tau/2, (1+\tau)/2\}$ for the ordinary (Klein) modular group, which produces three conjugates $j(\tau) \to \{j(2\tau), j(\tau/2), j((1+\tau)/2)\}$.

For simplicity, we begin by ignoring the mapping $a \to A$, leaving only $(x,y) \to (X,Y)$ (see Cohn [4]). We then start with a two-valued correspondence (to be resolved later by the choice of sign of a).

Given (x,y) the symmetrized (X,Y) are determined by an incomplete intersection with four *cylindrical* equations of co-dimension 2, (note the degrees of the variables):

(16.2a)
$$f_1(\overset{3}{x}, \overset{2}{y}; \overset{3}{X}) = X^3 x + (432 + 156y - xy)X^2 x + x(xy + 144y - 1728)^2$$
$$+ (4x^2 y^2 + 207x^2 y + 1152y^2 x + 19008yx + 62208x + 82944y^2)X = 0,$$

(16.2b)
$$(f_2 =) \; f_1(X,Y; x) = 0,$$

(16.2c)
$$f_3(\overset{2}{x}, \overset{3}{y}; \overset{3}{Y}) = Y^3 y^2(y+4) - (x - 4y + 48)y^2 Y^2 + (-5y + 108)xyY$$
$$+ x(xy + 144y - 1728) = 0,$$

(16.2d)
$$(f_4 =) \ f_3(X, Y; y) = 0.$$

The symmetry of $(x, y) \leftrightarrow (X, Y)$ is reminiscent of the symmetry of $x = j(\omega)$ and $X = j(2\omega)$ in $\Phi_2(x, X) = 0$, the (cubic) Klein modular equation of order 2.

17. Uniqueness problem. As we see from the three 2-isogenies, for every (x, y) there should be only three pairings of (X, Y) with (x, y) (not nine from the determination of three values of X from $f_1 = 0$ and three values of Y *independently* from $f_3 = 0$). The equations $f_2 = 0$ or $f_4 = 0$ (generically) resolve the three legitimate pairings (X, Y).

If we take discriminants of the cubics f_1 and f_3 separately,

(17.1a)
$$X\text{-disc}(f_1) = 4xy^2 (x^2 y - 1728x - 288xy - 1024y^2 + 4xy^2)$$
$$\cdot (3375x^2 - 746496y + 10368xy + 252x^2 y + x^3 y)^2,$$

(17.1b)
$$Y\text{-disc}(f_3) = 4(x^2 y - 1728x - 288xy - 1024y^2 + 4xy^2)$$
$$\cdot (432 - 72y + xy + 3y^2)^2 y^4 x.$$

The fields $\mathbf{Q}(x, y, X)$ and $\mathbf{Q}(x, y, Y)$ are the same as we show next, but for now we may note the discriminants agree to within square factors!

18. Algorithm for identifying field generators. We can explicitly link Y, the root of f_3, with X, the root of f_1 by a construction using the following:

18.1 MAIN ALGORITHM. *Let K/k be an extension field of degree n defined by ξ where*

(18.1a)
$$(A(\xi) =) \ A_n \xi^n + A_{n-1} \xi^{n-1} + \cdots + A_0 = 0, \ (A_j \in k).$$

Let some other element η be defined by the equation

(18.1b)
$$(B(\eta) =) \ B_n \eta^n + B_{n-1} \eta^{n-1} + \cdots + B_0 = 0, \ (B_j \in k).$$

It is required to discover if η lies in K, and if so to express η in terms of ξ as a polynomial of degree $n - 1$,

(18.1c)
$$\eta = C_{n-1} \xi^{n-1} + \cdots + C_0 \ (= C(\xi)), \ (C_j \in k).$$

We look for the $n - 1$ unknown coefficients C_j. Note the converse process, finding $B(\eta)$ given $A(\xi)$ and $C(\xi)$, is taken care of by

(18.1d)
$$B(\eta) = \xi\text{-resultant of } A(\xi) \text{ and } \eta - C(\xi).$$

So even with $C(\xi)$ unknown, the ξ-resultant of $A(\xi)$ and $\eta - C(\xi)$ is a polynomial $R(\eta)$ of degree n whose coefficients are to be proportional to $B(\eta)$. By comparing the coefficients of $R(\eta)$ and $B(\eta)$ we have $n-1$ equations $T_i = 0$ for $i = 1, \cdots, n-1$ (in the $n-1$ unknowns C_j and the knowns A_j, B_j). Here the T_i are of degree i in the unknowns C_j.

Then the unknowns C_j are found by the use of resultants for successive elimination.

This process is tolerably easy where $n = 3$. We restrict further remarks to this case for simplicity. For instance, T_1 is linear, so C_0 can be solved in terms of C_1 and C_2. Then T_2 is quadratic and T_3 is cubic in the unknowns C_1 and C_2. So the C_1-resultant is a sixth degree polynomial in C_2, which we factor over k. In fact, (note the degrees),

$$(18.1e) \quad C_1\text{-resultant}(T_2, T_3) \;=\; R(\overset{6}{C_2}; \overset{6}{A_0}, \overset{9}{A_1}, \overset{9}{A_2}, \overset{22}{A_3}, \overset{2}{B_0}, \overset{3}{B_1}, \overset{6}{B_2}, \overset{12}{B_3}) \;=\; 0 \quad (n = 3).$$

If a solution (C_0, C_1, C_2) does exist, there must be a linear factor giving C_2. Going back we have a linear factor for C_1 in T_1, and a formula for C_0, completing the solution.

For efficiency, it is desirable to store the resultants and equations T_i. For instance, for $n = 3$, we would solve the linear T_1 to eliminate (say) C_0 in T_2 and T_3. So we store T_1, T_2, and the C_1-resultant of T_2 and T_3, (a polynomial of degree 6 in C_2).

18.2 EXTRANEOUS SOLUTIONS. *This process will always produce extraneous solutions. For instance T_2 is quadratic so it has two roots if it has one. Actually, in the case of a normal equation, each of the n roots of $B(\eta)$ are expressed by some $C(\xi)$, but we still have n extraneous solutions. So verification of $C(\eta)$ is still necessary.*

18.3 MONIC EQUATIONS. *If we consider the fact that the degrees of A_3 and B_3 are exceptionally high in the C_1-resolvent (even for $n = 3$), it is advantageous to change variables before executing the algorithm so as to make ξ and η monic, (e.g., $\xi \to A_n\xi$ and $\eta \to B_n\eta$, so $A_n \to 1, B_n \to 1$). In practice this helps avoid overflow.*

19. Improved parameters. With this algorithm (using $k = Q(x, y)$) we determine $X \in Q(x, y, Y) = K$ from $f_3(= A(Y))$ and $f_1(= B(X))$ as

(19.1a)
$$\begin{aligned}
(X =) \quad X(Y; x, y) \;=\; & [-(yx - 1728 + 144y)(y^2 + 3y - 36)x]/d_X \\
& + Y[(x^2y + 2160x - 360yx - 20736y - 81xy^2 + 3456y^2 + 4y^3x - 144y^3)y]/d_X \\
& + Y^2[(yx + 3x + 432 - 36y)(y + 4)y^2]/d_X,
\end{aligned}$$
(19.1b)
$$d_X \;=\; (yx + 432 - 72y + 3y^2)x.$$

At the expense of much more complicated relations, we now use the single-valued notation, in which the extended modular function field is birationally parametrized as

$$(19.2a) \qquad\qquad Q(x, y, a) \;=\; Q(s, y)$$

by introducing the new parameter

$$(19.2\mathrm{b}) \qquad (s =)\ s(a,x,y)\ =\ \frac{a + 4(y+4)(35y+108)}{x + 200y + 864}.$$

Conversely, $x = x_s(s,y)$ and $a = a_s(s,y)$ where

(19.3a)
$$x_s(s,y)\ =\ \frac{16(25y+108)s^2 - 16(y+4)(35y+108)s + 4(y+4)(49y^2 + 288y + 432)}{y(y+4) - 2s^2},$$

(19.3b)
$$a_s(s,y)\ =\ \frac{-8(y+4)(35y+108)s^2 + 36(11y+12)(y+4)^2 s - 4y(y+4)^2(35y+108)}{y(y+4) - 2s^2}.$$

The derivation (see Müller [15]) comes from the interpretation of the equation (15.3a) for a^2 as a conic in a, x over $\mathbf{Q}(y)$. Then by the "method of Pythagoras," all we need for parametrization is a rational point in $\mathbf{Q}(y)$, namely

$$(19.4\mathrm{a}) \qquad a_0\ =\ -4(y+4)(35y+108), \quad x_0\ =\ -200y - 864.$$

Then the slope is the new parameter

$$(19.4\mathrm{b}) \qquad s\ =\ (a - a_0)/(x - x_0).$$

Now, the multiplication on (τ, τ') by $(2 + \sqrt{2}, 2 - \sqrt{2})$ (and conjugates) leads to a three-valued mapping of

$$(19.4\mathrm{c}) \qquad (x, y, a, s)^1 \to^3 (X, Y, A, S).$$

In terms of the one-valued parametrization we want just the mapping

$$(19.4\mathrm{d}) \qquad (s, y)^1 \to^3 (S, Y).$$

We pay for the "better" parametrization by more complicated relations. The cubic f_3 which defines Y now becomes

$$(19.5\mathrm{a}) \qquad f_{3s}(\overset{3}{Y}; \overset{4}{s}, \overset{7}{y})\ =\ Y_3 Y^3 + Y_2 Y^2 + Y_1 Y + Y_0\ =\ 0,$$

with the more formidable coefficients

(19.5b)
$$Y_3 = y^2(y+4)(y^2 + 4y - 2s^2)^2,$$

(19.5c)
$$Y_2 = -8y^2(y+4)(y^2+4y-2s^2)(24y^2+150y+216-70sy-216s+51s^2),$$

(19.5d)
$$Y_1 = -4y(5y-108)(y^2+4y-2s^2)$$
$$(49y^3-140sy^2+484y^2+1584y-992sy+100s^2y-1728s+1728+432s^2),$$

(19.5e)
$$Y_0 = 16(49y^4+520y^3-140y^3s+1296y^2+100y^2s^2-992sy^2-1728sy$$
$$+360s^2y+864s^2)(49y^3-140sy^2+484y^2+1584y-992sy$$
$$+100s^2y-1728s+1728+432s^2).$$

Note that the equation for $X(Y;x,y)$ can now be rewritten by substitution as $X_s(Y;s,y)$, with similar loss of simplicity.

Given s and y, for each of these three (generic) values of Y there must be one $S(s,y;Y)$, so we have the three-valued correspondence $(s,y)^1 \to^3 (S,Y)$ precisely formulated (generically). This is a much larger computation. The "total" equations are too long to be manageable (or even instructive), but we outline the component steps.

First, we define ρ as the ratio $A/a = a(X,Y)/a(x,y)$, i.e.,

(19.6a)
$$\rho^2 = \frac{(Y+4)(X^2Y-1728X-288XY-1024Y^2+4XY^2)}{(y+4)(x^2y-1728x-288xy-1024y^2+4xy^2)}$$

If we substitute $X = X(Y;x,y)$ then the Y-resultant of the above equation with f_3 yields a cubic in ρ^2. This equation *factors* into two cubics (trivially equivalent under $\rho \leftrightarrow -\rho$):

(19.6b)
$$0 = \rho_0 + \rho_1\rho + \rho_2\rho^2 + \rho_3\rho^3, \quad \text{where}$$

(19.6c)
$$\rho_0 = (xy-1728+144y)(x^2y^2+2x^2y-904xy^2-9920xy-16000x+165888y^2),$$

(19.6d)
$$\rho_1 = y(-746496y^4+2y^3x^3-1055y^3x^2+9289728y^2-7741440y^3,$$
$$-4032xy^4+8x^2y^4-14008x^2y^2-3365632xy^2-213184y^3x$$
$$-9040896x-23280x^2y-8294400x+4x^3y^2),$$

(19.6d)
$$\rho_2 = -y^2(-2xy+480+464y+54y^2-xy^2)(x^2y-1728x-288xy-1024y^2+4xy^2),$$

(19.6e)
$$\rho_3 = y^3(y+4)^2(x^2y-1728x-288xy-1024y^2+4xy^2).$$

We note (with some relief!) that no radicals (like $\sqrt{2}$) were needed for factorization leading to this equation!

By applying the Main Algorithm again, we express ρ directly in $\mathbf{Q}(Y, x, y)$ as

(19.7a)
$$(\rho =) \rho(Y; x, y) = [(yx - 1728 + 144y)(4y^3 x - 576y^3 + x^2 y^2 - 174xy^2$$
$$+ 6912y^2 + x^2 y + 576yx + 4320x)]/d_\rho$$
$$- Y[2y(-1488y^4 + 8xy^4 + 2y^3 x^2 - 643y^3 x + 38976y^3 - 292608y^2$$
$$- 51x^2 y^2 + 8252xy^2 + 470016y - 108x^2 y - 11664yx - 139968x)]/d_\rho$$
$$- Y^2[y^2 (x^2 y^2 + 2x^2 y - 432x - 235008 - 760yx - 12672y - 211xy^2$$
$$+ 2400y^2 + 4y^3 x - 744y^3)]/d_\rho,$$

(19.7b)
$$d_\rho = y(x^2 y - 1728x - 288yx - 1024y^2 + 4xy^2)(yx + 432 - 72y + 3y^2).$$

The s, y parameters can now be made basic. Since $\rho \in \mathbf{Q}(Y, x, y) \subset \mathbf{Q}(s, Y, y)$, the same can now be said for $A = a\rho$ and finally for

(19.8a)
$$(S =) S(Y; s, y) = \frac{A + 4(Y + 4)(35Y + 108)}{X + 200Y + 864}.$$

(In fact we use $a = a_s(s, y), X = X(Y; x, y)$ and $x = x_s(s, y)$ to obtain S as the ratio of two quadratics in Y with coefficients in $\mathbf{Q}[s, y]$).

The "final" result, by another use of the algorithm, is

(19.8b)
$$S = S_0(s, y) + Y S_1(s, y) + Y^2 S_2(s, y),$$

which would, unfortunately, involve unmanageable polynomials of $\mathbf{Q}(s, y)$ in the rational coefficients $S_i(s, y)$. We content ourselves with the implicit result.

20. Fixed points as illustrations. The fixed points of the system (16.2a-d), where $(x, y) = (X, Y)$ (see Cohn [4]), lead to the following table:

$x = X$	$y = Y$	$a = A$	$A/a\ (= \rho)$	$s = S$
576	12	$+3072$	1	$48/5$
"	"	-3072	1	8
$\frac{117+\sqrt{17}}{2}$	$\frac{-3-3\sqrt{17}}{2}$	$\frac{3}{\sqrt{2}}(\frac{5+\sqrt{17}}{2})^{16}(4 - \sqrt{17})^9$	1	$\frac{8-24\sqrt{17}+9\sqrt{2}+\sqrt{34}}{20}$
"	"	$-\frac{3}{\sqrt{2}}(\frac{5+\sqrt{17}}{2})^{16}(4 - \sqrt{17})^9$	1	$\frac{8-24\sqrt{17}-9\sqrt{2}-\sqrt{34}}{20}$
$\frac{117-\sqrt{17}}{2}$	$\frac{-3+3\sqrt{17}}{2}$	$\frac{3}{\sqrt{2}}(\frac{5-\sqrt{17}}{2})^{16}(4 + \sqrt{17})^9$	1	$\frac{8+24\sqrt{17}+9\sqrt{2}-\sqrt{34}}{20}$
"	"	$-\frac{3}{\sqrt{2}}(\frac{5-\sqrt{17}}{2})^{16}(4 + \sqrt{17})^9$	1	$\frac{8+24\sqrt{17}-9\sqrt{2}+\sqrt{34}}{20}$
-64	-4	0	-1	

There are generally two values for $s(= S)$ corresponding to the two roots $\pm a(x, y)$. Note the role of conjugates in $\mathbf{Q}(\sqrt{2}, \sqrt{17})$. Also note in the last case, that the locus $a = A = 0$ can be a point of indeterminacy for some parametrizations.

21. Remark on algebraic invariants. Something sorely lacking in this exposition is explicit algebraic expressions for the Hilbert modular invariants, (e.g., x, y, a, s, etc.,) in terms of coefficients of the hyperelliptic equation, analogously with classical elliptic formulas for j, such as

$$(21.1a) \qquad w^2 = z^3 + Az + B \;\to\; j = \frac{1728 \cdot 4A^3}{4A^3 + 27B^2}.$$

Thus j is seen directly as an elliptic curve parameter, e.g.,

$$(21.1b) \qquad B = \frac{-27j}{4(j - 1728)t^3}, \quad A = tB \quad (t \text{ arbitrary}).$$

Armand Brumer has suggested that corresponding relations should be known from specialization of the Siegel modular functions, but it would be helpful to have explicit results in palatable form.

REFERENCES

1. J.-B. Bost and J.-F. Mestre, *Moyenne arithmético-géometrique et périodes des courbes de genre 1 et 2*, Gazette des Mathematiciens (Soc. Math. de France) (1988), 36–64.

2. A. Brumer, *(Unpublished communication, March 1997)*.

3. J.W.S. Cassels and E.V. Flynn, "Prolegomena to a Middlebrow Arithmetic of Curves of Genus 2," Cambridge Univ. Press, 1996.

4. H. Cohn, *An explicit modular equation in two variables and Hilbert's twelfth problem*, Math. of Comput. **38** (1982), 227–236.

5. H. Cohn, *Introductory remarks on complex multiplication*, Internat. Journ. of Math. and Math. Sci. **5** (1982), 675–690.

6. K.B. Gundlach, *Die Bestimmung der Funktionen zu einigen Hilbertschen Modulgruppen*, Journ. für die reine und angewandte Math. **220** (1965), 109–153.

7. E. Hecke, *Höhere Modulfunktionen und ihre Anwendung an der Zahlentheorie*, Math. Annalen **71** (1912), 1–37.

8. F. Hirzebruch, *Hilbert modular surfaces*, Enseignement Mathématique **19** (1973), 183–281.

9. E. Howe, *Constructing distinct curves with isomorphic Jacobians*, Journ. of Number Theory **56** (1996), 381–390.

10. G. Humbert, *Sur les fonctions abéliennes singulières I*, Journ. de Math. (5) **5** (1899), 233–350.

11. J. Igusa, *Arithmetic variety of moduli for genus two*, Annals of Math. **72** (1960), 612–649.

12. L. Königsberger, *Über die Transformation der Abelschen Functionen erster Ordnung*, Journ. für die reine und angewandte Math. **64** (1865), 17–42.

13. J.-F. Mestre, *Familles de courbes hyperelliptiques à multiplications réelles,*, Arithmetic algebraic geometry (1989), 193–208; Prog. Math., 89, Birkhäuser, Boston 1991.

14. J.-F. Mestre, *Courbes hyperelliptiques à multiplications réelles*, Séminaire de Théorie des Nombres, Univ. Bordeaux I, Talence (1987–1988), Exp. 34, 6 pp..

15. R. Müller, *Hilbertsche Modulformen und Modulfunktionen zu* $Q(\sqrt{8})$, Math. Annalen **266** (1983), 83–103.

16. S. Nagaoka, *On Hilbert modular forms*, Proc. Japan Acad. Ser. A **59** (1983), 346–348.

17. B. Poonen, *Computational aspects of curves of genus at least 2*, Algorithmic Number Theory (ANTS II, H. Cohen, ed.) (1996), 283–306; Second International Symposium, Talence, France; Springer-Verlag (Lecture Notes in Computer Science **1122**).

18. H.L. Resnikoff, *Singular Kummer surfaces and Hilbert modular forms*, Rice Univ. Studies **59** (1973), 109–121.

19. K. Rohn, *Transformation der hyperelliptischen Functionen* $p = 2$ *und ihre Bedeutung für die Kummersche Fläche*, Math. Annalen **15** (1879), 315–354.

IDA Center for Computing Sciences
Bowie, MD 20715-4300
email: hcohn@super.org

HUMBERT'S CONIC MODEL AND THE KUMMER SURFACE

HARVEY COHN

Abstract. In his Cours d'Analyse in 1904, Georges Humbert used the parametrization of a pencil of conics through four points by the Weierstrass ℘-function to prove theorems of geometry and mechanics. This method is implicit in his earlier applications of Kummer surfaces, for instance his criterion for real multiplication by $\sqrt{2}$ uses the special "quarter-period" configuration in the pencil.

0. Introduction. This talk is intended to relate two topics found in work of Georges Humbert (1859-1921). The first is an easy but unconventional parametrization of the conic (Part I), and the second is a conventional parametrization of the Kummer surface (Part II). The interrelation of these two topics is thinly sketched for the special goal of real multiplication by $\sqrt{2}$ (Part III). All this is over the field C.

Some of the best work on the parametrization of the Kummer surface was done by Humbert [8] in 1895. It represented the highest state of art in the use of theta-functions (which has survived intact) and the algebraic geometry of surfaces (which has transformed itself almost beyond recognition). Indeed, more modern treatments (see [3]) tend to use purely algebraic tools altogether, rejecting theta-functions. Some very readable accounts of hyperelliptic (genus two) jacobian manifolds are given in [14] and [18], and valuable bibliographies appear in [3] and [16].

After his advanced work, Humbert published an elliptic parametrization of the family of conics through four given points [10]. This tool has remained largely neglected despite the combination of its simplistic appeal (see [4], [13]) and relevance to the more prized real multiplication (see [2], [19]).

PART I. ELLIPTICAL PARAMETERS ON A CONIC PENCIL

1. Two theorems. Humbert [10] displayed two enticing results:

1.1 EULER. *Given two circles in the plane, one containing the other. It is generally impossible to construct a triangle which is inscribed in the larger and circumscribes the smaller, but when this is possible there is a continuum of such triangles.*

The obvious trivial case is where the circles are concentric. Only when the ratio of radii is $\sqrt{3}$ will there be such a triangle, and of course the configuration rotates.

While Euler simply counted parameters, Poncelet constructed the solution by involutions of point sets. Jacobi used the elliptic parametrization we shall discuss in the next section. The next theorem provides his special application.

1.2 JACOBI. *Consider a weight following a circular pendulum with position $P(t)$ in the plane of the pendulum at time t. Then for any (real) constant z, the chord connecting $P(z)$ and $P(t + z)$ will have as envelope a circular arc joining the high points of the pendulum path.*

Mathematics Subject Classification. Primary 11G15, 51A05. *Key words and phrases.* Poncelet, Humbert, real multiplication.
Presented at the New York Number Theory Seminar 6 Nov. 1997.

Typeset by $\mathcal{A}_{\mathcal{M}}$S-TEX

2. Parametrizations of a conic. For simplicity we agree that we are in the field $\mathbb{C}$ and conics are understood to be nondegererate, (not consisting of lines).

In the projective plane, to begin with, all conics are the same and all possible birational parametrizations are equivalent (affinely) to the simplest imaginable,

(2.1a)
$$x = t, \; y = t^2.$$

Actually, the most general parametrization is

(2.1b)
$$y : x : 1$$
$$= (a_{11}t^2 + a_{12}t + a_{13}) : (a_{21}t^2 + a_{22}t + a_{23}) : (a_{31}t^2 + a_{32}t + a_{33}),$$

with matrix form ("tr" for transpose),

(2.1c)
$$(y, x, 1)^{tr} = M \, (t^2, t, 1), \quad M = (a_{ij}), \; \det M \neq 0$$

It is clear that if M^{-1} acts on both sides, we get the new coordinates for (2.1a),

(2.1d)
$$(y', x', 1)^{tr} = M^{-1} \, (y, x, 1) = (t^2, t, 1).$$

Now by contrast, we create a unirational (two-to-one) parametrization of an arbitrary conic section. We consider a (complex) elliptic curve in Weierstrass form,

(2.2a)
$$y^2 = 4(x - e_1)(x - e_2)(x - e_3), \; (x = \wp(u), y = \wp'(u)),$$

(2.2b)
$$e_i \neq e_j (i \neq j); \; e_1 + e_2 + e_3 = 0.$$

For uniform notation, we write the parameters

(2.2c)
$$\{t_0, t_1, t_2, t_3\} = \{\infty, e_1, e_2, e_3\}.$$

We recall the periods and values at the half-periods are

(2.2d)
$$\{2w_1, 2w_2, 2w_3 (= 2w_2 + 2w_1)\},$$

(2.2e)
$$\wp(w_j) = e_j, \; \wp'(w_j) = 0, \; (j = 1, 2, 3).$$

The period lattice is

(2.2f)
$$\Omega = \{2n_1 w_1 + 2n_2 w_2\}, \; (n_1, n_2 \in \mathbb{Z}).$$

Remembering that t is the birational parameter of the conic (2.1b) in the (projective complex) xy-plane, we replace the parameter by u (writing "$[u]$" symbolically for the point (x, y)) as follows:

(2.3a)
$$[u] : \; (x, y) = (x(t), y(t)) \quad \text{for} \quad t = \wp(u).$$

Since $\wp(u) = \wp(-u)$, we have the two-to-one correspondence

(2.3b)
$$[u] \to (x, y), \; [-u] \to (x, y); \; (x, y) \to \{[u], [-u]\}.$$

If two symbols $[u]$ and $[z]$ were "added" we would obtain only two chords from $[u]$ to $[\pm u \pm z]$, namely

(2.3c)
$$\{\text{chord } ([u][u + z]), \; \text{chord } ([u][u - z])\}.$$

(The term "chord" will also refer to the corresponding secant).

Above all, the values of $[u]$ are periodic (2.2c) which makes for even more interesting results. Hence the chords in (2.3c) are distinct unless u or z is a half-period, ($2u$ or $2z$ lies in Ω).

3. Pencils of conics.

3.1 MAIN THEOREM. CONSTRUCTION OF THE PENCIL OF CONICS THROUGH FOUR POINTS OF A GIVEN CONIC. *Given a conic $C(0)$ with four designated base points in parameter $t = t_0, t_1, t_2, t_3$. Let us introduce an elliptic function $t = \wp(u)$ such that those four points correspond (uniquely) to $u = 0, w_1, w_2, w_3$ (with the periods as described earlier in (2.2d)). Then each point of $C(0)$ is described generically as $[u]$ and $[-u]$.*

Let z be a fixed parameter and draw the chords (secants) connecting $[u]$ with $[u + z]$ and connecting $[u]$ with $[u - z]$, assuming z is not a half-period. Then the chords vary with u sweeping out a conic $C(z)$ as their envelope with points of contact P and $-P$ (see Figure 1). This is the most general conic of the pencil.

Finally, for z a half-period, w_1, w_2, w_3, $C(z)$ is one of three limiting degenerate conics, namely the diagonals of the complete quadrangle determined by the base points.

The essence of the construction is that the envelope of the chords is a curve from which we can draw two tangents from an external point, hence a conic $C(z)$. When $u \in \{0, w_1, w_2, w_3\}$, then there is only one tangent, which corresponds to the four base points (of intersection of $C(0)$ with $C(z)$).

Let us next consider the three degenerate conics of the pencil, $C(w_i)$ with intersecting lines at A_i (as in Figure 1). Here, A_i is the intersection of the chords (of $C(w_i)$) from $[0]$ to $[w_i]$ with (cyclically) the chord from $[w_{i+1}]$ to $[w_{i+2}]$. Also the points $[u]$ and $[u + w_i]$ are *in involution*. This involution is created from the requirement that the chord from $[u]$ to $[u + w_i]$ must go through A_i, and indeed $C(w_i)$ is the "degenerate envelope" of these chords.

According to this same involution, the lines from A_i which are tangent to $C(0)$, (not shown in Figure 1), must occur where $\pm u \equiv u + w_i$ mod Ω, i.e., at *quarter-periods*, viz., $\pm w_i/2, \pm(w_i + w_{i+1})/2$. There are six of them, identified with the tangencies on $C(0)$ from A_1, A_2 and the imaginary tangencies from A_3, (using the notation of Figure 1).

3.2 COROLLARY. *Any given pair of nontangent conics, (not equivalent to concentric circles), can be embedded in a pencil so as to be $C(0)$ and $C(z)$ for a proper choice of parameter z.*

The cases considered are those where the conics have four distinct points of intersection, which become the base points. Of course all points are projective-complex.

4. Proofs for §1. For Theorem 1.1, we consider two circles (or any other conics indeed) as $C(0)$ and $C(z)$ of a pencil. Then a triangle inscribed in $C(0)$ and circumscribing $C(z)$ must have on $C(0)$ the vertices

$$(4.1a) \qquad S = \{[u], [u + z], [u - z]\},$$

and these must be the same if we start from $u = u + z$ or $u = u - z$. Easily, $[u + 2z] = [u - z]$, etc., so

$$(4.1b) \qquad 3z \equiv 0 \mod \Omega.$$

Thus we can "rotate" the configuration by making u vary.

136

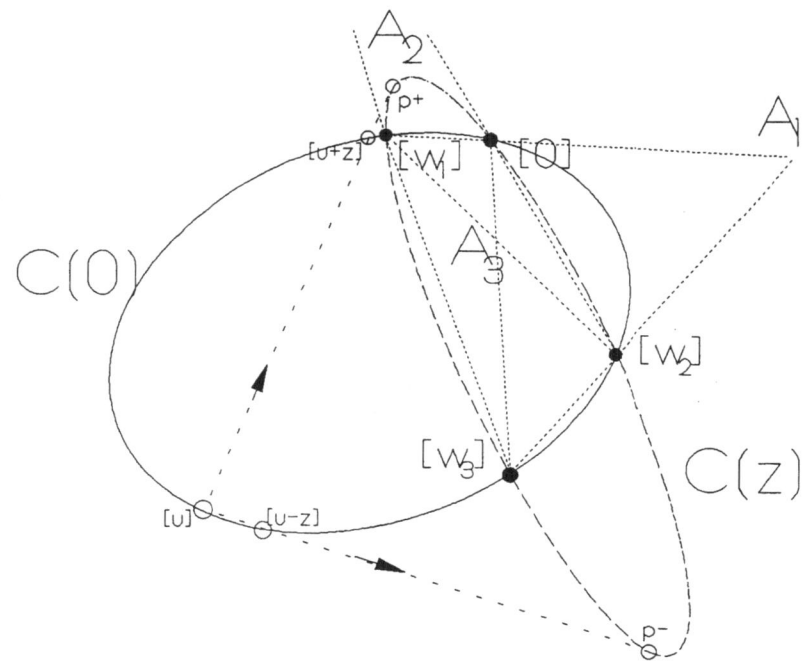

Figure 1. Generic real case

There is a similar result for an n-gon rather than a triangle. It corresponds to n-points of division of the period structure Ω. It would be dependent on the parity of n, and, worse, it *should* require a painful sorting out of reentrant polygons. We do not pursue this matter here (see [10]).

Theorem 1.2 is again relegated to the literature [10] with the remark that the spherical pendulum is parametrized by the $\wp$-function.

5. The quarter-period configuration of the pencil. We consider the special case of a primitive quarter-period

$$(5.1) \qquad\qquad 4z \in \Omega, \quad (2z \notin \Omega).$$

This is shown in Figure 2, where two base-points are real and two imaginary.

Here, $[u]$ creates its pair $[u \pm z]$ but $[u + 2z]$ creates the *same pair*. (Remember $3z \equiv -z \mod \Omega$). If we look at the four points just mentioned, the chords produce the following tangencies at $C(z)$:

chord($[u]$, $[u + z]$) is tangent at "5"

chord($[u]$, $[u - z]$) is tangent at "6"

chord($[u + 2z]$, $[u + z]$) is tangent at "4"

chord($[u + 2z]$, $[u - z]$) is tangent at "3."

As we saw earlier, since $2z$ is a half-period, the chords between $[u]$, $[u+2z]$ and $[u+z]$, $[u-z]$ go through A_2 (the center of the degenerate conic).

Additionally, if $u = z/2$ or $3z/2$, one chord joins them and another becomes tangent to $C(0)$. This produces two points of tangency of $C(z)$, written as "1" and "2."

These six points of tangency on $C(z)$ are a very valuable set. We first ask how many degrees of freedom are in the set. There is one module in the four base points on $C(0)$, essentially the j-invariant of the elliptic curve (2.2a). There is one other module, the value of u. Let s_i be the (biunique) t-parameter of the tangent point "i" on $C(z)$. Then Humbert [9] showed that the hyperelliptic curve

(5.2a) $$y^2 = (x - s_1)(x - s_2)(x - s_3)(x - s_4)(x - s_5)(x - s_6)$$

has modules creating real multiplication by $\sqrt{2}$, or otherwise expressed, modules transformed by the Hilbert modular function for $\sqrt{2}$ (see [6]).

In more convenient parameters (see [4]), the quintic (5.2a) becomes

(5.2b) $$y^2 = x(x - 1)(x - s^2)(x - a_1)(x - a_2),$$

(5.2c) $$a_1 = r\frac{r - rs + s^2 + s}{r + rs + s^2 - s}, \quad a_2 = r\frac{r + rs + s^2 - s}{r - rs + s^2 + s}.$$

6. The pencil as a symmetrized period manifold. The periodic structure C/Ω (the parallelogram) is the exact image of the elliptic curve (2.2a). Now $C(z)$ is by definition periodic in z, but also it is symmetric, so

(6.1a) $$C(z) = C(z + 2w_i) = C(-z).$$

So the usual parallogram is reduced by several identifications. Generally, from (6.1a),

(6.1b) $$C(w_i + z) = C(w_i - z).$$

So there are centers of symmetry at each of $\{0, w_1, w_2, w_3\}$. It is easy to see that under the symmetry the manifold for z namely $(C/\Omega)/\{\pm u\}$ is of genus 0. Correspondingly, the manifold of $[u]$ can be the (birationally parametrizable) conic.

PART II. THE KUMMER SURFACE

7. The symmetrized jacobian. We make the generalization from elliptic functions (of one variable and two independent periods) to *hyperelliptic* functions of two variables with four periods. We restrict ourselves to period systems reducible to the columns of the so-called Riemann matrix

(7.1a)
$$R = \begin{pmatrix} 1 & 0 & a & b \\ 0 & 1 & b & c \end{pmatrix},$$

with three complex parameters a, b, c whose imaginary parts form a positive definite matrix inside (7.1a),

(7.1b)
$$\Im(a\xi^2 + 2b\xi\eta + c\eta^2) > 0, \text{ for real } (\xi, \eta) \neq (0, 0).$$

Such period matrices come from the periods of hyperelliptic integrals (of genus 2). The quadruple periods form a lattice L in $\mathbf{C}^2$ (with complex variables u and v), viz.,

(7.1c)
$$\text{Periods of } \begin{pmatrix} u \\ v \end{pmatrix}: L = \left\{ G_1 \begin{pmatrix} 1 \\ 0 \end{pmatrix} + G_2 \begin{pmatrix} 0 \\ 1 \end{pmatrix} + H_1 \begin{pmatrix} a \\ b \end{pmatrix} + H_2 \begin{pmatrix} b \\ c \end{pmatrix} \right\};$$

(7.1d)
$$G_1, G_2, H_1, H_2 \in \mathbf{Z}.$$

The manifold of $\mathbf{C}^2/L$ is a period parallelopiped in two complex dimensions. It is called a "jacobian" manifold because of its original application (see [14]) where u and v are hyperelliptic abelian integrals of the first kind with the periods shown.

Functions defined on it are quadruply periodic (meromorphic) functions, which may be represented as the quotients of special holomorphic functions. These are not quite quadruply periodic but have exponential periodicity factors of a special form. They are called *normal* theta-functions of order N (see [5]) and are defined as a linear space (over $\mathbf{C}$) by the period relations

(7.2a)
$$\theta_N(u, v) = \theta_N(u + 1, v) = \theta_N(u, v + 1),$$

(7.2b)
$$\begin{aligned} \theta_N(u, v) &= \exp 2\pi[i(u + a/2)N] \, \theta_N(u + a, v + b) \\ &= \exp 2\pi[i(v + c/2)N] \, \theta_N(u + b, v + c). \end{aligned}$$

Special attention is given to *even* theta-functions defined by

(7.2c)
$$\theta_N(u, v) = \theta_N(-u, -v).$$

It can be shown, a priori, that the space of normal theta-functions is of dimension N^2 and the subspace of even functions is of dimension $(N^2 + 4)/2$ when N is even. (Odd theta-functions are the complementary subspace; they have a (-1) factor in (7.2c), but they are not needed here).

Analogously with elliptic functions, we say the jacobian manifold is $\mathbf{C}^2/L$. The symmetrized jacobian manifold is $(\mathbf{C}^2/L)/\{\pm(u,v)\}$; it is the object that corresponds to a Kummer surface.

8. Formulas for theta-functions. We give only the briefest survey, with minimal proofs.

The best known example of a normal theta function is of order $N = 1$:

$$(8.1a) \qquad \theta(u,v) = \sum_{m,n}^{-\infty,+\infty} \exp 2\pi i[(mu + nv) + \phi(m,n)/2]$$

$$(8.1b) \qquad \phi(m,n) = am^2 + 2bmn + cn^2.$$

Trivially, $\theta(u,v)^N$ is of order N but more sophisticated examples are needed.

We must construct 16 auxiliary theta-functions of (order one) but with *characteristics* defined by the quadruples:

$$(8.2a) \qquad g_1, g_2, h_1, h_2 \in \{0,1\}.$$

This introduces signs into the relations (7.2ab). We define

$$(8.2b) \qquad \Theta\begin{pmatrix} g_1 & g_2 \\ h_1 & h_2 \end{pmatrix}\begin{pmatrix} u \\ v \end{pmatrix}$$
$$= \sum_{m,n}^{-\infty,+\infty} \exp 2\pi i[(\frac{g_1}{2} + m)u + (\frac{g_2}{2} + n)v + \frac{mh_1 + nh_2}{2} + \frac{1}{2}\phi(\frac{g_1}{2} + m, \frac{g_2}{2} + n)].$$

If the characteristics are all 0, then of course we are back to (8.1a). (It is possible to define such functions as in (8.2b) with order n, but they are not needed here, compare [14]).

The modified period and symmetry relations are

$$(8.3a) \qquad \Theta\begin{pmatrix} g_1 & g_2 \\ h_1 & h_2 \end{pmatrix}\begin{pmatrix} u \\ v \end{pmatrix}$$
$$= (-1)^{g_1}\Theta\begin{pmatrix} g_1 & g_2 \\ h_1 & h_2 \end{pmatrix}\begin{pmatrix} u+1 \\ v \end{pmatrix} = (-1)^{g_2}\Theta\begin{pmatrix} g_1 & g_2 \\ h_1 & h_2 \end{pmatrix}\begin{pmatrix} u \\ v+1 \end{pmatrix},$$

$$(8.3b) \qquad \Theta\begin{pmatrix} g_1 & g_2 \\ h_1 & h_2 \end{pmatrix}\begin{pmatrix} u \\ v \end{pmatrix} = (-1)^{h_1}\exp 2\pi i(u + \frac{a}{2})\Theta\begin{pmatrix} g_1 & g_2 \\ h_1 & h_2 \end{pmatrix}\begin{pmatrix} u+a \\ v+b \end{pmatrix}$$

$$(8.3c) \qquad = (-1)^{h_2}\exp 2\pi i(v + \frac{c}{2})\Theta\begin{pmatrix} g_1 & g_2 \\ h_1 & h_2 \end{pmatrix}\begin{pmatrix} u+b \\ v+c \end{pmatrix}.$$

$$\Theta\begin{pmatrix} g_1 & g_2 \\ h_1 & h_2 \end{pmatrix}\begin{pmatrix} u \\ v \end{pmatrix} = (-1)^{g_1 h_1 + g_2 h_2}\Theta\begin{pmatrix} g_1 & g_2 \\ h_1 & h_2 \end{pmatrix}\begin{pmatrix} -u \\ -v \end{pmatrix}.$$

The squares of these functions are all normal even functions of order 2.

The important fact about these 16 functions is that they have as roots the 16 half-periods in some combination. From (7.1c) these 16 half-periods are $\lambda/2, \mu/2$ where

(8.4a)
$$\lambda = G_1 + H_1 a + H_2 b, \quad \mu = G_2 + H_1 b + H_2 c,$$

where, from now on,

(8.4b)
$$G_1, H_1, G_2, H_2 \in \{0,1\}.$$

By combining all the identities (8.3abc), we find

(8.4c)
$$\Theta \begin{pmatrix} g_1 & g_2 \\ h_1 & h_2 \end{pmatrix} \begin{pmatrix} u + \lambda \\ v + \mu \end{pmatrix} = \Theta \begin{pmatrix} g_1 & g_2 \\ h_1 & h_2 \end{pmatrix} \begin{pmatrix} u \\ v \end{pmatrix}$$
$$\times \exp 2\pi i [H_1(u + G_1 + \frac{a}{2}) + H_2(v + G_2 + \frac{c}{2}) + \frac{g_1 G_1 + g_2 G_2 + h_1 H_1 + h_2 H_2}{2}].$$

We look for roots of the theta-functions among the half-periods (u_0, v_0) with

(8.4d)
$$u_0 = -\lambda/2, \quad v_0 = -\mu/2.$$

By substituting these values in (8.4d) and using symmetry (8.3c), we find

(8.4e)
$$\Theta \begin{pmatrix} g_1 & g_2 \\ h_1 & h_2 \end{pmatrix} \begin{pmatrix} \lambda/2 \\ \mu/2 \end{pmatrix} = \Theta \begin{pmatrix} g_1 & g_2 \\ h_1 & h_2 \end{pmatrix} \begin{pmatrix} \lambda/2 \\ \mu/2 \end{pmatrix} \varepsilon,$$

with multiplier reminiscent of (8.3c),

(8.4f)
$$\varepsilon = \exp \pi i [(g_1 + H_1)(h_1 + G_1) + (g_2 + H_2)(h_2 + G_2)].$$

Of course both sides of (8.4e) vanish if $\varepsilon = -1$. So for each of the sixteen characteristics there corresponds of six roots (i.e., half-periods):

(8.5a)
$$(g_1, g_2, h_1, h_2) \rightarrow (G_1, G_2, H_1, H_2) \text{ (one-to-six)}$$

for which $\varepsilon = -1$. Actually, the six values follow modulo 2 from the columns:

(8.5b)
$$
\begin{array}{rrrrrrr}
g_1 + H_1 \equiv & 0 & 1 & 0 & 1 & 1 & 1 \\
g_2 + H_2 \equiv & 1 & 0 & 1 & 0 & 1 & 1 \\
h_1 + G_1 \equiv & 0 & 1 & 1 & 1 & 0 & 1 \\
h_2 + G_2 \equiv & 1 & 0 & 1 & 1 & 1 & 0
\end{array}
$$

We later verify that half-periods may occur as roots only from the above table. This is not a trivial observation, since a function of two variables has as its roots a submanifold of dimension one.

It is convenient to use the dyadic representation for the characteristics of each theta-function

(8.5c) $$j = 8g_1 + 4g_2 + 2h_1 + h_2, \quad (0 \le j \le 15),$$

and for each of the half-periods

(8.5d) $$J = 8H_1 + 4H_2 + 2G_1 + G_2, \quad (0 \le J \le 15).$$

So from (8.5b) each j engenders six values of J, denoted for later convenience by the set C_j:

(8.5e) $$C_j = \{J\} : j \to J, \text{ (one-to-six)}.$$

This relation is shown in the columns showing six values of J for each C_j:

8.6 THE KUMMER TABLE OF THETA-FUNCTIONS AND ROOTS.

C_0	C_1	C_2	C_3	C_4	C_5	C_6	C_7	C_8	C_9	C_{10}	C_{11}	C_{12}	C_{13}	C_{14}	C_{15}
5	4	7	6	1	0	3	2	13	12	15	14	9	8	11	10
10	11	8	9	14	15	12	13	2	3	0	1	6	7	4	5
7	6	5	4	3	2	1	0	15	14	13	12	11	10	9	8
11	10	9	8	15	14	13	12	3	2	1	0	7	6	5	4
13	12	15	14	9	8	11	10	5	4	7	6	1	0	3	2
14	15	12	13	10	11	8	9	6	7	4	5	2	3	0	1

8.7 GÖPEL QUADRUPLES OF THETA-FUNCTIONS. *These are unordered sets of four distinct (ordered) quadruples of characteristics*

(8.7a) $$\kappa^{(r)} = (g_1^{(r)}, g_2^{(r)}, h_1^{(r)}, h_2^{(r)}), \quad (r = 1, \cdots, 4)$$

such that (summing over r in each quadruple $\kappa^{(r)}$)

(8.7b) $$\sum g_1^{(r)} \equiv \sum g_2^{(r)} \equiv \sum h_1^{(r)} \equiv \sum h_2^{(r)}$$
$$\equiv \sum (g_1^{(r)} h_1^{(r)} + g_2^{(r)} h_2^{(r)}) \equiv 0 \bmod 2.$$

They have the property that sign factors of (8.3abc) cancel in the product of four theta-functions of these characteristics. (There are 60 such sets).

Thus for the product of elements of a Göpel quadruple, (say)

(8.7c) $$\Theta_{\alpha\beta\gamma\delta} = \Theta_\alpha \Theta_\beta \Theta_\gamma \Theta_\delta$$

is a normal even theta-function of order $N = 4$.

9. The Kummer surface. We start with a Göpel quadruple. Since there are only 10 linearly independent even normal functions of order four, some relation exists with constant coefficients among the 11 indicated functions:

(9.1a) $$\Theta_\alpha \Theta_\beta \Theta_\gamma \Theta_\delta = \sum_\alpha A_\alpha \Theta_\alpha^4 + \sum_{\alpha,\beta} B_{\alpha,\beta} \Theta_\alpha^2 \Theta_\beta^2.$$

If we introduce the variables

(9.1b) $$x = \Theta_\alpha^2, \; y = \Theta_\beta^2, \; z = \Theta_\gamma^2, \; w = \Theta_\delta^2,$$

then we square (9.1a) to obtain a homogeneous equation of degree four

(9.1c) $$xyzw = P(x, y, z, w)^2,$$

(with $P(x, y, z, w)$ the homogeneous form of degree two from (9.1a)). This equation represents the *Kummer surface,* which is important because it is *isomorphic to the jacobian modulo even symmetry* $(u, v) \to (-u, -v)$. This last result is essentially an easy consequence of the degree four (as we shall later note).

Actually, the Kummer surface has simple singularities, with the neighborhood of a point looking like a cone (rather than a plane). This condition leads to the form (see [17])

(9.1d) $$P(x, y, z, w) = [x^2 + y^2 + z^2 + w^2$$
$$+ 2f(yz + xw) + 2g(zx + yw) + 2h(xy + zw)]\text{const.}$$

The roots (8.4a) of the theta-functions, seen previously to be half-periods, are now also seen to be the singularities of the Kummer surface. Actually, this follows from the symmetry about the half-periods in (8.4d),

(9.1e) $f(u, v) = f(-u, -v) = f(u+\lambda, v+\mu) \implies f(\lambda/2+u, \mu/2+v) = f(\lambda/2-u, \mu/2-v)$.

This is more complicated than symmetry in one variable. It means that $f(\lambda/2+u, \mu/2+v)$ has an expansion in terms of u^2, v^2 and *(additionally)* uv. So a change of variable is not enough to remove the singularity (as in the case of a Riemann surface with a branch point).

9.2 THE KUMMER CONFIGURATION. *The sixteen singular points lie in sixteen conics (also denoted by C_j) with six points to a conic as shown in the Kummer table. Each point lies on six conics.*

Each conic lies in a different plane, so that each pair of conics intersects in two singularities. This creates a correspondence of each of the 120 pairs of singular points to a pair of (intersecting) conics.

Typical correspondences through intersections are shown here. Note from Kummer's table the reciprocity of the relations

(9.2a) $$C_0 \cap C_1 \to \{10, 11\}, \; \{0, 1\} \to C_{10} \cap C_{11}.$$

Also, for any conic (say C_0), each of its fifteen point-pairs (diagonals) will account for the intersections with the fifteen other conics. For instance, the point-pair $\{11, 13\}$ in C_0 accounts for C_6, i.e., $\{11, 13\} \to C_0 \cap C_6$.

The 60 Göpel (unordered) quadruples of characteristics (i.e., of theta-functions) become quadruples of conics $\{C_i, C_j, C_k, C_l\}$ for which, (taken in any or every order),

(9.2b) $$C_i = (C_i \cap C_j) \cup (C_i \cap C_k) \cup (C_i \cap C_l).$$

Thus $\{C_0, C_1, C_{10}, C_{11}\}$ is such an example:

$$C_0 \cap C_1 = \{10, 11\}, \quad C_0 \cap C_{10} = \{7, 13\}, \quad C_0 \cap C_{11} = \{5, 14\},$$
(9.2c) $\quad C_1 \cap C_{10} = \{4, 15\}, \quad C_1 \cap C_{11} = \{6, 12\}, \quad C_{10} \cap C_{11} = \{0, 1\}.$

Otherwise expressed, there are 6 singularities on each of the four conics, but in total there are only 12, as they each appear twice. (Compare [7] for further properties).

A Göpel quadruple of conics (9.2b) has the equivalent property that the singularities labelled $\{i, j, k, l\}$ have no subset of three lying on one conic.

10. Some typical proofs. The original theoretical development was highly intuitive from Riemann to Poincaré [15] and even beyond. Much of the flavor of the subject was lost by modern standards of rigor (see [11]).

10.1 POINCARÉ. *A theta-function of order N and one of order M have $2MN$ roots in common (or infinitely many).*

As a limiting case in (7.1a), when $b \to 0$, we have two periodic systems one for u of periods $(1, a)$ and one for v of periods $(1, c)$. The theta-functions break up into the product of two of the familiar elliptic type, and the simultaneous roots in the limiting case might come from

$$(10.1a) \qquad \theta_N^{(1)}(u)\theta_N^{(2)}(v) = \theta_M^{(3)}(u)\theta_M^{(4)}(v) = 0.$$

Each θ_N (or θ_M) has N (resp. M) roots, so we make the choice:

$$(10.1b) \qquad \theta_N^{(1)}(u) = \theta_M^{(4)}(v) = 0 \quad \text{versus} \quad \theta_N^{(2)}(v) = \theta_M^{(3)}(u) = 0.$$

With either choice we have MN common roots, hence the result. The counting formula must hold true for $b \neq 0$ by continuity.

As a consequence, any of the theta-functions of order one in (8.2b) indexed by j has only six of the roots indexed by J as shown in Kummer's table (no other half-periods). This is true because Kummer's table is an exact accounting of (half-period) roots, i.e., *two* such functions have only two roots in common, while any two columns share only two roots.

As a second consequence, the Kummer surface (9.1c) is an isomorphic image of the symmetrized jacobian, $(C^2/L)/\{\pm(u, v)\}$. If we make

$$(10.1c) \qquad x/w = \text{const}, \quad y/w = \text{const},$$

then by (9.1b), this means

$$(10.1d) \qquad \Theta_\alpha^2 - (x/w)\Theta_\delta^2 = \Theta_\beta^2 - (y/w)\Theta_\delta^2 = 0.$$

Thus a pair of theta-functions of order two vanishes from (10.1c). Such a pair is satisfied only by eight values (u, v) or (by symmetry) four values $\pm(u, v)$. Of course the line (10.1c) must have exactly four intersections with the Kummer surface (of degree four). This proves the desired isomorphism.

PART III. ELLIPTIC CONFIGURATIONS ON THE KUMMER SURFACE

11. Abelian functions and intermediate functions. The central object is the *abelian function*, a meromorphic function $f(u,v)$ (of two complex variables) which has a quadruple set of periods, so it is defined on a manifold like $\mathbf{C}^2/L$ in (7.1c). Specifically,

$$(11.1a) \qquad (\alpha_1, \beta_1) \in \{(1,0), (0,1), (a,b), (b,c)\},$$

$$(11.1b) \qquad f(u,v) = f(u + \alpha_j, v + \beta_j), \; j = 1, \cdots, 4.$$

By a theorem of Appell and Poincaré [1], such a meromorphic function is *globally* a quotient of holomorphic nonvanishing *intermediate functions* $\theta(u,v)$ with the property

$$(11.1c) \qquad \theta(u + \alpha_j, v + \beta_j) = \theta(u,v) \exp(\gamma_j u + \delta_j v + \epsilon_j), \; j = 1, \cdots, 4$$

with involved consistency conditions on the $\gamma_j, \delta_j, \epsilon_j$, (which come from the commutativity of the period addition, see[12]).

This structure comes from hyperelliptic curves, like that of (5.2a) if we set

$$(11.2a) \qquad u = \int_\infty^{(x,y)} dx/y + \int_\infty^{(x',y')} dx/y,$$

$$(11.2b) \qquad v = \int_\infty^{(x,y)} x\,dx/y + \int_\infty^{(x',y')} x\,dx/y.$$

Then the paths on the Riemann surface of H define four independent periods (on say $\mathbf{C}^2/L$) and symmetric functions such as

$$(11.2c) \qquad f_1 = xx', \; f_2 = x + x', \; f_3 = y + y', \; f_4 = yy',$$

are abelian functions. We can replace u, v by linear combinations of u, v and choose paths cleverly so that we return to the periods in

$$(11.2d) \qquad R = \begin{pmatrix} 1 & 0 & a & b \\ 0 & 1 & b & c \end{pmatrix},$$

with three complex parameters a, b, c (satisfying (7.1b)). It has been shown [14] that f_1, f_2, f_3 are generating functions of the field of abelian functions, while f_1, f_2, f_4 (as well as $x/w, y/w, z/w$ of (9.1b)) are generators of the subfield of even functions. (Note that the change $y \rightarrow -y$ through a different path leaves f_4 unchanged while $(u,v) \rightarrow -(u,v)$.)

Special intermediate functions are the theta-functions, which were used to construct the Kummer surface. *Since an algebraic curve on the Kummer surface is rational in the homogeneous coordinates x, y, z, w of (9.1b), it follows that such a curve is determined by the zero locus of an intermediate function.*

12. Singular abelian manifolds. The abelian manifold $\mathbf{C}^2/L$ is called *singular* when the lattice of periods L has a 2×2 nonscalar matrix ($S \notin E\mathbf{Z}$) as an endomorphism, called a "complex multiplication"

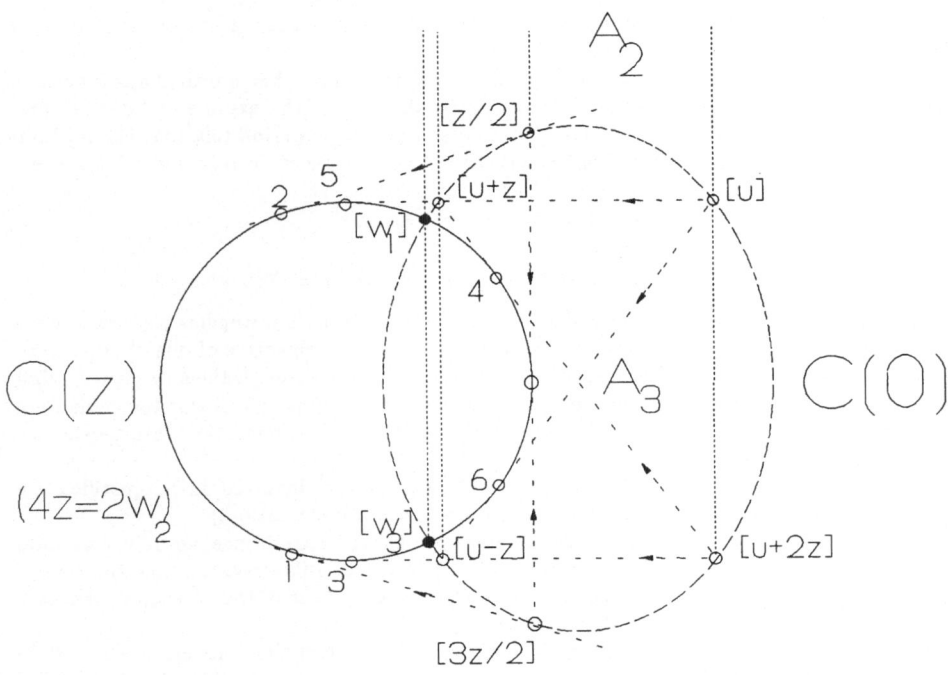

Figure 2. Quarter-period case

(12.1a)
$$SR = RM$$

for M an integral 4×4 matrix. Here S signals a change of basis for (u, v) and M a subbasis for the lattice of periods R.

Generically, (12.1a) is not possible if $\det(M) \neq \pm 1$. This is similar to complex multiplication which is *nongeneric* for elliptic curves over **C**. In other words there must be a loss of free parameters in (11.2d). For instance, we just take the case where a *singular relation* happens to be

(12.1b)
$$a = 2c.$$

Then the relation (12.1a) becomes valid with

(12.1c)
$$S = \begin{pmatrix} 0 & 2 \\ 1 & 0 \end{pmatrix}, \quad M = \begin{pmatrix} S & 0 \\ 0 & S^{tr} \end{pmatrix}.$$

This is called a "real" case [9] of complex multiplication since the eigenvalues of S are real $(\pm\sqrt{2})$.

To follow through on Humbert's approach, we note that the intermediate functions $\theta(u,v)$ of (1.1c) may have special consistency conditions on the exponential coefficients. Thus we define a *singular intermediate function* as one whose period relations (11.1c) hold only by virtue of a singular relation like (12.1b). An example of such period relations is

(12.2a)
$$\theta(u,v) = \theta(u+1,v) = \theta(u,v+1),$$
(12.2b)
$$\theta(u+a,v+b) = \theta(u,v)\exp 4\pi iv, \quad \theta(u+b,v+c) = \theta(u,v)\exp 2\pi iu.$$

Now the vanishing locus of a singular intermediate function is a *singular* algebraic curve on the Kummer surface. If everything is viewed from the perspective of one of the singularities (at the tangent cone) there will be six conics passing through the singularity, lying in six planes. These planes are viewed as six lines with 15 intersections (for the remaining 15 singularities). Some singular curve now will project into a configuration with two conics and six lines.

We shall describe the configuration and skip the details involved in its justification (which is unfortunately lengthier than the entire current presentation!)

We look again at Figure 2. Here the six lines are those dashed lines, actually tangents to $C(z)$ passing through the points numbered $1,2,3,4,5,6$. (The notation has the risk of confusing these points with singularities but, actually, only four of the 15 singularities are shown, at intersections labeled $[u],[u+z],[u-z],[u+2z]$).

Also $C[0]$ is the projection of some singular curve and the fact that the points $1,\cdots,6$ lie on six tangents from $C[0]$ causes the points to be special. (Only five constraints determine a conic). A further link in this arcane chain of reasoning is that the parameters of the hyperelliptic curve represent the six points of tangency. This explains the specialization of the hyperelliptic (5.2bc), (see [4] for explicit computation).

13. Concluding remarks. There is a larger literature for real mutiplication by $\sqrt{5}$ because the conditions on the hyperelliptic curve are more interesting from the standpoint of group theory. Here there is a special "fifth-period" configuration which we do not pursue, restricting ourselves to the easier case of $\sqrt{2}$. These are the only two cases which can be treated with any degree of satisfaction as of now (see [2], [9], [19]).

The author gratefully acknowledges helpful conversations with Armand Brumer.

REFERENCES

1. P. Appell, *Sur les fonctions périodiques de deux variables*, Journ. de Math. (4) **7** (1891), 157–219.

2. P. Bending, *Curves of genus 2 with $\sqrt{2}$ multiplication*; (unpublished dissertation) .

3. J.W.S. Cassels and E.V. Flynn, "Prolegomena to a Middlebrow Arithmetic of Curves of Genus 2," Cambridge Univ. Press, 1996.

4. H. Cohn, *A hyperelliptic curve with real multiplication of degree two*; (see Contents) .

5. J.D.Fay, "Theta Functions on Riemann Surfaces," Springer Verlag, 1973; Lect. Notes in Math 352 .

6. E. Hecke, *Höhere Modulfunktionen und ihre Anwendung an der Zahlentheorie*, Math. Annalen **71** (1912), 1–37.

7. M.R. Gonzalez-Dorrego, $(16,6)$ *configurations and geometry of Kummer surfaces in* $\mathbf{P}^3$, Memoirs of the AMS **512** (1994).

8. G. Humbert, *Théorie générale des surfaces hyperelliptiques*, Journ. de Math. (4) **9** (1893), 27–171, 361–475.

9. G. Humbert, *Sur les fonctions abéliennes singulières I*, Journ. de Math. (5) **5** (1899), 233–350.

10. G. Humbert, "Cours d'Analyse II," Gauthier-Villars, 1904, pp. 238–249.

11. H. Lange and C. Birkenhake, "Complex Abelian Varieties," Springer-Verlag, 1992.

12. J. Lewittes, *Riemann surfaces and the theta function*, Acta Mathematica **111** (1964), 37–61.

13. J.-F. Mestre, *Courbes hyperelliptiques à multiplications réelles*, C.R. Acad. Sci. Paris (1) **307** (1988), 721–724.

14. E. Picard, "Quelques Applications Analytiques de la Théorie des Courbes et des Surfaces Algébriques," Gauthier-Villars, 1931; (Notes taken by J. Dieudonné) .

15. H. Poincaré, *Sur les fonctions abéliennes*, Acta Math. **26** (1902), 43–98.

16. B. Poonen, *Computational aspects of curves of genus at least 2*, Algorithmic Number Theory (ANTS II, H. Cohen, ed.) (1996), 283–306; Second International Symposium, Talence, France; Springer-Verlag (Lecture Notes in Computer Science **1122**).

17. G. Salmon, "A Treatise on the Analytic Geometry of Three Dimensions II," 1914, pp. 50–51; Chelsea Reprint 1965 .

18. C.E. Traynard, "Fonctions Abéliennes et Fonctions Thêta de Deux Variables," Mém. de Sci. CLI Gauthier-Villars, 1962; (Lectures of P. Painlevé 1902)

19. J. Wilson, *Curves of genus 2 whose Jacobians have a $\sqrt{5}$ multiplication*; (unpublished dissertation) .

IDA Center for Computing Sciences
Bowie, MD 20715-4300
email: hcohn@super.org

ARITHMETICITY AND THETA CORRESPONDENCE
ON AN ORTHOGONAL GROUP

ZE-LI DOU

To Professor Kenneth B. Kramer

Introduction

In this paper we discuss several related topics in the theory of automorphic forms, on which the author's current research is focused. Roughly speaking, we shall deal with the questions of arithmeticity regarding special values of certain zeta and L-functions, as well as algebraicity results concerning periods and Fourier coefficients of certain automorphic forms.

The author has attempted to maintain a balance among several goals. He naturally wishes to bring the reader to his current research. On the other hand, he considers it an essential part of his task to explain how his work fits into the larger program concerning the same topics mentioned above, which was initiated by Shimura, and is being intensely pursued by Shimura himself and many others. From this point of view, it is not desirable to showcase only the results in one specific setting. Finally, as this paper is based on a talk given at the Graduate Center of the City University of New York, the author also wishes to address as large an audience as is practical. This puts a further constraint on the amount of technical material involved, at least initially. Consequently, he has adopted a "gradual generalization" approach—the article begins with a motivational section discussing the simplest possible setting only, and then each subsequent section introduces a generalization in at least one direction. Also, in each of the sections §§2–4, we concentrate on only one key concept. Thus in §2 we discuss the arithmeticity of automorphic forms; a nearly holomorphic arithmetic function (at critical points) is constructed in §3; and the main part of §4 is devoted to the discussion of a theta correspondence. These topics are all mutually related, and such relationships are pointed out in the discussions in the text. In general terms, we investigate, in the setting of a totally real quadratic extension of a totally real algebraic number field, a theta correspondence (see §4), and the algebraicity results on periods and L-values that follow (such as results of the type mentioned in §2), which can be derived with the aid of certain arithmeticity results of functions analogous to the $f(w, s)$ considered in §3.

With this approach, the author hopes to have given an adequate explanation of the essential points of that part of the program in which his research is engaged. Admittedly, this is neither the most efficient nor the most "logical" approach—one

Research for this article was partially supported by NSA Grant MDA904-97-1-0109 and by the Texas Christian University Research Foundation.

149

could have started from the setting of §4 and then proceeded backwards, thereby eliminating §1 altogether. Also, with this approach this paper cannot contain a full list of the author's results with all their technical precision. However, he views his results as ones complementing those of Shimura, in a naturally generalized setting. Therefore, an understanding of Shimura's methodology is crucial. For this reason the author considers the tradeoff more than reasonable, in the sense that the author's results can to a certain extent be "inferred" from a understanding of Shimura's theorems in this paper, but not the other way around. Finally, due to the limitation of space, no proof has been included in this paper. To compensate for these shortcomings, the author has inserted references throughout this paper, where precise statements of the latest results can be found, as well as their proofs.

I would like to thank Toni Bluher for several conversations concerning both the content and style of this paper; I have found them extremely helpful. It is also my most pleasant duty to thank the editors, Professor Melvyn Nathanson and Professors David and Gregory Chudnovsky, for their invitation to give a talk at the Graduate Center, and for their subsequent invitation to include this paper in their volume. In addition, Professor Josef Dodziuk's hospitality is responsible for some of the fondest memories of my visit to New York City in 1997.

This paper is dedicated to Professor Ken Kramer of Queens College, as a token of my sincere gratitude to him. I am very fortunate to have been a recipient of his generous and constant support for well over a decade, starting from my years as an undergraduate student.

1. Two examples of arithmeticity and near holomorphy

In this section we shall provide some motivation for the concept of arithmeticity by considering two examples in the case of the most classical modular forms. In so doing we follow the lead of Shimura. (See [Sh95].) We introduce some notation as follows:

$$H = \{z \in \mathbb{C} \mid \text{Im}(z) > 0\}, \qquad F = \mathbb{Q},$$

and

$$\Gamma_0(N) = \left\{ \alpha = \begin{pmatrix} a & b \\ c & d \end{pmatrix} \in \text{SL}_2(\mathbb{Z}) \ \middle| \ c \equiv 0 \ (\text{mod } N) \right\}.$$

A meromorphic function $f : H \longrightarrow \mathbb{C}$ is called an *automorphic form of weight k* for some $k \in \mathbb{Z}$ if it satisfies the following condition with respect to the usual action of elements $\alpha \in \Gamma_0(N)$ on $z \in H$, namely,

$$f(\alpha z) = j(\alpha, z)^k f(z), \quad \forall \alpha \in \Gamma_0(N),$$

where

$$\alpha z \overset{\text{def}}{=} \frac{az + b}{cz + d}, \qquad \text{and} \qquad j(\alpha, z) \overset{\text{def}}{=} cz + d.$$

We shall write

$$f\|_k \alpha(z) = j(\alpha, z)^{-k} f(\alpha z).$$

Then the above condition becomes simply

(1) $$f\|_k \alpha = f, \quad \forall \alpha \in \Gamma_0(N).$$

A *holomorphic* automorphic form f is one which is holomorphic on H and whose Fourier expansion at every cusp has the form $\sum_{n=0}^{\infty} a_n e^{2\pi i n z/h}$, where h is a positive integer. That is, we require f to be holomorphic at every cusp. (Note that this requirement is not needed in the case of $F \neq \mathbb{Q}$.) If in fact $a_0 = 0$, then we say f is a *cusp form*. The space of all such holomorphic automorphic forms is denoted by $\mathcal{M}_k(\Gamma_0(N))$; its subspace of cusp forms is denoted by $\mathcal{S}_k(\Gamma_0(N))$. Both of these spaces are finite dimensional vector spaces, whose dimensions can be easily computed via the Riemann-Roch Theorem. To avoid having to refer to a specific congruence subgroup such as $\Gamma_0(N)$, we take the union of all such spaces of holomorphic automorphic forms and denote it simply by $\mathcal{M}_k$. Similarly we have the definition of $\mathcal{S}_k$.

There are commutative, self-adjoint (with respect to the Petersson inner product) linear operators acting on the spaces of automorphic forms, which are called Hecke operators. As a consequence, there exist common eigenforms with respect to the Hecke operators. Also, we have the theory of newforms. These topics are very standard, and therefore we will not give any further explanation. A normalized Hecke eigenform will be referred to as a primitive form.

Let ϕ be a Dirichlet character defined modulo N. Then a character on $\Gamma_0(N)$ is given by $\phi(\alpha) \overset{\text{def}}{=} \phi(d_\alpha)$, where d_α is the $(2,2)$-entry of α. We may consider the elements f of $\mathcal{M}_k$ satisfying the condition

$$f\|_k \alpha = \phi(\alpha) f, \quad \forall \alpha \in \Gamma_0(N).$$

Suppose f is a cusp form satisfying this condition. Then we can write

$$f(z) = \sum_{n=1}^{\infty} a_n e^{2\pi i n z}.$$

For every primitive Dirichlet character χ, we define a Dirichlet series $D(s, f, \chi)$ as follows:

$$D(s, f, \chi) = \sum_{n=1}^{\infty} \chi(n) a_n n^{-s}.$$

Here s is a complex variable. This series is convergent on a half plane and can be analytically continued to the entire $\mathbb{C}$-plane.

We can now state the following theorem:

Theorem. *Suppose $k > 1$ and f is a primitive Hecke eigenform. There exist two periods p_+ and p_- such that*

$$\tau(\chi)^{-1}(\pi i)^{-m} D(m, f, \chi) \in p_\varepsilon K_\chi, \quad 0 < m < k,$$

where $\tau(\chi)$ is the Gaussian sum of χ, K_χ is the field generated over $\mathbb{Q}$ by the $\chi(n)$ and the Fourier coefficients a_n of f, and $\varepsilon = \pm 1$ is determined by the relation $\chi(-1)\varepsilon = (-1)^m$.

For a proof, see for example [Sh76]. Note that there are infinitely many possible characters χ and hence also infinitely many values $D(m, f, \chi)$ as described above.

Thus the existence of the periods p_+ and p_- in the above theorem is remarkable and is a manifestation of the *arithmeticity* of the values of the zeta function $D(s, f, \chi)$.

Our next example is an Eisenstein series, which in the simplest case is given by

$$E_k(z, s) = \sum_{\alpha \in (P \cap \Gamma) \backslash \Gamma} \phi(\alpha) j(\alpha, z)^{-k} (\mathrm{Im}(\alpha z))^s, \quad z \in H, s \in \mathbb{C}.$$

Here we have written $\Gamma_0(N)$ as Γ for brevity, and the meaning of k, α, ϕ, and $\phi(\alpha)$ are the same as above. The symbol P is defined by $P = \{\alpha \in \mathrm{SL}_2(\mathbb{Q}) \mid c_\alpha = 0\}$, where c_α is the $(2,1)$-entry of α. This Eisenstein series is convergent for $k + \mathrm{Re}(2s) > n + 1$. Also, after we multiply $E_k(z, s)$ by a non-zero entire function, the resulting product can be continued to a real analytic function on $H \times \mathbb{C}$ which is holomorphic in s. We are interested in the arithmetic nature of $E_k(z, m)$ for certain critical integers. As an example, let us take $m = 0, k = 2$, and the trivial character $\phi = 1$. Then we have

$$E_2(z, 0) = -\frac{3}{\pi \mathrm{Im}(z)} + 1 - 24 \sum_{n=1}^{\infty} \left(\sum_{0 < d | n} d \right) e^{2\pi i n z}.$$

Notice that this is not holomorphic. For this reason we introduce the notion of near holomorphy.[1] A function $f : H \longrightarrow \mathbb{C}$ is said to be *nearly holomorphic* if it can be written as a finite sum of expressions of the form $p(\mathrm{Im}(z)^{-1}) g(z)$, where p is a polynomial and g is a holomorphic function. We may also speak of nearly holomorphic automorphic forms if we add in the automorphy condition (1).[2] We then say that a nearly holomorphic automorphic form is *arithmetic*, or $\overline{\mathbb{Q}}$-*rational*, if the Fourier coefficients of such a form are algebraic. Theorems of Shimura, when specialized to our example at hand, then show that a modified version of $E_2(z, 0)$ is a nearly holomorphic and arithmetic automorphic form. In fact, Shimura showed that it is $\mathbb{Q}_{ab}$-rational. For an illuminating discourse on the concepts of arithmeticity and near holomorphy, Shimura's [Sh95] is strongly recommended. Later on in this paper, we shall present, in certain more general settings, the results mentioned in this section in precisely formulated forms. We shall also see the interconnection of such results.

2. Arithmetic automorphic forms with respect to a quaternion algebra

In this section we shall define arithmeticity more precisely. As always in this paper, we confine our discussion to the orthogonal case. Also, in accordance to the plan of "gradual generalization" we have adopted, we shall not operate in the most general setting possible. For the general development we refer the reader to Shimura's papers [Sh80], [Sh86], and [Sh87]. (See also §3 of this paper.) In this

---

[1] If we had chosen $k > 2$ in this case, the situation would have been simpler. But the occurrence of near holomorphy is natural and not sporadic. See [Sh86].

[2] In the setting of our example, we also need to assume that f has a Fourier expansion of the form $(f\|_k \alpha)(z) = \sum_{\nu=1}^{m_\alpha} \mathrm{Im}(z)^{-\nu} \sum_{n=0}^{\infty} c_{\nu \alpha n} \exp(2\pi i n z / N_\alpha)$. This condition is not needed if we take a number field different from $\mathbb{Q}$ or if the automorphic forms are defined on multiple copies of H.

section, we shall consider the setting of a quaternion algebra over a totally real algebraic number field. The main reference for this section is [Sh81I] and [Sh81II].

Let, then, F be a totally real algebraic number field with $[F : \mathbb{Q}] = n$, and let B be a quaternion algebra over F. Recall that this means B is a central simple algebra of dimension 4 over F. We have an isomorphism of the following type:

$$(2) \qquad B \otimes_{\mathbb{Q}} \mathbb{R} \cong M_2(\mathbb{R})^r \times \mathbb{H}^{n-r},$$

where $\mathbb{H}$ denotes the ring of Hamilton quaternions. We arrange the archimedean primes v_i in such a way that B is unramified at $v_1, v_2, \dots, v_r$ and ramified at $v_{r+1}, v_{r+2}, \dots, v_n$. We assume that $r \geq 1$. Notice that this development allows the possibility $B = M_2(F)$; therefore it includes the Hilbert modular forms as a special case.

Writing $\mathbb{A}$ for the adele ring of $\mathbb{Q}$, we let

$$B_{\mathbb{A}} = B \otimes_{\mathbb{Q}} \mathbb{A} \quad \text{and write} \quad G_{\mathbb{A}} = (B_{\mathbb{A}})^{\times}.$$

Here the symbol $R^{\times}$ denotes the group of invertible elements of any ring R. Then we have $G_{\mathbb{Q}} = B^{\times}$. The infinite part of $G_{\mathbb{A}}$ will be denoted by G_{∞}, which is isomorphic to $\mathrm{GL}_2(\mathbb{R})^r \times (\mathbb{H}^{\times})^{n-r}$. We also write the identity component of G_{∞} as $G_{\infty+}$. We then define

$$G_{\mathbb{A}+} = \{x \in G_{\mathbb{A}} \mid x_{\infty} \in G_{\infty+}\} \quad \text{and} \quad G_{\mathbb{Q}+} = G_{\mathbb{Q}} \cap G_{\mathbb{A}+}.$$

For every $0 \leq m \in \mathbb{Z}$, there is an $\mathbb{R}$-rational irreducible representation $\sigma_m : \mathbb{H}^{\times} \longrightarrow G_{m+1}(\mathbb{C})$ of degree m, which is unique up to equivalence. By fixing a suitable isomorphism (2), we may assume that the following properties are satisfied:

$$x_{v_i} \in M_2(\overline{\mathbb{Q}}), \qquad\qquad \forall 1 \leq i \leq r, \forall x \in B;$$
$$\sigma_m(x_{v_i}) \in M_{m+1}(\overline{\mathbb{Q}}), \qquad \forall r + 1 \leq i \leq n, \forall x \in B;$$
$$\sigma_m(x^*) = {}^t\overline{\sigma_m(x)}, \qquad\qquad \forall x \in \mathbb{H}.$$

Here $*$ denotes the main involution of B as well as its natural extension to $B_{\mathbb{A}}$. If $\kappa = (\kappa_{r+1}, \kappa_{r+2}, \dots, \kappa_n) \in \mathbb{Z}^{n-r}$ such that $\kappa_i \geq 0$, then we may define a representation σ_{κ} on $G_{\mathbb{A}}$ by

$$\sigma_{\kappa}(x) = \bigotimes_{i=r+1}^{n} \sigma_{\kappa_i}(x_i).$$

The representation space is $\bigotimes_{i=r+1}^{n} \mathbb{C}^{\kappa_i+1}$, which will be denoted by $\mathcal{X}$.

We can now define the action of elements of $G_{\mathbb{A}+}$ on H^r, the factor of holomorphy $j(x, z)$ (where $z \in H^r$), and congruence subgroups of $G_{\mathbb{Q}+}$ as natural generalizations of the corresponding concepts defined in the previous section. This done, then given $k \in \mathbb{Z}^r$ and a vector valued mapping $f : H^r \longrightarrow \mathbb{C}^d$, where $d = \prod_{i=r+1}^{n}(\kappa_i + 1)$, we define another mapping $f\|_{k,\kappa} x$ of the same type by the following formula:

$$(f\|_{k,\kappa} x)(z) = \prod_{i+1}^{n} \mathrm{N}(x_i)^{\kappa_i/2} j(x, z)^{-k} \sigma_{\kappa}(x)^{-1} f(x(z)).$$

If Γ is a congruence subgroup, then we denote by $\mathcal{A}_{k,\kappa}(\Gamma), \mathcal{M}_{k,\kappa}(\Gamma)$, and $\mathcal{S}_{k,\kappa}(\Gamma)$ the space of meromorphic, holomorphic, and cusp automorphic forms, respectively, with respect to Γ. That is, we require $f\|_{k,\kappa}x = f$ for all $x \in \Gamma$. If $B = \mathrm{M}_2(\mathbb{Q})$, then we also need to impose the usual condition at the cusps. As usual, the union of the spaces $\mathcal{A}_{k,\kappa}(\Gamma)$ over all congruence subgroups Γ is denoted simply by $\mathcal{A}_{k,\kappa}$. The symbols $\mathcal{M}_{k,\kappa}$ and $\mathcal{S}_{k,\kappa}$ are defined in the same way.

Furthermore, if ϕ is a character of Γ of finite order, then $\mathcal{M}_{k,\kappa}(\Gamma, \phi)$ is the subspace of $\mathcal{M}_{k,\kappa}$ such that $f\|_{k,\kappa}x = \phi(x)f$ for all $x \in \Gamma$. Similarly we have the definition of $\mathcal{S}_{k,\kappa}(\Gamma, \phi)$. Notice that the distinction between $\mathcal{M}_{k,\kappa}$ and $\mathcal{S}_{k,\kappa}$ is unnecessary if B is a division algebra, because in that case $\mathcal{M}_{k,\kappa} = \mathcal{S}_{k,\kappa}$.

Next we turn our attention to the meaning of arithmeticity for automorphic forms. Let K be a totally imaginary quadratic extension of F. Suppose h is an F-linear embedding of K into B. Then we have $h(K^\times) \subset G_{\mathbb{Q}+}$ and $h(K^\times)$ has a unique common fixed point on H^r. This point is called a *CM-point*. To define arithmeticity, we need to make use of a mapping $P_{k,\kappa}$, which takes values in $\mathrm{GL}(\mathcal{X})$, and is constructed by means of a bilinear mapping related to the *periods of abelian varieties*. Since we do not wish in this article to stress the technical details, we refer the reader to Shimura's [Sh80] and [Sh81I] for the definition of these mappings. An element $f \in \mathcal{A}_{k,\kappa}$ is said to be arithmetic at a CM-point w (at which f is holomorphic) if $P_{k,\kappa}(w)^{-1}f(w)$ has algebraic components. If this is so for all such CM-points, then f is called an *arithmetic automorphic form*. We shall use the symbol $\overline{\mathbb{Q}}$ to indicate the set of arithmetic elements. Therefore we speak of $\mathcal{S}_{k,\kappa}(\overline{\mathbb{Q}}), \mathcal{S}_{k,\kappa}(\Gamma, \overline{\mathbb{Q}}), \mathcal{S}_{k,\kappa}(\Gamma, \phi, \overline{\mathbb{Q}})$, and so forth.

We have the following identities:

$$\mathcal{M}_{k,\kappa} = \mathcal{M}_{k,\kappa}(\overline{\mathbb{Q}}) \otimes_{\overline{\mathbb{Q}}} \mathbb{C}, \quad \text{and} \quad \mathcal{S}_{k,\kappa} = \mathcal{S}_{k,\kappa}(\overline{\mathbb{Q}}) \otimes_{\overline{\mathbb{Q}}} \mathbb{C}.$$

For the remainder of this section, we shall describe an algebraicity result concerning a zeta function. In order to do so, we first need to consider adelic automorphic forms, which are defined on $G_\mathbb{A}$. Given an integral ideal $\mathfrak{c}$ of F and a character ψ, we can define the space of automorphic forms and its subspace of cusp forms of weight (k, κ), level $\mathfrak{c}$, and character ψ. These spaces will be denoted by $\mathcal{M}_{k,\kappa}(\mathfrak{c}, \psi)$ and $\mathcal{S}_{k,\kappa}(\mathfrak{c}, \psi)$. Elements of such spaces will denoted by bold face letters. In the interest of economy of space, we shall not develop the definition of such forms from scratch. It turns out that every element $\mathbf{f} \in \mathcal{M}_{k,\kappa}(\mathfrak{c}, \psi)$ can be identified as a vector $(f_1, \ldots, f_h) \in \prod_{\lambda=1}^h \mathcal{M}_{k,\kappa}(\Gamma_\lambda, \phi_\lambda)$ for suitably defined Γ_λ and $\phi_\lambda, \lambda = 1, \ldots, h$. We refer to, say, [D93] for a full explanation. The meaning of arithmeticity for such adelic automorphic forms is a natural generalization of what we have just defined— if $\mathbf{f} \in \mathcal{M}_{k,\kappa}(\mathfrak{c}, \psi)$ is identified with $(f_1, \ldots, f_h)$, then $\mathbf{f}$ is called arithmetic if all f_i are arithmetic for $i = 1, \ldots, h$.

Assume that F has a subfield E such that $[E : \mathbb{Q}] = r$ and that the restrictions of $v_1, \ldots, v_r$ to E are all different from one another. Our zeta function involves three main ingredients: a locally constant function η on F, which is $\overline{\mathbb{Q}}$-valued, a Hilbert modular form Ω of $\mathcal{M}_{l,0}(\mathrm{M}_2(E))$, and a primitive Hecke eigenform $0 \neq \mathbf{f} \in \mathcal{S}_{k,\kappa}(\mathfrak{c}, \psi)$. The automorphic forms are taken among the arithmetic elements. This is possible because of the identities mentioned above concerning $\mathcal{S}_{k,\kappa}$ and $\mathcal{S}_{k,\kappa}(\overline{\mathbb{Q}})$, etc. Recall, also, that a function η is *locally constant* if there exist two lattices L

and M in F such that $\eta(x) = 0$ for $x \notin L$ and $\eta(x)$ depends only on x modulo M. We can find a subgroup U of $\mathfrak{o}_F^\times$, where the symbol $\mathfrak{o}_F$ stands for the ring of integers in F, such that $\eta(ux) = \eta(x)$ for all $u \in U$. We then say that η is U-*invariant*.

Let us take such a locally constant function η and a Hilbert modular form Ω as above, and write

$$\Omega(z) = \sum_{a \in E} \omega(a) e_E(az),$$

where $e_E(az) \overset{\text{def}}{=} \exp(2\pi i \sum_{j=1}^r a^{v_j} z_j)$. Here we have denoted the restrictions of v_i to E by the same symbols for notational simplicity. Let U_E denote the group of totally positive units in E. We then take U to be a subgroup of U_E of finite index such that both η and Ω are U-invariant, and introduce the notation $[U] \overset{\text{def}}{=} [U_E : U]^{-1}$.

As for the primitive Hecke eigenform $\mathbf{f}$, suppose we have $\mathbf{f} \mid T_{\mathfrak{r}}(\mathfrak{a}) = \chi(\mathfrak{a})\mathbf{f}$ for all $\mathfrak{a}$, where the $T_{\mathfrak{a}}(\mathfrak{a})$ denote the Hecke operators. Then the set $\{\chi(\mathfrak{a})\}$ is called a primitive system of eigenvalues. The zeta function $Z(s)$ is defined by

$$Z(S) = [U] \sum_{0 \ll b \in F/U} \eta(b)\chi(b_{\mathfrak{r}})\omega_u(\text{Tr}_{F/E}(tb))b^{c-sE}.$$

Here $t \in F$ is such that $t^{v_i} > 0$ for $1 \le i \le r$ and $t^{v_i} < 0$ for $r + 1 \le i \le n$; $\mathfrak{r}$ is a fractional ideal of F; $c = \sum_{i=1}^n c_i\tau_i$, where the c_i are integers or half-integers; $b^{sE} = (b^{v_1}, \ldots, b^{v_r})^s$; and $\omega_u(a) = a^u\omega(a)$, where $u = u_1v_1 + \cdots + u_nv_n$ with $0 \le u_i \in \mathbb{Z}$.

We can now state a result of Shimura ([Sh81II]) concerning the algebraicity of certain critical values of the zeta function $Z(s)$ as follows:

Theorem. *Suppose* $l + 2u + 2Re_{F/E}(c) = \gamma(v_1 + \cdots + v_n)$ *with* $\gamma \in \mathbb{Z}$, $2c_i \equiv k_i$ (mod 2) *for* $i \le r$, $\kappa_i \le 2c_i \equiv \kappa_i$ (mod 2) *for* $i > r$. *Then for every integer satisfying*

(i) $\left(\dfrac{\gamma}{2}\right) - 1 + [F : E] < s_0 \le \left(\dfrac{k_i}{2}\right) + c_i,$ *if* $1 \le i \le r$; *and*

(ii) $2s_0 \ne \gamma + 2[F : E],$ *if* $E = \mathbb{Q}$,
we have $Z(s_0)$ *finite, and*

$$Z(s_0) \sim \pi^{|k|}\langle \mathbf{g}, \mathbf{g} \rangle.$$

Here $x \sim y$ *means* $\dfrac{x}{y} \in \overline{\mathbb{Q}}$, *and* $\mathbf{g}$ *is any arithmetic Hecke eigenform belonging to the system of eigenvalues* $\{\chi\}$.

Moreover, if in addition we have

$$2c_i + k_i + 2 = \gamma + 2[F : E], \qquad \forall 1 \le i \le r,$$

then the residue of Z *at* $s = \left(\dfrac{\gamma}{2}\right) + [F : E]$ *is an algebraic number times* $R_E\pi^{|k|}\langle \mathbf{g}, \mathbf{g} \rangle$, *where* R_E *is the regulator of* E.

The above theorem is certainly rather complicated technically; several more theorems of this nature can be stated in this setting, and can be found in [Sh81II], together with detailed proofs. Moreover, theorems of this type have been obtained by Shimura in much more general settings, rendering the theorem stated above only a special case. For example, we may replace F by a direct sum of totally real algebraic number fields. The function $f(w,s)$, which we define in the next section, will be given in this generality. In any case, for a fuller discussion we refer to [Sh83] and [Sh88]. In addition to the intrinsic interest of such results, which is evident, there are also important consequences concerning the L-functions attached to automorphic forms—indeed, they may be regarded as special cases of the theorems of the above type. (The Theorem in §1 is a simple example of this.) Furthermore, these results are also intimately related to the several period invariants, also investigated by Shimura. Shimura's article [Sh88] is one recent reference in which all these aspects are discussed in depth. Also in that paper, a list of far-reaching conjectures are stated on the precise nature of the periods of automorphic forms. Several of these conjectures have been settled by Shimura himself ([Sh90]), in the case where we have a division quaternion algebra. Parallel results for the non-division algebra case (i.e., the case of Hilbert modular forms) have been obtained by the author; see [D94]. M. Harris and H. Yoshida also made penetrating contributions in this direction. See especially Yoshida's paper [Y95].

We are interested in a further generalization of the type of results treated in this section. Namely, we wish to investigate zeta functions of the above type in the setting involving a field extension. In particular, we wish to examine the case where the field extension is a totally real quadratic extension of a totally real algebraic number field. A theta lifting for automorphic forms with respect to such a field extension will be constructed in the final section of this paper.

3. The function $f(w,s)$ in the orthogonal case

As we have explained, the Theorem in the previous section can be considered as a generalization of the Theorem we cited in §1. In this section we shall construct a mapping $f(w,s)$, which can be regarded as a generalized version of the Eisenstein series we discussed in §1. Here the variable w belongs to a certain bounded symmetric domain and s is a complex variable. We shall see that $f(w,s)$ is nearly holomorphic and arithmetic at certain critical points s. This does not merely round out the ideas about arithmeticity introduced in the first section, but, more importantly, the proof of the Theorem of §2 is aided by the consideration of such a function. (See [Sh81II] for the setting of §2 and [Sh88] for the general setting.) Moreover, we shall see that the mapping $f(w,s)$ is defined via a theta correspondence, which leads naturally to the main topic of the next section.

The main references for this section are Shimura's [Sh86] and [Sh87], as well as the work of A. W. Bluher, [B90], [B94], and [B98]. We shall largely follow the notation in [Sh86] in this section. The setting of this section includes that of §2 as a special case, as we shall see.

Let F denote a *direct sum* of totally real algebraic number fields $F_1, \ldots, F_t$. Since many of the concepts in this setting are to be defined as natural generalizations of corresponding ones in §2, we shall denote them by the same symbols. For example, the symbol $\mathbf{e}_F(z)$ will now have the meaning

$$\mathbf{e}_F(z) = \mathbf{e}\left(\sum_{\sigma \in J_F} z_\sigma\right) = \mathbf{e}\left(\sum_{\sigma \in \cup_{i=1}^t J(F_i)} z_\sigma\right) = \exp\left(2\pi i\left(\sum_{\sigma \in \cup_{i=1}^t J(F_i)} z_\sigma\right)\right).$$

The symbol $J(F_i)$ here means the set of Archimedean primes of F_i. Such modifications in meaning will not be explicitly mentioned, as so doing will increase the length of this article unnecessarily. Instead, we refer the reader to the papers referred to above for precise definitions, whenever there is a doubt.

We assume that E is a common subfield of $F_1, \ldots, F_t$, and that J_F contains a subset ε such that the restrictions of places of ε to E yields a bijection of ε onto $J(E)$, with $\varepsilon \cap J(F_i) \neq \emptyset$ for all $1 \leq i \leq t$. Thus our E is analogous to the field in §2 denoted by the same symbol. We write $\varepsilon' = J_F - \varepsilon$.

We now recall some general principles concerning the action of a certain orthogonal group on a bounded symmetric domain. See [Sh80] and [B90] for details.

Let n be a positive integer and V be a vector space over $\mathbb{R}$ of dimension $n + 2$, and let S be a symmetric bilinear form on V (over $\mathbb{R}$) of signature $(n, 2)$. Then S can be naturally extended to a $\mathbb{C}$-valued symmetric form on $V \otimes_{\mathbb{R}} \mathbb{C}$. Let us consider the set

$$\mathcal{N}(S) = \{v \in V \otimes_{\mathbb{R}} \mathbb{C} \mid S[v] = 0, S(v, \bar{v}) < 0\},$$

where $S[v] \overset{\text{def}}{=} S(v, v)$. Denote by $\mathcal{B}$ the set of ordered bases (x, y) of two-dimensional totally negative subspaces of the quadratic space (V, S) such that $S[x] = S[y] = -1$. Then we have a bijection $\mathcal{B} \longleftrightarrow \mathcal{N}(S)/\mathbb{R}_+$, which is induced by $(x, y) \leftrightarrow x + iy$.

We define the block diagonal matrices

$$Q = \operatorname{diag}[I_n, -I_2], \qquad R = \operatorname{diag}\left[I_n, \begin{pmatrix} 0 & -1 \\ -1 & 0 \end{pmatrix}\right],$$

where I_k denotes the identity matrix of order k. Then a matrix $A \in \operatorname{GL}_{n+2}(\mathbb{C})$ (or, strictly speaking, a $\mathbb{C}$-isomorphism $V \otimes_{\mathbb{R}} \mathbb{C} \to \mathbb{C}^{n+2}$) can be found, such that the following identities hold:

$$S(x, y) = {}^t(Ax)R(Ay), \quad \text{and} \quad S(\bar{x}, y) = {}^t(\overline{Ax})Q(Ay), \quad \forall x, y \in V \otimes_{\mathbb{R}} \mathbb{C}.$$

Therefore the mapping $v \mapsto Av$ gives a $\mathbb{C}$-linear isomorphism

$$A: \quad \mathcal{N}(S) \longrightarrow \mathcal{N}(Q, R) \overset{\text{def}}{=} \{u \in \mathbb{C}^{n+2} \mid {}^tuRu = 0, {}^t\bar{u}Qu < 0\}.$$

The space $\mathcal{N}(Q, R)$ has two connected components, which may be described as follows. Let

(3) $$\mathcal{Z}_n = \{w \in \mathbb{C}^n \mid {}^t\bar{w}w < 1 + \frac{1}{4}|{}^tww|^2 < 2\}.$$

We define two mappings $p : \mathcal{Z}_n \to \mathcal{N}(Q, R)$ and $p' : \mathcal{Z}_n \to \mathcal{N}(Q, R)$ by

$$p(w) = \begin{pmatrix} w \\ {}^t\overline{w}w \\ \frac{2}{} \\ 1 \end{pmatrix}, \quad \text{and} \quad p'(w) = \begin{pmatrix} \overline{w} \\ 1 \\ \frac{{}^t\overline{w}w}{2} \end{pmatrix}.$$

Then p and p' naturally give rise to bijective mappings of $\mathbb{C} \times \mathcal{Z}_n$ onto the components of $\mathcal{N}(Q, R)$ alluded to above. Indeed, we may simply define $(c, w) \mapsto cp(w)$ and $(c, w) \mapsto cp'(w)$, respectively. Clearly the two components are homeomorphic to each other. We denote these components by $\mathcal{N}_0(Q, R)$ and $\mathcal{N}_0'(Q, R)$, and the corresponding components of $\mathcal{N}(S)$ by $\mathcal{N}_0(S)$ and $\mathcal{N}_0'(S)$.

Summarizing, we have the following sequence of bijections:

(4)

$$\mathcal{Z}_n \xrightarrow{p} \mathcal{N}_0(Q, R)/\mathbb{C}^\times \xrightarrow{A^{-1}} \mathcal{N}_0(S)/\mathbb{C}^\times \longrightarrow$$

$$\{\text{two-dimensional totally negative subspaces of } (V, S)\}.$$

We introduce here a majorizing form of S which will be needed in the next section. If $w \in \mathcal{Z}_n$, then by (4) we see that w corresponds to a subspace W spanned by the real and imaginary parts of $w' = A^{-1}p(w)$ over $\mathbb{R}$. Let $W^\perp$ be the orthogonal complement of W. Then S is negative definite on W and positive definite on $W^\perp$. Our (positive definite) majorizing form P_w is then defined by $P_w = -S$ on W and $P_w = S$ on $W^\perp$.

For notational clarity, we shall from now on write

$$P[v; w] = P_w(v).$$

We then have the following formula:

(5) $$P[v; w] - S[v] = -4S(w', \overline{w'})^{-1}|S(v, w')|^2 \geq 0.$$

Returning to the preparation for the definition of $f(w, s)$, let V_i be a vector space of dimension $n_i + 2$ over F_i for each $1 \leq i \leq t$ with $n_i \geq 0$, and consider an F_i-bilinear symmetric form $S_i : V_i \times V_i \longrightarrow F_i$. We adopt the following notation:

$$V = \prod_{i=1}^{t} V_i, \qquad G(S_i) = \{\alpha \in \mathrm{SL}(V_i) \mid S_i(\alpha x, \alpha x) = S_i(x, x)\},$$

$$S = \{S_1, \ldots, S_t\}, \qquad G(S) = \prod_{i=1}^{t} G(S_i).$$

If $\tau \in J_F$, then $\tau \in J(F_i)$ for some i, and we denote by V_τ the τ-completion of V_i and by S_τ the natural extension of S_i to $V_\tau \otimes_\mathbb{R} \mathbb{C}$. We also write $n_\tau = n_i$. We then have the identification

$$V \otimes_\mathbb{Q} \mathbb{C} = \prod_{\tau \in J_F} V_\tau \otimes_\mathbb{R} \mathbb{C}.$$

In the following, we assume that S_τ has signature $(n_\tau, 2)$ on V_τ if $\tau \in \varepsilon$, and that S_τ has signature $(n_\tau + 2, 0)$ on V_τ if $\tau \in \varepsilon'$.

We can define $\mathcal{Z}_\tau$ as in (3) for each τ, and let

$$\mathcal{Z} = \prod_{\tau \in \delta} \mathcal{Z}_\tau.$$

The symbol δ in the definition above is defined as follows. Arrange indices in such a manner that we have $n_i > 0$ for $1 \leq i \leq r$ and $n_i = 0$ for $r + 1 \leq i \leq t$. We assume $0 < r \leq t$. We then put

$$\delta = \varepsilon \cap J(F_1 \times \cdots F_r), \quad \text{and} \quad \delta' = \varepsilon' \cap J(F_1 \times \cdots \times F_r).$$

Given $0 \leq \kappa \in \mathbf{Z}^{\delta'}$, we define $\mathcal{P}_\kappa(S)$ to be the vector space of $\mathbb{C}$-valued polynomial functions on $\prod_{\tau \in \delta'} V_\tau$ which are homogeneous of degree κ_τ and S_τ-harmonic on V_τ. (See [Sh80] for the definition of S_τ-harmonic functions.) We now take $\Omega(z) = \sum_a \omega(a) e_E(az)$ to be an element of $\mathcal{M}_{l,0}(E)$ with $0 \leq l$. (Therefore Ω is allowed to be a constant.) Let U_E and U be defined in the same way as in the previous section. Then we define

$$f(w, s) = [U] \sum_{0 \neq v \in V/U} c(v)\omega\big(\mathrm{Tr}_{F/E}(-S[v])\big)S[v]^j q(v)v^\phi$$
$$\cdot \mu[v, w]^{-k}\eta(w)^{s\eta}|\mu[v, w]^\delta v^\psi|^{-2s}.$$

Here $c \in \mathcal{L}(V)$ is a locally constant function, and $q \in \mathcal{P}_\kappa(S)$. Also, $0 \leq k \in \mathbf{Z}^\delta, 0 \leq \kappa \in \mathbf{Z}^{\delta'}, 0 \leq j \in \mathbf{Z}^F, \phi \in \mathbf{Z}^\kappa$, and ψ is a subset of J_K such that

$$\mathrm{Res}_{V_i/F_i}(\psi \cap J(V_i)) = \varepsilon \cap J(F_i), \quad \forall i = r + 1, \ldots, t,$$

where $K \stackrel{\mathrm{def}}{=} V_{r+1} \times \cdots \times V_t$. Finally, μ is a factor of automorphy defined by

$$\mu_\tau[x, w] = S_\tau\big(x_\tau, A_\tau^{-1} p(w_\tau)\big) = {}^t(A_\tau x^\tau)Rp(w_\tau), \quad x \in \prod V_\tau, w \in \mathcal{Z}.$$

We have already explained that consideration of mappings of type $f(w, s)$ is indispensable for algebraicity results such as the Theorem in the previous section, which, in turn, yields further results in the study of arithmetic. Here we note that *the setting of the previous section corresponds to the case $t = r = 1$ here.*

We now consider the question of near holomorphy and arithmeticity. The meaning of near holomorphy and arithmeticity are natural generalizations of the definitions given in the previous sections. The precise definitions can be found in [Sh86]. (See also [B90] and [B98].) Let us state a result of Shimura ([Sh86]) as follows.

Theorem. *The series $f(w, s)$ can be continued as a meromorphic function to the whole s-plane; more precisely, there exist a non-zero entire function $A(s)$ and a real analytic function $B(w, s)$ on $\mathcal{Z} \times \mathbb{C}$, which is holomorphic in s, such that*

$$A(s)f(w, s) = B(w, s)$$

160

for large Re(s). Furthermore, under certain conditions, $f(w,s)$ is finite, nearly holomorphic, and arithmetic. This is more explicitly explained as follows. We assume that

(i) $k_\tau > 0$ for every $\tau \in \delta$ and $\phi_\sigma \geq 0$ for $\sigma \notin \psi \cup \psi\rho$;

(ii) $Res_{F/E}(k - \kappa - 2j) - Res_{K/E}(\phi) - l = a_f \iota_E$, where $a_f \in 2^{-1}\mathbb{Z}$; and

(iii) $\phi_\sigma \geq \phi_{\sigma\rho}$, where ρ is the complex conjugation.

Then we put

$$b_f = a_f + 2 - \sum_{i=1}^{t}(n_i + 2)[F_i : E]/2.$$

If $b_f \geq 1$, then $f(w,0)$ is finite; in fact it is holomorphic in w with a few exceptions which can be explicitly identified. Furthermore, with some additional conditions, $\pi^{\#}p_K(-\phi, 2\psi)f(w,o)$ is arithmetic, where the power $\#$ can be explicitly given, and p_K is defined in [Sh80] as we alluded to in §2.

If $b_f = 0$ or $1/2$, then $f(w,s)$ has at most a simple pole at $s = 1 - b_f$. Let h be the residue there. Then, under some additional conditions, $R_E^{-1}\pi^{\#}p_K(-\phi, 2\psi)h$ is arithmetic, where the power $\#$ can be explicitly computed, though not always equal to the power above.

The conditions omitted above can be found in [Sh86]. The proof can be found in [Sh88]. Shimura's techniques have been extended by A. W. Bluher to show that such results hold for a certain range of other integers and half integers as well. Moreover, by imposing somewhat more stringent conditions on the locally constant function c as well as the Hilbert modular form Ω, Bluher showed that similar results hold for an even larger range of critical points for another series $\tilde{f}(w,s)$, which is defined as a product of $f(w,s)$ with a certain L-series. These results can be found in the papers [B90] and [B94].

Another feature of the theory we wish to mention here is the fact that $f(w,s)$ has an integral representation of the type

$$f(w,s) = \int \Omega(z)\Theta(z,w;C)y^{*}E(z,s)\{dz\},$$

where $E(z,s)$ is an Eisenstein series and $\Theta(z,w;C)$ is a certain theta function. Therefore, we may say that $f(w,s)$ is defined via a *theta correspondence*. In this regard, Bluher has proved that to a certain extent, arithmeticity is preserved by theta correspondence. (See [B98].) This result is relevant in the discussion of the arithmeticity of $\tilde{f}(w,s)$ above. Although we do not wish to discuss this particular result in detail, the idea of theta correspondence will be our focus in the next section.

4. Theta correspondence with respect to a quadratic extension

In this final section of the paper, we shall present a theta correspondence of automorphic forms with respect to a totally real quadratic extension of a totally real algebraic number field. Thus our setting is again a generalization of that of §2, but in a different direction from that of §3. In this setting, one may again hope to derive results analogous to those stated in §2, and to do so is the main focus of the author's current research. Rather than entering into a discussion on the generalities, we shall devote most of this section to a description of the correspondence in our specific setting, together with some of the properties.

Let us explain our new setting. Happily, most of the ground work has already been laid in the last two sections, and consequently we can afford to be quite efficient. Let F be a totally real algebraic number field, and E/F a totally real quadratic extension.[3] Let B_E be a quaternion algebra over E which is equipped with an F-linear automorphism τ such that $\tau^2 = \mathrm{id}_{B_E}$, but $\tau|_E \neq \mathrm{id}_E$. The main involution of B_E will be denoted by $*$.

We shall need to consider the following subsets of B_E:

$$B = \{x \in B_E \mid x^\tau = x\} \qquad \text{and} \qquad V = \{x \in B_E \mid x^\tau = -x^*\}.$$

It can easily be shown that B is a quaternion algebra over F, and V is a vector space over F of dimension 4.

Regarding the adele rings, we fix notation by setting

$$B_\infty \cong \mathrm{M}_2(\mathbb{R})^\delta \times \mathbb{H}^{\delta'} \qquad \text{and} \qquad (B_E)_\infty \cong \mathrm{M}_2(\mathbb{R})^\zeta \times \mathbb{H}^{\zeta'}.$$

Let $\mathbf{a}$ be the set of Archimedean primes of F. For each $v \in \mathbf{a}$, we fix, once and for all, an extension $u \in J(E)$. The collection of these primes is then written ι. Further, we denote by η and η', respectively, the subsets of ι corresponding to δ and δ'. To summarize our notation, we have the following identities:

$$\mathbf{a} = \delta \sqcup \delta', \quad \text{and} \quad J(E) = \zeta \sqcup \zeta';$$
$$J(E) = \iota \sqcup \tau\iota, \quad \text{and} \quad \iota = \eta \sqcup \eta';$$
$$\zeta = \eta \sqcup \tau\eta, \quad \text{and} \quad \zeta' = \eta' \sqcup \tau\eta'.$$

The meaning of the notation $\tau\iota$, etc., is self-evident. Throughout this paper, we shall assume (as in an analogous situation in §2) that $\zeta \neq \emptyset$.

We now define holomorphic automorphic forms on H^ζ. This time they are matrix valued mappings taking values in $\mathrm{End}(\mathcal{X})$, where $\mathcal{X}$ is defined in the same manner as the representation space in §2 denoted by the same symbol, and have weight $k + \tau k$, where $k \in \mathbb{Z}^\iota$.

Given a mapping $f : H^\zeta \to \mathrm{End}(\mathcal{X})$, we define another mapping of the same kind, $f\|_{k+\tau k}\alpha$, by the following formula:

$$(f\|_{k+\tau k}\alpha)(w) = j(\alpha, w)^{-k\eta - \tau(k\eta)} \sigma(\mathrm{N}(\alpha)^{\frac{1}{2}} \alpha^{-1}) f(\alpha w) \sigma(\mathrm{N}(\alpha)^{-\frac{\tau}{2}} \alpha^\tau).$$

We may then define $\mathcal{S}_{k+\tau k}(\Gamma)$ and $\mathcal{S}_{k+\tau k}(B_E)$ by generalizing the definition in §2 in a natural way. The detailed definitions can be found in [D97a].

To define our theta function, we take $2\mathrm{N}$, where N is the norm, to be the symmetric F-bilinear form S (on V) in §3. Thus we have

$$S(x, y) = \mathrm{Tr}(xy^*), \qquad \forall x, y \in V.$$

---

[3]Note that E is no longer a *subfield* of F!

In this case, the majorizing form of S_v for $v \in \delta$ can be explicitly computed. It turns out that we have

$$P_v[\xi; w] = 2N(\xi) + \text{Im}(w_1)^{-1}\text{Im}(w_2)^{-1}|[\xi; w_1, w_2]|^2,$$

$$\forall v \in \delta, \forall \xi \in V_v, \forall w = (w_1, w_2) \in H \times H.$$

We note that here the discussion in §2 is relevant, since the computations can be essentially reduced to that case.

If $C \in \mathcal{L}(V)$ is a locally constant function and an element $r \in F$ is chosen such that $r_v > 0$ for $v \in \delta$ and $r_v < 0$ for $v \in \delta'$, then the theta function is defined to be an $\text{End}(\mathcal{X})$-valued function $\theta(z, w; C, r)$ on $H^a \times H^\varsigma$ as follows:

$$(6) \qquad \theta(z, w; C, r) = \text{Im}(z)^\delta \text{Im}(w)^{-k\eta - \tau(k\eta)}$$

$$\cdot \sum_{\xi \in V} C(\xi)[\xi, \overline{w}]^{k\eta}\sigma(\xi)\mathbf{e}_F(rR[\xi, z, w]).$$

Here

$$R[\xi, z, w]_v = \begin{cases} N(\xi_u)\overline{z_v}, & \text{if } u|_F = v \in \delta', \\ N(\xi_u)z_v + i\text{Im}(z_v)|[\xi, w]_v|^2(\text{Im}(w_u)\text{Im}(w_{\tau u}))^{-1}, & \text{if } u|_F = v \in \delta. \end{cases}$$

In order to give the theta lift and its properties explicitly, which is crucial for algebraicity results, we must also derive explicit transformation formulas for the theta function.

Let us first consider the action of $(B_E)_+^\times$ on the variable w. If $\alpha \in (B_E)_+^\times$ and $N(\alpha) \in F$, then

$$\theta(z, w; C, r)\|_{k+\tau k}\alpha = \theta(z, w; C_\alpha, r),$$

where $C_\alpha(\xi) \stackrel{\text{def}}{=} C(\alpha\xi\alpha^{-\tau})$. Also, if $p \in F^\times$, then

$$\theta(z, w; C, r) = \theta(z, w; C', p^2 r),$$

where $C'(\xi) \stackrel{\text{def}}{=} p^{\frac{k}{2} - \eta'}C(p\xi)$.

We now consider the action of elements of $SL_2(F)$ on the variable z. In this case we can specialize certain results of Shimura to our setting. See [Sh93] for a discussion in the general setting. Let $SL_2(F)$ be the G, and the form $2r\,N$ be the S, in §3 of [Sh93]. (Notice that, therefore, the usage of the symbol S differs in this paragraph from the rest of the paper.) Thus S is of signature $(2, 2)$ for $v \in \delta$ and signature $(0, 4)$ for $v \in \delta'$. Then the q loc. cit. equals 4. We then have $F(\det(S)^{\frac{1}{2}}) = E$ and a straightforward calculation shows that

$$J^S(\alpha, z) = |j(\alpha, z)|^{2\delta}j(\alpha, \bar{z})^k,$$

where $J^S(\alpha, z)$ is Shimura's notation, except that here we have written z in place of Shimura's $\mathfrak{z}$. We further note the following notational correspondences (in our specific setting), writing the notations of Shimura in [Sh93] on the right-hand side:

$$G_{\mathbb{A}} = \mathfrak{M}_q; \qquad \mathcal{L}(V) = \mathcal{S}(V_{\mathfrak{h}}^n);$$

$$C \in \mathcal{L}(V) \leftrightarrow \lambda \in \mathcal{S}(V_{\mathbf{h}}^n); \qquad \gamma C \leftrightarrow {}^{\sigma}\lambda \text{ for } \gamma = \sigma \in G_{\mathbf{A}}.$$

In such a case we have the following results. Notations as above, every $\gamma \in G_{\mathbf{A}}$ gives rise to a $\mathbb{C}$-linear automorphism of $\mathcal{L}(V)$, (this action depends on the choice of r!) which we denote by $(\gamma, C) \mapsto \gamma C$, such that the following properties hold:

(a) $j(\gamma, \bar{z})^{-k} \theta(\gamma z, w; \gamma C, r) = \theta(z, w; C, r)$, $\forall \gamma \in G$;

(b) $(\gamma \delta) C = \gamma(\delta C)$, $\forall \gamma, \delta \in G_{\mathbf{A}}$.

(c) For every C, there exists a congruence subgroup Γ of G, such that

$$\gamma C = C, \qquad \forall \gamma \in \Gamma.$$

(d) If $(2) \in G_{\mathbf{h}}$ and $c_{(2)} = 0$, then

$$((2)C)(x) = |a_{(2)}|_{\mathbf{A}}^2 \omega_{\mathbf{h}}(a_{(2)}) \mathbf{e}_{\mathbf{h}}(r \, \mathrm{N}(x) a_{(2)} b_{(2)}) C(x a_{(2)}),$$

where ω denotes the Hecke character of F corresponding to E, and $\mathbf{e}_{\mathbf{h}}(y) \overset{\text{def}}{=} \mathbf{e}_{\mathbf{A}}(y_{\mathbf{h}})$.

We also have, for $p \in F$, $p \gg 0$,

$$p^{\frac{k}{2}} \theta(pz, w; C, r) = \theta(z, w; p^{\delta + \frac{k}{2}} C, pr).$$

We now define the theta lifting as follows. Given $h \in \mathcal{S}_{k+\tau k}(B_E)$, we define

$$I(z; C, r; h) = \langle \theta(z, w; C, r), h(w) \rangle.$$

Theorem. *We have*

$$I(z; C, r; h) = vol(D)^{-1} \int_D Tr\left(\overline{{}^t \theta(z, w; C, r)} h(w)\right) Im(w)^{k\eta + \tau(k\eta)} d_H^\varsigma w,$$

where $D = \Delta \backslash H^\varsigma$ and Δ is a suitable congruence subgroup. $I(z; C, r; h)$ is a Hilbert modular form belonging to $\mathcal{S}_k(\mathrm{SL}_2(F))$, and its Fourier coefficients are related to the periods of h. More precisely, we have an expression of the form

$$vol(D) I(z; C, r; h) = \kappa \sum_\alpha 2^{||\delta||} |r \, \mathrm{N}(\alpha)|^{-\delta} \overline{C(\alpha)} P(h, \alpha, \Delta) \mathbf{e}_F(-r \, \mathrm{N}(\alpha) z),$$

where $P(h, \alpha, \Delta)$ is a certain period of h, which is defined by

$$P(h, \alpha, \Delta) = \int_{\Delta_\alpha \backslash J_\alpha} [\alpha, pw]^{k\eta} Tr\left(\overline{{}^t \sigma(\alpha)} h(pw)\right) d(\Delta_\alpha p).$$

Here α is a totally negative element of V, and $w \in H^\varsigma$ can be arbitrarily chosen. In fact, the above integral is independent of the choice of w, and vanishes for all α which are not totally negative. Finally the symbols Δ_α and J_α are suitably chosen subgroups of $\mathrm{SL}_2(\mathbb{R})^\varsigma$.

For a proof of the above theorem, see [D97a]. This theorem generalizes a theorem of Shimura (see [Sh82]). Its significance is partly due to the relation to the period

164

conjectures of Shimura; for a discussion of that connection we refer the reader to [Sh88]. On a more concrete level, we see that several of the key concepts discussed in §2 have already made an appearance in the Theorem above. In any case, this clearly provides a setting in which suitable generalizations of theorems of the type mentioned in §§2–3 can be sought. In fact, if we retrace our development through those two sections, we may to a certain extent anticipate such results. Therefore, the consideration of the setting of this section is natural. Also, in view of Yoshida's work (see[Y95]), our setting is relevant. Without engaging ourselves in any speculations, we simply mention that the author has obtained several results in this direction, which can be found in the papers [D97a] and [D97b]. A third work is currently under preparation, which he hopes to complete in the near future. Since these results require a considerable amount of further technical preparation, their precise statements are omitted here.

REFERENCES

[B90]   A. W. Bluher, *Near Holomorphy of some automorphic forms at critical points*, Invent. Math. **102** (1990), 335–376.

[B94]   _____, *Arithmeticity of some automorphic forms at critical points*, Amer. J. Math. **116** (1994), 1283–1335.

[B98]   _____, *Near holomorphy, arithmeticity, and the theta correspondence*, To appear in [DDG98].

[DDG98] R. Doran, Z.-L. Dou, G. Gilbert, eds., Proceedings of Symposia in Pure Mathematics, to appear 1998.

[D93]   Z.-L. Dou, *Fundamental periods of certain arithmetic cusp forms*, Thesis, Princeton University (1993).

[D94]   _____, *On the fundamental periods of Hilbert modular forms*, Trans. Amer. Math. Soc. **346** (1994), 147–158.

[D97a] _____, *On a theta correspondence with respect to a quadratic extension*, Preprint.

[D97b] _____, *Theta correspondence and Hecke operators relative to a quadratic extension*, Preprint.

[Sh76] G. Shimura, *The special values of the zeta functions associated with cusp forms*, Comm. Pure and Appl. Math. **29** (1976), 783–804.

[Sh80] _____, *The arithmetic of certain zeta functions and automorphic forms on orthogonal groups*, Ann. Math. **111** (1980), 313–375.

[Sh81I] _____, *On certain zeta functions attached to two Hilbert modular forms, I*, Ann. Math. **114** (1981), 127–164.

[Sh81II] _____, *On certain zeta functions attached to two Hilbert modular forms, II*, Ann. Math. **114** (1981), 569–607.

[Sh82] _____, *The periods of certain automorphic forms of arithmetic type*, J. Fac. Sci. Univ. Tokyo (Sect. IA) **28** (1982), 605–632.

[Sh83] _____, *Algebraic relations between critical values of zeta functions and inner products*, Amer. J. Math. **104** (1983), 253–285.

[Sh86] _____, *On a class of nearly holomorphic automorphic forms*, Ann. of Math. **123** (1986), 347–406.

[Sh87] _____, *Nearly holomorphic functions on Hermitian symmetric spaces*, Math. Ann. **278** (1987), 1–28.

[Sh88] _____, *On the critical values of certain Dirichlet series and the periods of automorphic forms*, Invent.. Math. **94** (1988), 245–305.

[Sh90] _____, *On the fundamental periods of automorphic forms of arithmetic type*, Invent. Math. **102** (1990), 399–428.

[Sh93] _____, *On the transformation formulas of theta series*, Amer. J. Math. **115** (1993), 1011–1052.

[Sh95]   _____, *Arithmeticity of the special values of various zeta functions and the periods of abelian integrals*, Sugaku Expositions **8** (1995), 17–38.

[Sh97]   _____, *Euler products and Eisenstein series*, CBMS Regional Conference Series in Mathematics **93** (1997).

[Y95]   H. Yoshida, *On a conjecture of Shimura concerning periods of Hilbert modular forms*, Amer. J. Math. **117** (1995), 1019–1038.

MATHEMATICS DEPARTMENT, P. O. BOX 298900, TEXAS CHRISTIAN UNIVERSITY, FORT WORTH, TX 76129

E-mail address: z.dou@tcu.edu

MORPHIC HEIGHTS AND PERIODIC POINTS

M. EINSIEDLER, G. EVEREST, AND T. WARD

June 13, 2000

ABSTRACT. An approach to the calculation of local canonical morphic heights is described, motivated by the analogy between the classical height in Diophantine geometry and entropy in algebraic dynamics. We consider cases where the local morphic height is expressed as an integral average of the logarithmic distance to the closure of the periodic points of the underlying morphism. The results may be thought of as a kind of morphic Jensen formula.

1. INTRODUCTION

Let $\phi : \mathbb{P}^1(\overline{\mathbb{Q}}) \to \mathbb{P}^1(\overline{\mathbb{Q}})$ be a morphism of degree d, defined over the rationals. Call, Goldstine and Silverman (see [3],[4]) have associated to ϕ a canonical global *morphic height* $\hat{\lambda}_\phi$ on $\overline{\mathbb{Q}}$, with the properties that

1. $\hat{\lambda}_\phi(\phi(q)) = d\hat{\lambda}_\phi(q)$ for any $q \in \mathbb{P}^1(\overline{\mathbb{Q}})$;
2. q is pre-periodic if and only if $\hat{\lambda}_\phi(q) = 0$.

A point q is called *pre-periodic* under ϕ if the orbit $\{\phi^{(n)}(q)\}$ is finite (write $f^{(n)}$ for the nth iterate of a map f). The global height decomposes into a sum of local canonical morphic heights $\lambda_{\phi,v}$:

$$\hat{\lambda}_\phi(q) = \sum_v n_v \lambda_{\phi,v}(q).$$

Here v runs over all the valuations (both finite and infinite) of the number field generated by q and the n_v denote the usual normalising constants. In the special case that ϕ takes the form $\phi[x,y] = [y^d f(x/y), y^d]$ for a polynomial f of degree d, Call and Goldstine [3] prove that

$$\lambda_{\phi,v}(q) = \lim_{n \to \infty} \frac{1}{d^n} \lambda_v(\phi^{(n)}(q)), \tag{1}$$

where λ_v is the local projective height $\lambda_v(q) = \log^+ |q|_v$, and that the local height $\lambda_{\phi,v}(q)$ vanishes if and only if $|\phi^{(n)}(q)|_v$ is bounded for all

1991 *Mathematics Subject Classification.* 11S05, 37F10.

The first author acknowledges the support of EPSRC postdoctoral award GR/M49588, the second thanks Jonathan Lubin and Joe Silverman for the AMS Sectional meeting on Arithmetic Dynamics at Providence, RI, 1999.

n, and finally, that q is pre-periodic if and only if

$$\lambda_{\phi,v}(q) = 0 \text{ for all } v. \tag{2}$$

Example 1.1.   1. Let $f(z) = z^d$ with $d > 1$. Here the canonical morphic heights and the projective heights agree. Jensen's formula ([13, Theorem 15.18]) gives

$$\int_{\mathbb{S}^1} \log |y - q| dm(y) = \log^+ |q|,$$

where m is the Haar measure on the circle $\mathbb{S}^1$. The circle is also the Julia set for this morphism on $\mathbb{C}$, and it is the closure of the set of non-zero periodic points, which are all roots of unity. In the p-adic case, the Julia set is empty but it is still true that the local height is the Shnirel'man integral of the logarithmic distance from the closure of the set of periodic points. For all v, finite and infinite, the following holds:

$$\log^+ |q|_v = \lim_{n \to \infty} \frac{1}{d^n} \sum_{\zeta \neq q: \zeta^{d^n-1}=1} \log |\zeta - q|_v.$$

For finite v this may be seen, for instance, using Section 3. Alternatively, it follows from the Diophantine estimate

$$|q^{d^n} - q|_v > C(q)|n|_v$$

provided the left hand side is non-zero (cf. Remark 3.6). For $v|\infty$ a result from transcendence theory (in this case, Baker's theorem) is needed.

2. Suppose $a, b \in \mathbb{Q}$ with $4a^3 + 27b^2 \neq 0$ and let

$$f(z) = \frac{z^4 - 2az^2 - 8bz + a^2}{z^3 + az + b}.$$

Then f gives rise to a morphism of degree 4 which describes the duplication map on an elliptic curve. The global and local morphic heights coincide with the usual notions of height on the curve. For the infinite valuations, the local height is again the integral average of the logarithmic distance to points on the Julia set, which is the closure of the set of periodic points. At the singular reduction primes it is still true that the local height is the integral average of the logarithmic distance to the periodic points. Both of these assertions are proved in [6] where this morphism was used to construct a dynamical system which interprets these heights in dynamical terms. The proofs require elliptic transcendence theory to show that a rational point cannot approximate a periodic point too closely (cf. Proposition 3.8). This is part of a much broader analogy between heights in Diophantine geometry and entropy in algebraic dynamics (see [5], [6], [7], [10]).

In this paper, our purpose is to describe a family of examples where the local canonical morphic height can be expressed as a limiting integral over periodic points of the underlying morphism. The finite and infinite cases require different approaches. In both, we consider the special class of morphisms corresponding to affine polynomial maps and in the former case, we assume good reduction in the sense of Morton and Silverman [12].

There are two directions in which this work can be made more sophisticated that are not pursued here. The first is to give a more formal interpretation of the limiting process using Shnirel'man integrals (see [14], or [8] for a modern treatment); the second is to extend the arguments to other morphisms.

2. COMPLEX CASE

Assume that $\phi : \mathbb{P}^1(\mathbb{C}) \to \mathbb{P}^1(\mathbb{C})$ is a morphism of degree d, with $\phi[x, y] = [y^d f(x/y), y^d]$ for a polynomial f of degree d. For basic definitions of complex dynamics, consult [1]. The following theorem expresses the local morphic height as an integral over the Julia set $J(f)$ of the polynomial f, and standard results from complex dynamics show that this is in turn a limiting integral over periodic points.

Theorem 2.1. *If $f(z) = az^d + \cdots + a_0$ is a polynomial, then for any $q \in \mathbb{C}$,*

$$\lambda_{\phi,\infty}(q) = \frac{1}{d-1} \log |a| + \int_{J(f)} \log |x - q| dm(x), \qquad (3)$$

where m is the maximal invariant measure for f on $J(f)$.

Proof. Assume first that q is in the domain of attraction of ∞ for f. The zeros of the polynomial $f_n(x) = f^{(n)}(x) - x$ are precisely the solutions of the equation $f^{(n)}(x) = x$. Note that $d_n = \deg(f_n) = d^n$, where $d = \deg(f)$. Since $|f^{(n)}(q)| \to \infty$, $\frac{1}{d_n} \log |f_n(q)|$ is approximately $\frac{1}{d^n} \log |f^{(n)}(q)|$, which converges to $\lambda_{\phi,\infty}(q)$. Since q lies in the open Fatou set, $\log |x - q|$ is continuous on $J(f)$. Now

$$\frac{1}{d_n} \log |f_n(q)| = \frac{1}{d_n} \sum_{f^{(n)}(x)=x} \log |x - q| + \frac{1}{d_n} \log |B_n|, \qquad (4)$$

where the sum is over the nth 'division points' and

$$B_n = a^{1+d+d^2+\cdots+d^{(n-1)}}$$

is the leading coefficient of $f^{(n)}(x)$. Thus

$$\frac{1}{d_n}\log|B_n| = \frac{1}{d^n}\left(\frac{d^n-1}{d-1}\right)\log|a| \to \frac{1}{d-1}\log|a|. \tag{5}$$

Now it is known that

$$\frac{1}{d_n}\sum_{f^{(n)}(x)=x}\log|x-q| \to \int_{J(f)}\log|x-q|dm(x),$$

where m is the maximal invariant measure for f restricted to the Julia set (see [11]; [9]).

In the following it is convenient to assume that a=1, we can ensure this by conjugating by a linear map. Now $|x-f(q)| = \prod_{f(t)=x}|t-q|$, so

$$\int_{J(f)}\log|x-f(q)|dm(x) = \int_{J(f)}\sum_{f(t)=x}\log|t-q|dm(x)$$

$$= d\int_{J(f)}\log|x-q|dm(x). \tag{6}$$

(The last equality follows from [11] or [9, Theorem (d)]).

Let now $q \notin J(f)$ have bounded orbit. Since $J(f)$ is closed, we can find $\epsilon > 0$ such that $B_\epsilon(q) \cap J(f) = \emptyset$. If $|f^n(q) - q| > \epsilon/2$ for almost all n, then

$$\frac{1}{d_n}\log|f^n(q) - q| \to 0.$$

We can argue as in the first case to get

$$\int_{J(f)}\log|x-q| = 0 = \lambda_{\phi,\infty}(q).$$

Assume now $|f^{n_j}(q) - q| \le \epsilon/2$ for some sequence $n_j \to \infty$. Then $|f^{n_j}(q) - x| > \epsilon/2$ for $x \in J(f)$. However, since $J(f)$ and $f^{n_j}(q)$ are bounded, we have also an upper bound

$$\log\frac{\epsilon}{2} \le \int_{J(f)}\log|x-f^{n_j}(q)| \le M.$$

Together with Equation (6) we get

$$1/d^{n_j}\log\frac{\epsilon}{2} \le \int_{J(f)}\log|x-q| \le 1/d^{n_j}M$$

which concludes the proof in this case.

It remains to show that the formula holds for $q \in J(f)$. Since $J(f)$ has no interior, there is a sequence $q_n \to q$ with $q_n \notin J(f)$. Then $\log|x-q_n| \to \log|x-q|$ for all $x \in J(f)\backslash\{q\}$. Since $J(f)$ is bounded,

$\log|x - q_n|$ and $\log|x - q|$ are uniformly bounded above by M say for $x \in J(f)\backslash\{q\}$. So by Fatou's lemma

$$0 = \lim_{n\to\infty} \int_{J(f)} \log|x - q_n|dm(x) \leq \int_{J(f)} \log|x - q|dm(x) \leq M. \tag{7}$$

This shows that $x \mapsto \log|x - q|$ is in $L^1(m)$. If

$$\int_{J(f)} \log|x - q|dm(x) > 0,$$

then Equation (6) contradicts (7).   □

3. The p-adic case

Let $\mathbb{C}_p$ denote the usual completion of the algebraic closure of the p-adic numbers $\mathbb{Q}_p$, and use $|\cdot|$ to denote the extension of the p-adic norm to $\mathbb{C}_p$. Write $\mathcal{O}_p$ for the ring of integral elements in $\mathbb{C}_p$, and $\mathcal{P}$ for the maximal ideal of $\mathcal{O}_p$. In this section we assume that $\phi : \mathbb{P}^1(\mathbb{C}_p) \to \mathbb{P}^1(\mathbb{C}_p)$ is a morphism of degree d corresponding to an affine polynomial f of degree d with coefficients in $\mathcal{O}_p$ and leading coefficient in $\mathcal{O}_p^*$. Notice that these assumptions are, for polynomials, equivalent to the assumption that the map ϕ has good reduction in the sense of [12]: ϕ induces a morphism of schemes over $\mathrm{Spec}(\mathcal{O}_p)$. The Julia set is empty in this setting (see [2], [12]), so a direct analogue of (3) is not possible.

The main result expresses the local canonical morphic height as a limiting integral over periodic points for the polynomial f.

Theorem 3.1. *If ϕ has good reduction and is defined by a polynomial f of degree d, then*

$$\lambda_{\phi,p}(q) = \log^+|q| = \lim_{n\to\infty} \frac{1}{d^n} \sum_{\xi \neq q : f^{(n)}(\xi)=\xi} \log|\xi - q|$$

where the sum is taken with multiplicities.

Notice first that for $q \in \mathbb{C}_p\backslash\mathcal{O}_p$ this is clear, so from now on we assume that $q \in \mathcal{O}_p$. Despite the simple resulting value of the height, the convergence involved requires an argument. The main issue is to produce lower bounds on the size of $|\xi - \zeta|$ for distinct periodic points ξ and ζ.

The *least period* of a periodic point ξ is the cardinality of the orbit of ξ. The points of period n are the solutions to the polynomial equation

$$f^{(n)}(x) - x = 0, \tag{8}$$

and are therefore all elements of $\mathcal{O}_p$. Following [12], for a periodic point ξ define $a_n(\xi)$ to be the multiplicity of ξ in (8), with the obvious convention that ξ has multiplicity zero in an equation that it does not satisfy. Notice that $a_n(\xi) \neq 0$ if and only if n is a multiple of the least period of ξ. Define $a_n^*(\xi)$ by

$$a_n^*(\xi) = \sum_{d|n} \mu\left(\frac{n}{d}\right) a_n(\xi),$$

where μ is the Möbius function. Increases in the multiplicity of the periodic point ξ along the sequence of multiples of its least period are recorded by $a_n^*(\xi)$. The periodic point ξ is an *essential n-periodic point* if $a_n^*(\xi) > 0$.

The following proposition is a special case of [12, Prop. 3.2].

Proposition 3.2. *Let K be an algebraically closed field of characteristic $p \geq 0$, and f a polynomial over K with degree $d \geq 2$. Fix a periodic point $\zeta \in K$ with least period m, and let r denote the multiplicative order of $(f^{(m)})'(\zeta)$ in K^* or ∞ if $(f^{(m)})'$ is not a root of unity. Then for $n \geq 1$, $a_n^*(\zeta) \geq 1$ if and only if one of the following conditions hold.*

1. *$n = m$;*
2. *$n = mr$;*
3. *$p > 0$ and $n = p^e mr$ for some $e \geq 1$.*

When $K = \mathbb{C}_p$, this proposition will also be applied to the polynomial $\bar{f}$ induced by reduction mod $\mathcal{P}$. Notice that the sum of the multiplicities of the points of period n under f lying in one residue class gives the multiplicity of the image point as a point of period n under $\bar{f}$.

For the proof of Theorem 3.1 the following proposition is needed; this will be proved later.

Proposition 3.3. *Suppose ξ_n is a periodic point with least period n for a polynomial f of good reduction. Then for any fixed $q \in \mathcal{O}_p$, $|q - \xi_n| \to 1$ as $n \to \infty$.*

Proof. (of Theorem 3.1) By Proposition 3.2 applied to the field $\mathbb{C}_p$ with characteristic zero, if ξ is a periodic point with least period m, then the multiplicity of ξ viewed as a periodic point of period $m, 2m, 3m, \ldots$ is uniformly bounded. Fix $q \in \mathcal{O}_p$ and $s \in (0,1)$. Proposition 3.3 says that the number of periodic points in the metric ball $D_s(q)$ is finite. It follows that

$$\liminf_{n \to \infty} \frac{1}{d^n} \sum_{\xi \neq q: f^{(n)}(\xi) = \xi} \log|\xi - q| \geq \log s.$$

On the other hand, each term in the sum is non-positive, so letting $s \to 1$ proves the theorem. $\qquad\square$

All that remains is to prove Proposition 3.3, for which we need some lemmas.

Lemma 3.4. *Assume that $f(0) = 0$, and let ζ be a periodic point of f. Then $|\zeta| = |f^{(n)}(\zeta)|$ for all $n \geq 1$.*

Proof. The spherical metric used in [12] coincides with the usual metric in $\mathcal{O}_p$, and f has good reduction. So by [12, Prop. 5.2],

$$|f(x) - f(y)| \leq |x - y|$$

for $x, y \in \mathcal{O}_p$. The lemma follows at once. $\qquad\square$

Lemma 3.5. *Assume that $f(0) = 0$ and $n > 1$ is fixed.*

1. *If $|f'(0)| < 1$ then $\prod_{\xi \neq 0} |\xi|^{a_n^*(\xi)} = 1$.*
2. *If $|f'(0)-1| < p^{-1}$ then $\prod_{\xi \neq 0} |\xi|^{a_n^*(\xi)} = \begin{cases} 1/p & \text{if } n \text{ is a power of } p, \\ 1 & \text{if not.} \end{cases}$*

Proof. In case 1., $\bar{f}'(0) = 0$ in the algebraically closed field $\mathcal{O}_p/\mathcal{P}$, so Proposition 3.2 may be applied with $\zeta = 0 + \mathcal{P}$, $m = 1$ and $r = \infty$. It follows that for $n > 1$ $a_n^*(0 + \mathcal{P}) = 0$, so there cannot be an essential n-periodic point ξ for f with $|\xi| < 1$.

In case 2., $m = 1$ and $r = 1$ for the point $\zeta = 0 + \mathcal{P}$ in Proposition 3.2. It follows that only values of n of the form p^k are relevant. Notice that

$$\prod_{\xi \neq 0} |\xi|^{a_{p^k}^*(\xi)} = \left| \left(\frac{f^{(p^k)}(x) - x}{f^{(p^{k-1})}(x) - x} \right)_{x=0} \right| \tag{9}$$

If $f'(0) \neq 1$, then the right-hand side of (9) is given by

$$\left| \frac{(f'(0))^{p^k} - 1}{(f'(0))^{p^{k-1}} - 1} \right| = \frac{1}{p}$$

by the binomial theorem. If $f'(0) = 1$, write $f(x) = x + x^e g(x)$ with $e > 1$ and $g(0) \neq 0$, then a simple induction argument shows that

$$f^{(k)}(x) = x + kx^e g(x) + O(x^{2e-1}).$$

It follows that (9) is equal to p^{-1} again. $\qquad\square$

Proof. (of Proposition 3.3) Let ζ be any periodic point, with least period ℓ. The first step is to prove the proposition for $q = \zeta$. Let ξ have least period n under f. The multiplicity of $\zeta + \mathcal{P}$, which has least period m for some $m|\ell$ must increase at ℓ (because $a_\ell^*(\zeta) > 1$). It follows by Proposition 3.2 that ℓ is equal to m, mr, or mrp^e for some $e \geq 1$. Assume first that n is not of one of those forms; then $|\xi - \zeta| = 1$ because the multiplicity of $\zeta + \mathcal{P}$ cannot increase at n in $\mathcal{O}_p/\mathcal{P}$.

In the remaining cases, we may assume for large n that $\ell | n$. Then ξ is a periodic point with least period n/ℓ under $f^{(\ell)}$. Applying the conjugation $x \mapsto x - \zeta$ means that 0 is a fixed point of g, where $g(x) = f^{(\ell)}(x + \zeta) - \zeta$.

If $|g'(0)| < 1$, then by Lemma 3.5 applied to g, $|\xi - \zeta| = 1$.

If $|g'(0)| = 1$, let t be the order of $g'(0) + \mathcal{P}$ in $\mathcal{O}_p/\mathcal{P}$. Then

$$|(g'(0))^t - 1| < 1.$$

There exists a $c \geq 1$ such that

$$|(g'(0))^{tp^c} - 1| < 1/p.$$

As before, we may assume that $tp^c\ell | n$, so Lemma 3.5 may be applied to the map $h = g^{(tp^c)}$ and the periodic point $\xi - \zeta$ of least period $n/tp^c\ell$ to give

$$\prod_{j=1,\ldots,n/tp^c\ell} |h^{(j)}(\xi - \zeta)| \geq 1/p.$$

It follows by Lemma 3.4 that $|\xi - \zeta| \geq p^{-tp^c\ell/n}$. Since t, c, ℓ depend only on ζ, $|\xi - \zeta| \to 1$ as $n \to \infty$. The ultrametric inequality in $\mathbb{C}_p$ now gives the result for any $q \in \mathcal{O}_p$. $\qquad\square$

Remark 3.6. Notice that the discussion above also gives a quantitative version of Proposition 3.3. This Diophantine result may be of independent interest. If f is a polynomial of good reduction, then

$$|f^{(n)}(q) - q| > C(f, q)|n|$$

for all $n \geq 1$, provided the left hand side is non-zero.

Example 3.7. To see the different cases that are possible in Proposition 3.3, consider the following examples.

1. Let $f(x) = g(x^p) + ph(x)$ be a monic polynomial with coefficients in $\mathcal{O}_p$. Then $|f'(q)| < 1$ for any $q \in \mathcal{O}_p$, so in Lemma 3.5 only the first case is ever used. It follows that in Proposition 3.3, $|\zeta - \xi| = 1$ for any distinct periodic points ζ, ξ.

2. Let $f(x) = x^2 - (1+a)x$ for some small a. Then 0 is a fixed point, and $|f'(0)| = 1$. Now

$$f^{(2)}(x) - x = x^4 - (2a + 2)x^3 + (a^2 + a)x^2 + (a^2 + 2a)x,$$

so $(f^{(2)}(x) - x)/(f(x) - x)$ has constant term $(a^2 + 2a)/(-a - 2) = -a$. Therefore there must be two non-zero points of period 2 that are close to the fixed point 0.

If the polynomial f has coefficients outside $\mathcal{O}_p$, then in contrast to Proposition 3.3, there may be sequences of periodic points converging to a periodic point. For example, $f(x) = x^2 + \frac{1}{2}x$ on $\mathbb{C}_2$ has this property. Therefore, to recover Theorem 3.1 in greater generality (for polynomials of bad reduction or rational functions) some kind of Diophantine approximation results are needed. In Example 1.1.2 these tools are provided by elliptic transcendence theory.

Proposition 3.8. *Let ζ be a periodic point with least period ℓ under f. Assume that $|(f^{(\ell)})'(\zeta)| > 1$. Then there are periodic points $\xi \neq \zeta$ arbitrarily close to ζ.*

Proof. Define $g = f^{(\ell)}$ and $a = (f^{(\ell)})'(\zeta)$. Without loss of generality we can assume that $\zeta = 0$. Then

$$g(x) = ax + bx^e + O(x^{e+1}) \text{ with } b \neq 0$$

and

$$g^{(2)}(x) = a^2 x + (ab + a^e b)x^e + O(x^{e+1}).$$

Define $b_2 = (ab + a^e b)$, then $|b_2| = |b||a|^e$. By induction one can see that

$$g^{(k)}(x) = a^k x + b_k x^e + O(x^{e+1})$$

with $|b_k| = |b||a|^{ke}$.

Therefore the Newton polygon of $g^{(k)}(x) - x$ starts with a line with slope $s \leq -k + c$ for a fixed c (depending on b and e). So there exists a periodic point ξ with $|\xi| = p^s \leq p^{-k+c}$. $\qquad\square$

4. TCHEBYCHEFF POLYNOMIALS

Example 4.1. Consider the Tchebycheff polynomial of degree d, $f(z) = T_d(z) = \cos(d \arccos(z))$. The Julia set is the interval $J(f) = [-1, 1]$. The map $\phi : \mathbb{C} \to \mathbb{C}$ given by $\phi(z) = \frac{1}{2}(z + z^{-1})$ is a semi-conjugacy from $g : z \mapsto z^d$ onto $z \mapsto f(z)$, in other words, $f(\phi(z)) = \phi(z^d)$. Write ψ for the branch of the inverse of ϕ defined on $\{z \in \mathbb{C} \mid |z| > 1\}$. The canonical morphic height at the infinite place is (for $q \notin J(f)$)

$$
\begin{aligned}
\lambda_{\phi,\infty}(q) &= \lim_{n \to \infty} \frac{1}{d^n} \log^+ |f^{(n)}(q)| \\
&= \lim_{n \to \infty} \frac{1}{d^n} \log^+ |\phi g^{(n)} \psi(q)| \\
&= \lim_{n \to \infty} \frac{1}{d^n} \log^+ |\frac{1}{2}\left(g^{(n)}\psi(q) + \frac{1}{g^{(n)}\psi(q)} \right)| \\
&= \lim_{n \to \infty} \max\left\{ 0, \frac{1}{d^n} \log |g^{(n)}\psi(q)| \right\} \\
&= \log^+ |\psi(q)|.
\end{aligned}
$$

For $q \in J(f)$, the same formula holds since $\lambda_{\phi,\infty}(q) = 0$ there by [3] and $\log^+ |\psi(q)| = 0$ there by a direct calculation.

Now by Jensen's formula, for any $q \in \mathbb{C}$,

$$\log^+ |\psi(q)| = \log 2 + \int_{\mathbb{S}^1} |\phi(y) - q| dy$$
$$= \log 2 + \int_{J(f)} \log |t - q| dm(t)$$

since m is the image under ϕ of the maximal measure (Lebesgue) on the circle. That is,

$$\lambda_{\phi,\infty}(q) = \log 2 + \int_{J(f)} \log |t - q| dm(t). \tag{10}$$

The constant $\log 2$ in $\lambda_\infty(q)$ may be explained in accordance with Theorem 2.1. The leading coefficient of T_d is 2^{d-1}, so $\frac{1}{d-1} \log |a|$ in this case is exactly $\log 2$.

A similar approach can be adopted in the case of polynomials with connected Julia sets. There the local conjugacy near ∞ extends to the whole domain of attraction of ∞, which is the complement of the filled Julia set.

Example 4.2. As before, let $f(x) = T_d(x) = \cos(d \arccos(x))$ be the Tchebycheff polynomial of degree d and let ϕ be the corresponding morphism. We would like to use Theorem 3.1, but f does not satisfy the assumptions since it is not monic.

Let $g(x) = 2f(\frac{x}{2})$. Notice that f is defined uniquely by the property $f(\frac{1}{2}(z + z^{-1})) = \frac{1}{2}(z^d + z^{-d})$. It follows that g is characterized by the property $g(z + z^{-1}) = (z^d + z^{-d})$, which shows that $g \in \mathbb{Z}[x]$ is a monic polynomial. Let ψ be the morphism defined by g, then by Theorem 3.1 we have for $q \in \mathbb{C}_p$

$$\lambda_{\psi,p}(q) = \log^+ |q| = \lim_{n \to \infty} \frac{1}{d^n} \sum_{\xi \neq q : g^{(n)}(\xi) = \xi} \log |\xi - q|. \tag{11}$$

Since $g(x) = 2f(\frac{x}{2})$, we have that $\lambda_{\phi,p}(q) = \lambda_{\psi,p}(2q)$ and on the right hand side of (11) that $f^{(n)}(\xi) = \xi$ if and only if $g^{(n)}(2\xi) = 2\xi$. Therefore

$$\lambda_{\phi,p}(q) = \log^+ |2q| = \log |2| + \lim_{n \to \infty} \frac{1}{d^n} \sum_{\xi \neq q : f^{(n)}(\xi) = \xi} \log |\xi - q|,$$

which is again analogous to Equation (10) in Example 4.1.

Example 4.2 works because the Tchebycheff polynomial can be conjugated to a polynomial of good reduction; a similar approach can be adopted for any polynomial that is conjugate to one of good reduction.

REFERENCES

1. A. Beardon, *Iteration of Rational Functions*, Springer, New York, 1991.
2. R. Benedetto, *Reduction, dynamics, and Julia sets of rational functions*, J. Number Theory, to appear.
3. G.S. Call and S.W. Goldstine, *Canonical heights on projective space*, J. Number Theory **63** (1997), 211–243.
4. G.S. Call and J.H. Silverman, *Canonical heights on varieties with morphisms*, Compositio Math. **89** (1993), 163–205.
5. P. D'Ambros, G. Everest, R. Miles, and T. Ward, *Dynamical systems arising from elliptic curves*, Colloq. Math. (to appear) (2000).
6. M. Einsiedler, G. Everest, and T. Ward, *Entropy and the canonical height*, Preprint.
7. G. Everest and T. Ward, *A dynamical interpretation of the global canonical height on an elliptic curve*, Experiment. Math. **7** (1998), 305–316.
8. N. Koblitz, *p-adic analysis: a short course on recent work*, LMS Lecture Notes 46, Cambridge Univ. Press, 1980.
9. A. Freire, A. Lopes, and R. Mañé, *An invariant measure for rational maps*, Bol. Soc. Brasil. Mat. **14** (1983), no. 1, 45–62.
10. D.A. Lind and T. Ward, *Automorphisms of solenoids and p-adic entropy*, Ergodic Theory Dynam. Systems **8** (1988), 411–419.
11. M.Y. Lyubich, *Entropy of analytic endomorphisms of the Riemann sphere*, Funktsional. Anal. i Prilozhen. **15** (1981), 83–84.
12. P. Morton and J.H. Silverman, *Periodic points, multiplicities, and dynamical units*, J. Reine Angew. Math. **461** (1995), 81–122.
13. W. Rudin, *Real and Complex Analysis*, McGraw–Hill, New York, 1974.
14. L.G. Shirel'man, *On functions in normed algebraically closed division rings*, Izv. Akad. Nauk. SSSR Ser. Mat. **2** (1938), 487–498 (Russian).

(M.E.) MATHEMATICAL INSTITUTE, UNIVERSITY OF VIENNA, STRUDLHOF-GASSE 4, A-1090 WIEN, AUSTRIA.
E-mail address: manfred@mat.univie.ac.at

(G.E. & T.W.) SCHOOL OF MATHEMATICS, UNIVERSITY OF EAST ANGLIA, NORWICH NR4 7TJ, UK.
E-mail address: g.everest@uea.ac.uk
E-mail address: t.ward@uea.ac.uk

THE ELEMENTARY PROOF OF THE PRIME NUMBER THEOREM:
AN HISTORICAL PERSPECTIVE

(by D. Goldfeld)

The study of the distribution of prime numbers has fascinated mathematicians since antiquity. It is only in modern times, however, that a precise asymptotic law for the number of primes in arbitrarily long intervals has been obtained. For a real number $x > 1$, let $\pi(x)$ denote the number of primes less than x. The prime number theorem is the assertion that

$$\lim_{x \to \infty} \pi(x) \bigg/ \frac{x}{\log(x)} = 1.$$

This theorem was conjectured independently by Legendre and Gauss.

The approximation

$$\pi(x) = \frac{x}{A \log(x) + B}$$

was formulated by Legendre in 1798 [Le1] and made more precise in [Le2] where he provided the values $A = 1, B = -1.08366$. On August 4, 1823 (see [La1], page 6) Abel, in a letter to Holmboe, characterizes the prime number theorem (referring to Legendre) as perhaps the most remarkable theorem in all mathematics.

Gauss, in his well known letter to the astronomer Encke, (see [La1], page 37) written on Christmas eve 1849 remarks that his attention to the problem of finding an asymptotic formula for $\pi(x)$ dates back to 1792 or 1793 (when he was fifteen or sixteen), and at that time noticed that the density of primes in a chiliad (i.e. $[x, x + 1000]$) decreased approximately as $1/\log(x)$ leading to the approximation

$$\pi(x) \approx \text{Li}(x) = \int_2^x \frac{dt}{\log(t)}.$$

The remarkable part is the continuation of this letter, in which he said (referring to Legendre's $\frac{x}{\log(x) - A(x)}$ approximation and Legendre's value $A(x) = 1.08366$) that whether the quantity $A(x)$ tends to 1 or to a limit close to 1, he does not dare conjecture.

The first paper in which something was proved at all regarding the asymptotic distribution of primes was Tchebychef's first memoir ([Tch1]) which was read before the Imperial Academy of St. Petersburg in 1848. In that paper Tchebychef proved that if any approximation to $\pi(x)$ held to order $x/\log(x)^N$ (with some fixed large positive integer N) then that approximation had to be $\text{Li}(x)$. It followed from this that Legendre's conjecture that $\lim_{x \to \infty} A(x) = 1.08366$ was false, and that if the limit existed it had to be 1.

The first person to show that $\pi(x)$ has the order of magnitude $\frac{x}{\log(x)}$ was Tchebychef in 1852 [Tch2]. His argument was entirely elementary and made use of properties of factorials. It is easy to see that the highest power of a prime p which divides $x!$ (we assume x is an integer) is simply

$$\left[\frac{x}{p}\right] + \left[\frac{x}{p^2}\right] + \left[\frac{x}{p^3}\right] + \cdots$$

179

where $[t]$ denotes the greatest integer less than or equal to t. It immediately follows that

$$x! = \prod_{p \le x} p^{[x/p]+[x/p^2]+\cdots}$$

and

$$\log(x!) = \sum_{p \le x} \left(\left[\frac{x}{p} \right] + \left[\frac{x}{p^2} \right] + \left[\frac{x}{p^3} \right] + \cdots \right) \log(p).$$

Now $\log(x!)$ is asymptotic to $x \log(x)$ by Stirling's asymptotic formula, and, since squares. cubes, ... of primes are comparatively rare, and $[x/p]$ is almost the same as x/p, one may easily infer that

$$x \sum_{p \le x} \frac{\log(p)}{p} = x \log(x) + O(x)$$

from which one can deduce that $\pi(x)$ is of order $\frac{x}{\log(x)}$. This was essentially the method of Tchebychef, who actually proved that [Tch2]

$$B < \pi(x) \Big/ \frac{x}{\log(x)} < \frac{6B}{5}$$

for all sufficiently large numbers x, where

$$B = \frac{\log 2}{2} + \frac{\log 3}{3} + \frac{\log 5}{5} - \frac{\log 30}{30} \approx 0.92129$$

and

$$\frac{6B}{5} \approx 1.10555.$$

Unfortunately, however, he was unable to prove the prime number theorem itself this way, and the question remained as to whether an elementary proof of the prime number theorem could be found.

Over the years there were various improvements on Tchebychef's bound, and in 1892 Sylvester [Syl1], [Syl2] was able to show that

$$0.956 < \pi(x) \Big/ \frac{x}{\log(x)} < 1.045$$

for all sufficiently large x. We quote from Harold Diamond's excellent survey article [D]:

> The approach of Sylvester was *ad hoc* and computationally complex; it offered no hope of leading to a proof of the P.N.T. Indeed, Sylvester concluded in his article with the lament that "...we shall probably have to wait [for a proof of the P.N.T.] until someone is born into the world so far surpassing Tchebychef in insight and penetration as Tchebychef has proved himself superior in these qualities to the ordinary run of mankind."

The first proof of the prime number theorem was given by Hadamard [H1], [H2] and de la Vallée Poussin [VP] in 1896. The proof was not elementary and made use of Hadamard's theory of integral functions applied to the Riemann zeta function $\zeta(s)$ which is defined by the absolutely convergent series

$$\zeta(s) = \sum_{n=1}^{\infty} n^{-s},$$

for $Re(s) > 1$. A second component of the proof was a simple trigonometric identity (actually, Hadamard used the doubling formula for the cosine function) applied in an extremely clever manner to show that the zeta function didn't vanish on the line $Re(s) = 1$. Later, several simplified proofs were given, in particular by Landau [L] and Wiener [W1], [W2], which avoided the Hadamard theory.

In 1921 Hardy (see [B]) delivered a lecture to the Mathematical Society of Copenhagen. He asked:

"No elementary proof of the prime number theorem is known, and one may ask whether it is reasonable to expect one. Now we know that the theorem is roughly equivalent to a theorem about an analytic function, the theorem that Riemann's zeta function has no roots on a certain line. A proof of such a theorem, not fundamentally dependent on the theory of functions, seems to me extraordinarily unlikely. It is rash to assert that a mathematical theorem *cannot* be proved in a particular way; but one thing seems quite clear. We have certain views about the logic of the theory; we think that some theorems, as we say 'lie deep' and others nearer to the surface. If anyone produces an elementary proof of the prime number theorem, he will show that these views are wrong, that the subject does not hang together in the way we have supposed, and that it is time for the books to be cast aside and for the theory to be rewritten."

In the year 1948 the mathematical world was stunned when Paul Erdős announced that he and Atle Selberg had found a truly elementary proof of the prime number theorem which used only the simplest properties of the logarithm function. Unfortunately, this announcement and subsequent events led to a bitter dispute between these two mathematicians. The actual details of what transpired in 1948 have become distorted over time. A short paper, "The elementary proof of the prime number theorem," by E.G. Straus has been circulating for many years and has been the basis for numerous assertions over what actually happened. In 1987 I wrote a letter to the editors of the Atlantic Monthly (which was published) in response to an article about Erdős [Ho] which discussed the history of the elementary proof of the prime number theorem. At that time Selberg sent me his file of documents and letters (this is now part of [G]). Having been a close and personal friend of Erdős and also Selberg, having heard both sides of the story, and finally having a large collection of letters and documents in hand, I felt the time had come to simply present the facts of the matter with supporting documentation.

Let me begin by noting that in 1949, with regard to Paul Erdős's paper, "On a new method in elementary number theory which leads to an elementary proof of the prime number theorem," the Bulletin of the American Mathematical Society informed Erdős that the referee does not recommend the paper for publication. Erdős immediately withdrew the paper and had it published in the Proceedings of the National Academy of Sciences [E] . At the same time Atle Selberg published his paper, "An elementary proof of the prime–number theorem," in the Annals of Mathematics [S]. These papers were brilliantly reviewed by A.E. Ingham [I].

The elementary proof of the prime number theorem was quite a sensation at the time. For his work on the elementary proof of the PNT, the zeros of the Riemann zeta function (showing that a positive proportion lie on the line $\frac{1}{2}$), and the development of the Selberg sieve method, Selberg received the Fields Medal [B] in 1950. Erdős received the Cole Prize in 1952 [C]. The Selberg sieve method, a cornerstone in elementary number theory, is the basis for Chen's [Ch] spectacular proof that every positive even integer is the sum of a prime and a number having at most two prime factors. Selberg is now recognized as one of the leading mathematicians of this century for his introduction of spectral theory into number theory culminating in his discovery of the trace formula [A-B-G] which classifies all arithmetic zeta functions. Erdős has also left an indelible mark on mathematics. His work provided the foundations for graph and hypergraph theory [C–G] and the probabilistic method [A–S] with applications in combinatorics and elementary number theory. At his death in 1996 he had more than 1500 published papers with many coauthored papers yet to appear. It is clear that he has founded a unique school of mathematical research, international in scope, and highly visible to the world at large.

Acknowledgment: The author would like to thank Enrico Bombieri, Melvyn Nathanson, and Atle Selberg for many clarifying discussions on historical detail. In addition I received a wide variety of helpful comments from Michael Anshel, Harold Diamond, Ron Graham, Dennis Hejhal, Jeff Lagarias, Attila Mate, Janos Pach, and Carl Pomerance.

March 1948: Let $\vartheta(x) = \sum_{p \leq x} \log(p)$ denote the sum over primes $p \leq x$. The prime number theorem is equivalent to the assertion that

$$\lim_{x \to \infty} \frac{\vartheta(x)}{x} = 1.$$

In March 1948 Selberg proved the asymptotic formula

$$\vartheta(x)\log(x) + \sum_{p \leq x} \log(p)\vartheta\left(\frac{x}{p}\right) = 2x\log(x) + O(x).$$

He called this the fundamental formula.

We quote from Erdős's paper, "On a new method in elementary number theory which leads to an elementary proof of the prime number theorem," Proc. Nat. Acad. Scis. 1949:

> "Selberg proved some months ago the above asymptotic formula, ... the ingenious proof is completely elementary ... Thus it can be used as a starting point for elementary proofs of various theorems which previously seemed inaccessible by elementary methods."

Quote from Selberg: Letter to H. Weyl Sept. 16, 1948

"I found the fundamental formula ... in March this year ... I had found a more complicated formula with similar properties still earlier."

April 1948: Recall that $\vartheta(x) = \sum_{p \leq x} \log(p)$. Define

$$a = \liminf \frac{\vartheta(x)}{x}, \qquad A = \limsup \frac{\vartheta(x)}{x}.$$

Sylvester's estimates guarantee that

$$0.956 \leq a \leq A \leq 1.045.$$

In his letter to H. Weyl, Sept. 16, 1948, Selberg writes:

"I got rather early the result that $a + A = 2$,"

The proof that $a + A = 2$ is given as follows. Choose x large so that

$$\vartheta(x) = ax + o(x).$$

Then since $\vartheta(x) \leq Ax + o(x)$ it follows from Selberg's fundamental formula that

$$ax \log(x) + \sum_{p \leq x} A \frac{x}{p} \log(p) \geq 2x \log(x) + o(x \log(x)).$$

Using Tchebychef's result that

$$\sum_{p \leq x} \frac{\log(p)}{p} \sim \log(x)$$

it is immediate that $a + A \geq 2$. On the other hand, we can choose x large so that

$$\vartheta(x) = Ax + o(x).$$

Then since $\vartheta(x) \geq ax + o(x)$ it immediately follows as before that

$$Ax \log(x) + \sum_{p \leq x} a \frac{x}{p} \log(p) \leq 2x \log(x) + o(x \log(x)),$$

from which we get $a + A \leq 2$. Thus

$$a + A = 2.$$

Remark: Selberg was aware of the fact (already in April 1948) that $a + A = 2$, and that the prime number theorem would immediately follow if one could prove either $a = 1$ or $A = 1$.

May–July 1948: We again quote from Selberg's letter to H. Weyl of Sept. 16, 1948.

"In May I wrote down a sketch to the paper on Dirichlet's theorem, during June I did nothing except preparations to the trip to Canada. Then around the beginning of July, Turán asked me if I could give him my notes on the Dirichlet theorem so he could see it, he was going away soon, and probably would have left when I returned from Canada. I not only agreed to do this, but as I felt very much attached to Turán I spent some days going through the proof with him. In this connection I mentioned the *fundamental formula* to him, ... However, I did not tell him the proof of the formula, nor about the consequences it might have and my ideas in this connection... I then left for Canada and returned after 9 days just as Turán was leaving. It turned out that Turán had given a seminar on my proof of the Dirichlet theorem where Erdős, Chowla, and Straus had been present, I had of course no objection to this, since it concerned something that was already finished from my side, though it was not published. In connection with this Turán had also mentioned, at least to Erdős, the *fundamental formula*, this I don't object to either, since I had not asked him not to tell this further."

July 1948: Quote from E.G. Straus' paper, "The elementary proof of the prime number theorem."

"Turán who was eager to catch up with the mathematical developments that had happened during the war, talked with Selberg about his sieve method and now famous inequality (Fundamental Formula). He tried to talk Selberg into giving a seminar ... Selberg suggested Turán give the seminar.

This Turán did for a small group of us, including Chowla, Erdős and myself, ... After the lecture ... there followed a brief discussion of the unexpected power of Selberg's inequality."

"Erdős said,

I think you can also derive

$$\lim_{n \to \infty} \frac{p_{n+1}}{p_n} = 1$$

from this inequality.

In any case within an hour or two Erdős had discovered an ingenious derivation from Selberg's inequality. After presenting an outline of the proof to the Turán Seminar group, Erdős met Selberg in the hall and told him he could derive $\frac{p_{n+1}}{p_n} \to 1$ from Selberg's inequality."

"Selberg responded something like this:

You must have made a mistake because with this result I can get an elementary proof of the prime number theorem, and I have convinced myself that my inequality is not powerful enough for that."

Quote from Weyl's letter to Selberg August 31, 1948

"Is it not true that you were in possession of what Erdős calls the fundamental inequality and of the equation $a + A = 2$ for several months but could not prove $a = A = 1$ until Erdős deduced $\frac{p_{n+1}}{p_n} \to 1$ from your inequality?"

Here is Selberg's response in his letter to Weyl, Sept. 16, 1948.

"Turán had mentioned to Erdős after my return from Montreal he told me he was trying to prove $\frac{p_{n+1}}{p_n} \to 1$ from my formula.

Actually, I didn't like that somebody else started working on my unpublished results before I considered myself through with them."

"But though I felt rather unhappy about the situation, I didn't say anything since after all Erdős was trying to do something different from what I was interested in.

In spite of this, I became ... rather concerned that Erdős was working on these things . . .

I, therefore, started very feverishly to work on my own ideas. On Friday evening Erdős had his proof ready (that $\frac{p_{n+1}}{p_n} \to 1$) and he told it to me.

On Sunday afternoon I got my first proof of the prime number theorem. I was rather unsatisfied with the first proof because it was long and indirect. After a few days (my wife says two) I succeeded in giving a different proof."

Quote from Erdős's paper, "On a new method in elementary number theory which leads to an elementary proof of the prime number theorem," Proc. Nat. Acad. Scis. 1949:

" Using (1) (fundamental formula) I proved that $\frac{p_{n+1}}{p_n} \to 1$ as $n \to \infty$. In fact, I proved the following slightly stronger result: To every ϵ there exists a positive $\delta(\epsilon)$ so that for x sufficiently large we have

$$\pi\big(x(1 + \epsilon)\big) - \pi(x) > \delta(\epsilon)x/\log(x)$$

where $\pi(x)$ is the number of primes not exceeding x.

I communicated this proof to Selberg, who, two days later . . . deduced the prime number theorem."

Recently, Selberg sent me a letter which more precisely specifies the actual dates of events.

Quote from Selberg's letter to D. Goldfeld, January 6, 1998:

"July 14, 1948 was a Wednesday, and on Thursday, July 15 I met Erdős and heard that he was trying to prove $\frac{p_{n+1}}{p_n} \to 1$. I believe Turán left the next day (Friday, July 16), at any rate whatever lecture he had given (and I had not asked him to give one!) he had given before my return, and he was not present nor played any part in later events. Friday evening or it may have been Saturday morning, Erdős had his proof ready and told me about it. Sunday afternoon (July 18) I used his result (which was stronger than just $\frac{p_{n+1}}{p_n} \to 1$, he had proved that between x and $x(1+\delta)$ there are more than $c(\delta)\frac{x}{\log(x)}$ primes for $x > x_0(\delta)$, the weaker result would not have been sufficient for me) to get my first proof of the PNT. I told Erdős about it the next morning (Monday, July 19). He then suggested that we should talk about it that evening in the seminar room in Fuld Hall (as I thought, to a small informal group of Chowla, Straus and a few others who might be interested)."

In the same letter Selberg goes on to dispute Straus' recollection of the events.

"Turán's lecture (probably a quite informal thing considering the small group) could not have been later than July 14, since it was before my return. Straus has speeded up events; Erdős told me he was trying to prove $\frac{p_{n+1}}{p_n} \to 1$ on July 15. He told me he had a proof only late on July 16 or possibly earlier the next day. Straus' quote is also clearly wrong for the following reasons; first, I needed more than just $\frac{p_{n+1}}{p_n} \to 1$ for my first proof of the PNT, second, I only saw how to do it on Sunday, July 18.

It is true, however, as Erdős' and Straus' stories indicate, that when I first was told by Erdős that he was trying to prove $\frac{p_{n+1}}{p_n} \to 1$ from my formula, I tried to discourage him, by saying that I doubted whether the formula alone implied these things. I also said I had constructed a counterexample showing that the relation in the form

$$f(x)\log x + \int_1^x f\left(\frac{x}{t}\right) df(t) = 2x\log x + (O(x))$$

by itself does not imply that $f(x) \sim x$. It was true, I did have such an example. What I neglected to tell that in this example $f(x)$ (though positive and tending to infinity with x) was not monotonic! This conversation took place either in the corridor of Fuld Hall or just outside Fuld Hall so without access to a blackboard. This attempt to throw Erdős off the track (clearly not succeeding!) is somewhat understandable given my mood at the time.

Quote from Selberg's Paper, "An elementary proof of the prime–number theorem," Annals Math. 1949

"From the Fundamental Inequality there are several ways to deduce the prime number theorem ... The original proof made use of the following result of Erdős $\frac{p_{n+1}}{p_n} \to 1$. Erdős's result was obtained entirely independent of my work."

Selberg's first proof that the prime number theorem followed from the fundamental formula is given both in [**E**] and [**S**]. The crux of the matter goes something like this. We may write the fundamental formula in the form

$$\frac{\vartheta(x)}{x} + \sum_{p \leq x} \frac{\vartheta(x/p)}{x/p} \frac{\log(p)}{p \log(x)} = 2 + O\left(\frac{1}{\log(x)}\right).$$

Recall that a and A are the limit inferior and limit superior, respectively, of $\frac{\vartheta(x)}{x}$.

Now, choose x large so that $\frac{\vartheta(x)}{x}$ is near A. Since $a + A = 2$, it follows from the fundamental formula and

$$\sum_{p \leq x} \frac{\log(p)}{p \log(x)} \sim 1,$$

that $\frac{\vartheta(x/p)}{x/p}$ must be near a for most primes $p \leq x$. If S denotes the set of exceptional primes, then we have

$$\sum_{\substack{p \leq x \\ p \in S}} \frac{\log(p)}{p} \Big/ \sum_{p \leq x} \frac{\log(p)}{p} \approx 0.$$

Now, choose a small prime $q \notin S$ such that $\frac{\vartheta(x/q)}{x/q}$ is near a. Rewriting the fundamental formula with x replaced by x/q, the same argument as above leads one to conclude that $\frac{\vartheta(x/pq)}{x/pq}$ is near A for most primes $p \leq x/q$. It follows that $\vartheta(x/p) \approx ax/p$ for most primes $p \leq x$ and that $\vartheta(x/pq) \approx Ax/pq$ for most $p \leq x/q$. A contradiction is obtained (using Erdős's idea of nonoverlapping intervals) unless $a = A = 1$.

The Erdős-Selberg dispute arose over the question of whether a joint paper (on the entire proof) or seperate papers (on each individual contribution) should appear on the elementary proof of the PNT.

August 20, 1948: Quote from a letter of Selberg to Erdős.

"What I propose is the only fair thing: each of us can publish what he has actually done and get the credit for that, and not for what the other has done.

You proved that

$$\lim_{n \to \infty} \frac{p_{n+1}}{p_n} = 1.$$

I would never have dreamed of forcing you to write a joint paper on this in spite of the fact that the essential thing in the proof of the result was mine."

"Since there can be no reason for a joint paper, I am going to publish my proof as it now is. I have the opinion, . . . that I do you full justice by telling in the paper that my original proof depended on your result.

In addition to this I offered you to withhold my proof so your theorem could be published earlier (of course then without mentioning PNT).

I still offer you this. . .

If you don't accept this I publish my proof anyway."

Sept. 16, 1948: Quote from Selberg's letter to Weyl.

"when I came to Syracuse I discovered gradually through various sources that there had been made quite a publicity around the proof of the PNT. I have myself actually mentioned it only in one letter to one of my brothers ...

Almost all the people whom the news had reached seemed to attribute the proof entirely or at least essentially to Erdős, this was even the case with people who knew my name and previous work quite well."

Quote from E.G. Straus', "The elementary proof of the prime number theorem."

"In fact I was told this story (I forget by whom) which may well not be true . . . When Selberg arrived in Syracuse he was met by a faculty member with the greeting: "

Have you heard the exciting news of what Erdős and some Scandinavian mathematician have just done?"

Quote from Selberg's letter to D. Goldfeld, January 6, 1998:

"This is not true. What I did hear shortly after my arrival were some reports (originating from the Boston–Cambridge area) where only Erdős was mentioned. Later there were more such reports from abroad."

Sept. 20, 1948: We quote from a second letter of Selberg to Erdős.

"I hope also that we will get some kind of agreement. But I cannot accept any agreement with a joint paper.

How about the following thing. You publish your result, I publish my newest proof, but with a satisfactory sketch of the ideas of the first proof in the introduction, and referring to your result. I could make a thorough sketch on 2 pages, I think, and this would not make the paper much longer. If you like, I could send you a sketch of the introduction.

I have thought to send my paper to the Annals of Math., they will certainly agree to take your paper earlier."

Sept. 27, 1948: Quote from Erdős's letter to Selberg.

"I have to state that when I started to work on $\frac{p_{n+1}}{p_n} \to 1$ you were very doubtful about success, in fact stated that you believe to be able to show that the FUND. LEMMA does not imply the PNT (prime number theorem).

If you would have told me about what you know about a and A, I would have finished the proof of PNT on the spot.

Does it occur to you that if I would have kept the proof of $\frac{p_{n+1}}{p_n} \to 1$ to myself (as you did with $a + A = 2$) and continued to work on PNT ... I would soon have succeeded and then your share of PNT would have only been the beautiful FUNDAMENTAL LEMMA.

Sept. 27, 1948: Quote from Erdős's letter to Selberg.

"I completely reject the idea of publishing only

$$\lim_{n \to \infty} \frac{p_{n+1}}{p_n} = 1.$$

and feel just as strongly as before that I am fully entitled to a joint paper. So if you insist on publishing your new proof all I can do is to publish our simplified proof, giving you of course full credit for your share (stating that you first obtained the PNT, using some of my ideas and my theorem).

Also, I will of course gladly submit the paper to Weyl first, if he is willing to take the trouble of seeing that I am scrupulously fair to you.

Quote from E.G. Straus: "The elementary proof of the prime number theorem."

"It was Weyl who caused the Annals to reject Erdős's paper and published only a version by Selberg which circumvented Erdős's contribution, without mentioning the vital part played by Erdős in the first elementary proof, or even the discovery of the fact that such a proof was possible."

Quote from Selberg's letter to D. Goldfeld, January 6, 1998:

"This is wrong on several points, my paper mentioned and sketched in some detail how Erdős's result played a part and was used in the first elementary proof of PNT, but that first proof was mine as surely as Erdős' result was his. Also the discovery that such a proof was possible was surely mine. After all, you don't know that it is possible to prove something until you have done so!"

Excerpt: Handwritten Note by Erdős:

"It was agreed that Selberg's proof should be in the Annals of Math., mine in the Bulletin. Weyl was supposed to be the referee. To my great surprise Jacobson the referee. . . The Bulletin wrote that the referee does not recommend my paper for publication.

. . .

I immediately withdrew the paper and planned to publish it in the JLMS but . . . had it published in the Proc. Nat. Acad."

Feb. 15, 1949: Quote from H. Weyl's letter to Jacobson

"I had questioned whether Erdős has the right to publish things which are admittedly Selberg's. . . . I really think that Erdős's behavior is quite unreasonable, and if I were the responsible editor I think I would not be afraid of rejecting his paper in this form.

But there is another aspect of the matter. It is probably not as easy as Erdős imagines to have his paper published in time in this country if the Bulletin rejects it. . . So it may be better to let Erdős have his way. No great harm can be done by that. Selberg may feel offended and protest (and that would be his right), but I am quite sure that the two papers – Selberg's and Erdős's together – will speak in unmistakable language, and that the one who has really done harm to himself will be Erdős."

Quote from E.G. Straus: "The elementary proof of the prime number theorem."

"The elementary proof has so far not produced the exciting innovations in number theory that many of us expected to follow. So, what we witnessed in 1948, may in the course of time prove to have been a brilliant but somewhat incidental achievement without the historic significance it then appeared to have."

Quote from Selberg's letter to D. Goldfeld, January 6, 1998:

"With this last quote from Straus, I am in agreement (actually I did not myself expect any revolution from this). The idea of the local sieve, however, has produced many things that have not been done by other methods."

Remark: To this date, there have been no results obtained from the elementary proof of the PNT that cannot be obtained in stronger form by other methods. Other elementary methods introduced by both Selberg and Erdős have, however, led to many important results in number theory not attainable by any other technique.

Dec. 4, 1997: Letter from Selberg to D. Goldfeld.

"The material I have is nearly all from Herman Weyl's files, and was given to me probably in 1952 or 1953 as he was cleaning out much of his stuff in Princeton, taking some to Zurich and probably discarding some. The letters from Weyl to myself was all that I kept when I left Syracuse in 1949, all the rest I discarded. Thus there are gaps. Missing is my first letter to Erdős as well as his reply to it. . .

I did not save anything except letters from Weyl because I was rather disgusted with the whole thing. I never lectured on the elementary proof of the PNT after the lecture in Syracuse, mentioned in the first letter to Herman Weyl. However, I did at Cornell U. early in 1949 and later at an AMS meeting in Baltimore gave a lecture with an elementary proof of (using the notation of Beurling generalized primes & integers) the fact that if

$$N(x) = Ax + o\left(\frac{x}{\log^2(x)}\right)$$

then

$$\Pi(x) = \frac{x}{\log(x)} + o\left(\frac{x}{\log(x)}\right).$$

Beurling has the same conclusion if

$$N(x) = Ax + O\left(\frac{x}{(\log(x))^\alpha}\right),$$

with $\alpha > \frac{3}{2}$. I never published this.

Erdős of course lectured extensively in Amsterdam, Paris, and other places in Europe. After his lecture in Amsterdam, Oct. 30, 1948, v.d. Corput wrote up a paper, Scriptum 1, Mathematisch Centrum, which was the first published version!"

References

[A–B–G] K.E. Aubert, E. Bombieri, D. Goldfeld, *Number theory, trace formulas and discrete groups, symposium in honor of Atle Selberg,* Academic Press Inc. Boston (1989).

[A–S] N. Alon, J. Spencer, *The probabilistic method,* John Wiley & Sons Inc., New York (1992).

[B] H. Bohr, *Address of Professor Harold Bohr,* Proc. Internat. Congr. Math. (Cambridge, 1950) vol 1, Amer. Math. Soc., Providence, R.I., 1952, 127–134.

[Ch] J. Chen, *On the representation of a large even integer as the sum of a prime and the product of at most two primes,* Sci. Sinica **16** (1973), 157–176.

[C–G] F. Chung, R. Graham, *Erdős on graphs: his legacy of unsolved problems,* A.K. Peters, Ltd., Wellesley, Massachusetts (1998).

[C] L.W. Cohen, *The annual meeting of the society,* Bull. Amer. Math. Soc **58** (1952), 159–160.

[D] H.G. Diamond, *Elementary methods in the study of the distribution of prime numbers,* Bull. Amer. Math. Soc. vol. 7 number 3 (1982), 553–589.

[E] P. Erdős, *On a new method in elementary number theory which leads to an elementary proof of the prime number theorem,* Proc. Nat. Acad. Scis. U.S.A. **35** (1949), 374–384.

[G] D. Goldfeld, *The Erdős–Selberg dispute: file of letters and documents,* to appear.

[H1] J. Hadamard, *Étude sur les propriétés des fonctions entiéres et en particulier d'une fonction considérée par Riemann,* J. de Math. Pures Appl. (4) **9** (1893), 171–215; reprinted in Oeuvres de Jacques Hadamard, C.N.R.S., Paris, 1968, vol 1, 103–147.

[H2] J. Hadamard, *Sur la distribution des zéros de la fonction $\zeta(s)$ et ses conséquences arithmétiques,* Bull. Soc. Math. France **24** (1896), 199–220; reprinted in Oeuvres de Jacques Hadamard, C.N.R.S., Paris, 1968, vol 1, 189–210.

[Ho] P. Hoffman, *The man who loves only numbers,* The Atlantic, November (1987).

[I] A.E. Ingham, Review of the two papers: *An elementary proof of the prime–number theorem,* by A. Selberg and *On a new method in elementary number theory which leads to an elementary proof of the prime number theorem,* by P. Erdős. Reviews in Number Theory as printed in Mathematical Reviews 1940-1972, Amer. Math. Soc. Providence, RI (1974). See N20-3, Vol. 4, 191–193.

[La1] E. Landau, *Handbuch der Lehre von der Verteilung der Primzahlen,* Teubner, Leipzig (1909), 2 volumes, reprinted by Chelsea Publishing Company, New York (1953).

[La2] E. Landau, *Über den Wienerschen neuen Weg zum Primzahlsatz,* Sitzber. Preuss. Akad. Wiss., 1932, 514–521.

[Le1] A.M. Legendre, *Essai sur la théorie des nombres,* 1. Aufl. Paris (Duprat) (1798).

[Le2] A.M. Legendre, *Essai sur la théorie des nombres,* 2. Aufl. Paris (Courcier) (1808).

[S] A. Selberg, *An elementary proof of the prime–number theorem,* Ann. of Math. (2) **50** (1949), 305–313; reprinted in Atle Selberg Collected Papers, Springer–Verlag, Berlin Heidelberg New York, 1989, vol 1, 379–387.

[Syl1] J.J. Sylvester, *On Tchebycheff's theorem of the totality of prime numbers comprised within given limits,* Amer. J. Math. **4** (1881), 230–247.

[Syl2] J.J. Sylvester, *On arithmetical series,* Messenger of Math. (2) **21** (1892), 1–19 and 87–120.

[Tch1] P.L. Tchebychef, *Sur la fonction qui détermine la totalité des nombres premiers inférieurs à une limite donnée,* Mémoires présentés à l'Académie Impériale des Sciences de St.-Pétersbourg par divers Savants et lus dans ses Assemblées, Bd. 6, S. (1851), 141–157.

[Tch2] P.L. Tchebychef, *Mémoire sur les nombres premiers,* J. de Math. Pures Appl. (1) **17** (1852), 366–390; reprinted in Oeuvres **1** (1899), 49–70.

[VP] C.J. de la Vallée Poussin, *Recherches analytiques sur la théorie des nombres premiers,* Ann. Soc. Sci. Bruxelles **20** (1896), 183–256.

[W1] N. Wiener, *A new method in Tauberian theorems,* J. Math. Physics M.I.T. **7** (1927–28), 161–184.

[W2] N. Wiener, *Tauberian theorems,* Ann. of Math. (2) **33** (1932), 1–100.

ADDITIVE BASES REPRESENTATIONS
AND THE ERDŐS-TURÁN CONJECTURE

G. GREKOS, L. HADDAD, C. HELOU* , J. PIHKO

ABSTRACT. We give a lower bound to the maximal number of representations by an additive basis of the natural numbers, in conjunction with a celebrated conjecture of Erdős and Turán.

INTRODUCTION

Denote by $\mathbb{N} = \{0, 1, 2, \dots\}$ the set of natural numbers and consider a subset A of $\mathbb{N}$. The number $r(A, n)$ of representations of an element n of $\mathbb{N}$ by A is the number of ordered pairs $(a, b) \in A \times A$ such that $a + b = n$. We will say that A is a basis of $\mathbb{N}$ if $r(A, n) \geq 1$ for all $n \in \mathbb{N}$. Our main objective is the exploration of the following conjecture of P. Erdős and P. Turán [3], dating back to 1941.

(ET): If A is a basis of $\mathbb{N}$, then $r(A, n)$ is unbounded, as n ranges through $\mathbb{N}$.

There seems to be relatively little work concerned with this conjecture. Originally, Erdős and Turán used some deep function theory to show that $r(A, n)$ does not become constant for large n. But, in 1951, G. Dirac [1] proved this result by an elementary argument concerning the parity of $r(A, n)$. In 1956, P. Erdős and W. Fuchs [4] established, using Fourier series, that it is impossible to have $\sum_{k=0}^{n} r(A, k) = cn + o(n^{1/4}(\log n)^{-1/2})$. They also showed that if $A = \{a_1 < a_2 < \cdots < a_n < \dots\}$ satisfies the condition $a_n \leq Kn^2$ for some constant $K > 0$ and all $n \in \mathbb{N}$ (which is true of every basis of $\mathbb{N}$), then for any $c \geq 0$, one has $\limsup_{n \to \infty} \frac{1}{n} \sum_{k=0}^{n} (r(A, k) - c)^2 > 0$. They further asserted the existence of a sequence A satisfying the same condition, for which $\limsup_{n \to \infty} \frac{1}{n} \sum_{k=0}^{n} r(A, k)^2 < \infty$. In 1990, I. Ruzsa [8] confirmed this assertion, by constructing a basis A of $\mathbb{N}$ for which $\sum_{k=0}^{n} r(A, k)^2 = O(n)$. In 1988, M. Dowd [2] gave a finite form of (ET) in $\mathbb{N}$, equivalent to one of our formulations. His proof, using graph theory, was recently clarified and generalized by M. Nathanson [7]. Dowd also indicated that the validity of (ET) in $\mathbb{Z}$ implies its validity in $\mathbb{N}$. But, in 2002, Nathanson [6] showed that (ET) is not valid in $\mathbb{Z}$, by constructing arbitrarily sparse bases A of $\mathbb{Z}$ for which $r(A, n)$ is at most 2 for all $n \in \mathbb{Z}$.

Here, we will introduce some functions, defined in terms of a variable bound x in $\mathbb{N}$ and of the traces of all bases A of $\mathbb{N}$ on the interval $[0, x]$, involving the values of $r(A, n)$. This allows for equivalent formulations of the Erdős-Turán conjecture susceptible of partial

---

2000 *Mathematics Subject Classification.* 11B13.
*Presenter

Typeset by $\mathcal{A}\mathcal{M}\mathcal{S}$-TEX

quantitative verifications. In particular, we deduce that for any basis A of $\mathbb{N}$, the numbers $r(A, n)$ must at least take values ≥ 6. For more ample details, we refer to [5].

§1 Some functions describing finite traces of bases

For $x \in \mathbb{N}$ and $P \subset \mathbb{N}$, we write $P[x] = P \cap [0, x] = \{p \in P : p \leq x\}$, and we set $\rho(P, x) = \max\{r(P, n) : n \in \mathbb{N}[x]\}$. We also set $s(P) = \sup\{r(P, n) : n \in \mathbb{N}\}$; this is an element of $\overline{\mathbb{N}} = \mathbb{N} \cup \{\infty\}$.

The following properties are immediate but useful: If $y \in \mathbb{N}$ is such that $x \leq y$, then $\rho(P, x) \leq \rho(P, y)$; and if $A \subset P$, then $\rho(A, x) \leq \rho(P, x)$. Moreover, $\rho(P, x) = \rho(P[x], x)$. Also, $s(P) = \lim_{x \to \infty} \rho(P, x)$.

A set P will be called a basis of $\mathbb{N}[x]$ if $P \subset \mathbb{N}[x] \subset P + P$, where $P + P = \{p + q : (p, q) \in P \times P\}$; whereas P is a basis of $\mathbb{N}$ if $P + P = \mathbb{N}$. The set of all bases of $\mathbb{N}[x]$ will be denoted by $\mathcal{B}(x)$, and that of all bases of $\mathbb{N}$ by $\mathcal{B}(\mathbb{N})$. Naturally, $P \in \mathcal{B}(\mathbb{N})$ if and only if $P[x] \in \mathcal{B}(x)$ for every $x \in \mathbb{N}$.

The functions ρ and τ, from $\mathbb{N}$ into $\mathbb{N}$, are defined by $\rho(x) = \min\{\rho(P, x) : P \in \mathcal{B}(x)\}$ and $\tau(x) = \min\{s(P) : P \in \mathcal{B}(x)\}$.

Lemma 1.

(1) *The functions ρ and τ are increasing.*

(2) *For every $x \in \mathbb{N}$, we have $\rho(x) \leq \tau(x) \leq \rho(2x)$.*

Proof.

(1) Let $x, y \in \mathbb{N}$ be such that $x \leq y$. For any $Q \in \mathcal{B}(y)$, we have $Q[x] \in \mathcal{B}(x)$ and therefore $\rho(Q, y) \geq \rho(Q, x) = \rho(Q[x], x) \geq \rho(x)$ and $s(Q) \geq s(Q[x]) \geq \tau(x)$. Hence $\rho(y) = \min\{\rho(Q, y) : Q \in \mathcal{B}(y)\} \geq \rho(x)$ and $\tau(y) = \min\{s(Q) : Q \in \mathcal{B}(y)\} \geq \tau(x)$.

(2) The inequality $\rho(x) \leq \tau(x)$ follows from the obvious fact that $\rho(P, x) \leq \rho(P, 2x) = s(P)$, for all $P \in \mathcal{B}(x)$. Moreover, for any $Q \in \mathcal{B}(2x)$, since $Q[x] \in \mathcal{B}(x)$, we have $\tau(x) \leq s(Q[x]) = \rho(Q[x], 2x) \leq \rho(Q, 2x)$. Hence $\tau(x) \leq \min\{\rho(Q, 2x) : Q \in \mathcal{B}(2x)\} = \rho(2x)$. $\square$

A third significant function is $\sigma : \mathbb{N}^* \to \mathbb{N}^*$, where $\mathbb{N}^* = \mathbb{N} \setminus \{0\}$. It is defined by $\sigma(n) = \min\{s(P) : P \in \mathcal{B}(\#n)\}$, where $\mathcal{B}(\#n) = \{P \subset \mathbb{N} : |P| = n \text{ and } P \in \mathcal{B}(\max P)\}$. A finite, non-empty subset P of $\mathbb{N}$ such that $P \in \mathcal{B}(\max P)$ will simply be called a finite basis, so that $\mathcal{B}(\#n)$ is the set of all finite bases having exactly n elements. Furthermore, we will call successor of a finite basis P every finite basis Q obtained by adjoining to P an element $q > \max P$, so that $Q = P \cup \{q\}$ and $Q \in \mathcal{B}(q)$. Clearly, $Q = P \cup \{q\}$ is a successor of P if and only if $\max P + 1 \leq q \leq h$, where $h = \min(\mathbb{N} \setminus (P + P))$. It is also easy to see that $\mathcal{B}(\#(n + 1))$ consists exactly of the successors of the elements of $\mathcal{B}(\#n)$.

Lemma 2.

(1) *The function σ is increasing.*

(2) *For any $n \in \mathbb{N}^*$, we have $1 \leq \sigma(n) \leq n$.*

Proof.

(1) For any $n \in \mathbb{N}^*$ and any $Q \in \mathcal{B}(\#(n + 1))$, there is some $P \in \mathcal{B}(\#n)$ such that Q is a successor of P, namely $P = Q \setminus \{\max Q\}$. Then $s(Q) \geq s(P) \geq \sigma(n)$. Hence $\sigma(n + 1) = \min\{s(Q) : Q \in \mathcal{B}(\#(n + 1))\} \geq \sigma(n)$.

(2) For any finite subset P of $\mathbb{N}$ and any $x \in \mathbb{N}$, the number $r(P, n)$, of ordered pairs $(p, n - p)$ such that $p \in P$, cannot exceed the cardinality $|P|$ of P. Therefore $s(P) \leq |P|$. In particular, if $P \in \mathcal{B}(\#n)$, then $s(P) \leq |P| = n$. Hence $\sigma(n) \leq n$. Moreover, by (1) above, $\sigma(n) \geq \sigma(1) = 1$. $\square$

Lemma 3. *For any $x \in \mathbb{N}$, if $P \in \mathcal{B}(x)$, then $|P| \geq (\sqrt{8x + 9} - 1)/2$.*

Proof. Let $n = |P|$ and $P = \{p_1, \ldots, p_n\}$, where $p_1 < \cdots < p_n$. Since $\mathbb{N}[x] \subset P + P$, we have $|\mathbb{N}[x]| = x + 1 \leq |P + P|$. Moreover, $P + P = \{p_i + p_j : 1 \leq i \leq j \leq n\}$ and therefore $|P + P| \leq \sum_{j=1}^{n} j = n(n+1)/2$. Thus $x + 1 \leq n(n+1)/2$, i.e. $n^2 + n - 2(x+1) \geq 0$, which implies that $n \geq (-1 + \sqrt{8x + 9})/2$ (the positive root of the quadratic equation). $\square$

Corollary. *For any $n \in \mathbb{N}^*$, we have $\tau(n - 1) \leq \sigma(n) \leq \tau(n(n + 1)/2 - 1)$.*

Proof. For any $P \in \mathcal{B}(\#n)$, since $P \in \mathcal{B}(\max P)$, we have $s(P) \geq \tau(\max P)$; and since P has n elements, we have $\max P \geq n - 1$, so that, τ being an increasing function, $s(P) \geq \tau(\max P) \geq \tau(n - 1)$. Hence $\sigma(n) = \min\{s(P) : P \in \mathcal{B}(\#n)\} \geq \tau(n - 1)$.

On the other hand, if $x = n(n + 1)/2 - 1$ then, by Lemma 3, for any $P \in \mathcal{B}(x)$, we have $|P| \geq n$. Thus, σ being an increasing function, $s(P) \geq \sigma(|P|) \geq \sigma(n)$. Hence $\tau(n(n + 1)/2 - 1) = \min\{s(P) : P \in \mathcal{B}(x)\} \geq \sigma(n)$. $\square$

The problem considered here can be more generally stated as the determination of the element $\Lambda = \inf\{s(P) : P \in \mathcal{B}(\mathbb{N})\}$ of $\overline{\mathbb{N}}$. Indeed, the Erdős-Turán conjecture amounts to the assertion that $\Lambda = \infty$.

Lemma 4. *We have $\lim_{x \to \infty} \rho(x) = \lim_{x \to \infty} \tau(x) = \lim_{x \to \infty} \sigma(x) \leq \Lambda$.*

Proof. Since the functions ρ, τ and σ are increasing, they all have limits in $\overline{\mathbb{N}}$, as $x \to \infty$. The first equality then follows from Lemma 1, (2), and the second equality follows from the above Corollary. Moreover, for any $x \in \mathbb{N}$ and any $P \in \mathcal{B}(\mathbb{N})$, since $P[x] \in \mathcal{B}(x)$, we have $\tau(x) \leq s(P[x]) \leq s(P)$. Hence $\tau(x) \leq \Lambda$, for all $x \in \mathbb{N}$, and thus $\lim_{x \to \infty} \tau(x) \leq \Lambda$. $\square$

§2 EQUIVALENT FORMULATIONS OF THE ERDŐS-TURÁN CONJECTURE

We will need the following important set-theoretic notions and results.

Definition. *By an infinite family of subsets of $\mathbb{N}$, we will mean any family $\mathcal{P} = (P_i)_{i \in I}$ of subsets of $\mathbb{N}$ whose index set I is infinite. Furthermore, a subset A of $\mathbb{N}$ will be called a diagonal of such a family $\mathcal{P}$, if for any $n \in \mathbb{N}$, there are infinitely many indices $i \in I$ such that P_i has the same trace as A on the interval $[0, n]$, i.e. such that $A[n] = P_i[n]$.*

For instance, if $\mathcal{P} = (P_i)_{i \in I}$ is an infinite increasing, for inclusion, sequence of subsets of $\mathbb{N}$ (i.e. $I \subset \mathbb{N}$ and for $i, j \in I$, if $i \leq j$ then $P_i \subset P_j$), then $\bigcup_{i \in I} P_i$ is the only diagonal of $\mathcal{P}$. But in general, it is not obvious whether an arbitrary infinite family $\mathcal{P}$ has a diagonal.

As a matter of notation, for any set X, we denote by $\mathcal{S}(X)$ the set of all subsets of X.

Lemma 5 (The Diagonal Lemma). *Every infinite family $\mathcal{P} = (P_i)_{i \in I}$ of subsets of $\mathbb{N}$ has at least one diagonal $A \subset \mathbb{N}$.*

Proof. We construct, by induction on n, an increasing sequence of subsets A_n of $\mathbb{N}$ and a decreasing sequence of infinite subsets I_n of I such that for any $n \in \mathbb{N}$ and any $i \in I_n$, we have $A_n = P_i[n]$. Then we let $A = \bigcup_{n \in \mathbb{N}} A_n$ and we verify that A is a diagonal of the family $\mathcal{P}$.

For $n = 0$, let $F_0 : I \longrightarrow \mathcal{S}(\{0\})$ be the map defined by $F_0(i) = P_i[0]$, for all $i \in I$. Since $I = F_0^{-1}(\emptyset) \cup F_0^{-1}(\{0\})$ is an infinite set, then, one at least of the two sets $\emptyset$ or $\{0\}$, that we call A_0, has an infinite preimage $I_0 = F_0^{-1}(A_0)$. For $n \in \mathbb{N}^*$, we assume constructed $2n$ subsets $A_0 \subset \cdots \subset A_{n-1} \subset \mathbb{N}$ and $I_{n-1} \subset \cdots \subset I_0 \subset I$, such that I_{n-1} is infinite and $A_j = P_i[j]$ for all $i \in I_j$ and $0 \le j \le n-1$. Let $F_n : I_{n-1} \longrightarrow \mathcal{S}(\mathbb{N}[n])$ be the map defined by $F_n(i) = P_i[n]$, for all $i \in I_{n-1}$. Since $\mathcal{S}(\mathbb{N}[n])$ is finite and $I_{n-1} = \bigcup_{X \in \mathcal{S}(\mathbb{N}[n])} F_n^{-1}(X)$ is infinite, there exists an element $A_n \in \mathcal{S}(\mathbb{N}[n])$ such that $I_n = F^{-1}(A_n)$ is an infinite subset of I_{n-1}. Since $I_n \subset I_{n-1}$, then, for $n \in I_n$, we have $A_{n-1} = P_i[n-1] \subset P_i[n] = A_n$. This completes the construction by induction.

Now, letting $A = \bigcup_{k \in \mathbb{N}} A_k$, we have $A[n] = \bigcup_{k \in \mathbb{N}} A_k[n]$, for all $n \in \mathbb{N}$. But since the sequence (A_k) is increasing, for $0 \le k \le n$, we have $A_k[n] \subset A_n[n] = A_n$. Moreover, for $k \ge n$, since $I_k \subset I_n$, for any $i \in I_k$, we have $A_k[n] = (P_i[k])[n] = P_i[n] = A_n$. Hence $A[n] = A_n = P_i[n]$, for all n in $\mathbb{N}$ and all i in the infinite set I_n. Thus A is a diagonal of the family $\mathcal{P}$. $\square$

Corollary 1. *Let $\mathcal{P} = (P_i)_{i \in I}$ be an infinite family of subsets of $\mathbb{N}$, with $I \subset \mathbb{N}$, and let A be a diagonal of this family.*

(1) *If $P_i \in \mathcal{B}(i)$ for all $i \in I$, then $A \in \mathcal{B}(\mathbb{N})$.*

(2) *If $s(P_i) \le s$, for some $s \in \overline{\mathbb{N}}$ and all $i \in I$, then $s(A) \le s$.*

(3) *If $P_i \in \mathcal{B}(i)$ and $s(P_i) = \tau(i)$ for all $i \in I$, then $A \in \mathcal{B}(\mathbb{N})$ and we have $s(A) = \lim\limits_{x \to \infty} \tau(x) = \Lambda$.*

Proof. (1) and (2). For every $n \in \mathbb{N}$, there is an infinite subset I_n of I such that for all $i \in I_n$, we have $A[n] = P_i[n]$ and therefore $r(A, n) = r(P_i, n)$. Thus if $P_i \in \mathcal{B}(i)$ (resp. $s(P_i) \le s$) for all $i \in I$, then, choosing $i \ge n$ in I_n, we get $r(A, n) = r(P_i, n) \ge 1$ (resp. $r(A, n) = r(P_i, n) \le s(P_i) \le s$). Since this holds for every $n \in \mathbb{N}$, we conclude that $A \in \mathcal{B}(\mathbb{N})$ (resp. $s(A) \le s$).

(3) If $P_i \in \mathcal{B}(i)$ and $s(P_i) = \tau(i)$, for all $i \in I$, then, by (1), $A \in \mathcal{B}(\mathbb{N})$. Moreover, for any $n \in \mathbb{N}$, there is an infinite subset I_n of I such that for all $i \in I_n$, we have $A[n] = P_i[n]$ and therefore $r(A, n) = r(P_i, n) \le s(P_i) = \tau(i) \le \lim\limits_{x \to \infty} \tau(x)$, since τ is an increasing function. It follows that $s(A) \le \lim\limits_{x \to \infty} \tau(x)$. On the other hand, by Lemma 4 and the definition of Λ, we have $\lim\limits_{x \to \infty} \tau(x) \le \Lambda \le s(A)$. Hence the equalities. $\square$

Corollary 2. *We have $\lim\limits_{x \to \infty} \rho(x) = \lim\limits_{x \to \infty} \tau(x) = \lim\limits_{x \to \infty} \sigma(x) = \Lambda$.*

Proof. For every $i \in \mathbb{N}$, choose from the finite set $\mathcal{B}(i)$ an element P_i at which the map $P \mapsto s(P)$ attains its minimum $\tau(i)$. Then the family $\mathcal{P} = (P_i)_{i \in \mathbb{N}}$ satisfies the conditions $P_i \in \mathcal{B}(i)$ and $s(P_i) = \tau(i)$ for all $i \in \mathbb{N}$. Hence, by Corollary 1, (3), $\lim\limits_{x \to \infty} \tau(x) = \Lambda$. The other equalities result from Lemma 4. $\square$

Theorem 1. *The following statements are equivalent:*

(ET): If A is a basis of $\mathbb{N}$, then $s(A) = \infty$; i.e. $\Lambda = \infty$.

(ETρ): $\lim\limits_{x \to \infty} \rho(x) = \infty$.

$(ET\tau)$: $\lim\limits_{x\to\infty} \tau(x) = \infty$.

$(ET\sigma)$: $\lim\limits_{x\to\infty} \sigma(x) = \infty$.

Proof. This is an immediate consequence of the above Corollary 2. $\square$

§3 ON THE GROWTH OF THE FUNCTIONS ρ, τ AND σ

We will need an auxiliary function $\alpha : \mathbb{N} \to \mathbb{N}$. For $x \in \mathbb{N}$ and for a subset P of $\mathbb{N}$, we first set $\alpha(P, x) = \max\{r(P, n) + r(P, n + x + 1) : n \in \mathbb{N}[x]\}$; we then define $\alpha(x) = \min\{\alpha(P, x) : P \in \mathcal{B}(x)\}$.

Lemma 6. *For any $x \in \mathbb{N}$, we have $\rho(x) \leq \tau(x) \leq \alpha(x) \leq \min(2\tau(x)\,,\, [(x+3)/2])$, where $[r]$ denotes the integral part of the real number r.*

Proof. For any $P \subset \mathbb{N}$, we have $\alpha(P, x) = \max\{r(P, n) + r(P, n + x + 1) : n \in \mathbb{N}[x]\} \geq r(P, k)$, for $0 \leq k \leq 2x + 1$. It follows that $\alpha(P, x) \geq \rho(P, 2x + 1)$. Moreover, $\alpha(P, x) \leq \max\{r(P, n) : n \in \mathbb{N}[x]\} + \max\{r(P, n+x+1) : n \in \mathbb{N}[x]\} \leq \rho(P, x) + \rho(P, 2x+1) \leq 2s(P)$. In particular, if $P \subset \mathbb{N}[x]$, then $\rho(P, 2x + 1) = s(P)$ and therefore $s(P) \leq \alpha(P, x) \leq 2s(P)$. Taking the minimum, as P ranges through $\mathcal{B}(x)$, of all sides in the latter inequalities, we get $\tau(x) \leq \alpha(x) \leq 2\tau(x)$. In addition, by Lemma 1, $\rho(x) \leq \tau(x)$. This yields all the desired inequalities except one; so there only remains to show that $\alpha(x) \leq [(x + 3)/2]$.

Now, let $m = [(x + 1)/2]$ and $A = \mathbb{N}[m]$. Then $A \in \mathcal{B}(x)$ and thus $\alpha(x) \leq \alpha(A, x)$. Moreover, $\alpha(A, x) \leq \max\{r(A, n) : n \in \mathbb{N}[x]\} = \rho(A, x)$. Indeed, if $n \geq 1$, then $n + x + 1 > 2m$ and thus $r(A, n + x + 1) = 0$; while if $n = 0$, then $r(A, 0) + r(A, x + 1) = r(A, x)$, since both sides are equal to 1 or 2 according as $x = 2m$ or $x = 2m - 1$ respectively. Furthermore, $s(A) = r(A, m) = m + 1$, since it can be easily checked that $r(A, n) = n + 1$ if $0 \leq n \leq m$ and $r(A, n) = \max(2m - n + 1, 0)$ if $n \geq m + 1$. Therefore $\alpha(x) \leq \alpha(A, x) \leq \rho(A, x) \leq s(A) = m + 1 = [(x + 3)/2]$, which completes the chain of inequalities. $\square$

Another concept that we need is that of generating power series. For every subset P of $\mathbb{N}$, there is an associated formal power series $f_P(X) = \sum_{p \in P} X^p = \sum_{n=0}^{\infty} \chi(P, n) X^n$, where $\chi(P, .)$ is the characteristic function of P, defined by $\chi(P, n) = 1$ if $n \in P$ and $\chi(P, n) = 0$ if $n \in \mathbb{N} \setminus P$. The square of this series $g_P(X) = f_P(X)^2 = \sum_{n=0}^{\infty} r(P, n) X^n$ is the generating series of the sequence $(r(P, n))_{n \in \mathbb{N}}$, since $r(P, n) = \sum_{k=0}^{n} \chi(P, k) \chi(P, n - k)$, for any $n \in \mathbb{N}$. For instance, if $A = \mathbb{N}[m]$, where $m \in \mathbb{N}$, then $f_A(X) = \sum_{k=0}^{m} X^k = (1 - X^{m+1})/(1 - X)$ and therefore $g_A(X) = (1 - X^{m+1})^2/(1 - X)^2 = (1 - 2X^{m+1} + X^{2m+2}) \sum_{n=0}^{\infty} (n+1) X^n = \sum_{n=0}^{m} (n+1) X^n + \sum_{n=m+1}^{2m} (2m - n + 1) X^n$, which gives the values of $r(A, n)$ for all $n \in \mathbb{N}$ as noted in the proof of Lemma 6.

Lemma 7. *Let A, B be two subsets of $\mathbb{N}$ and $d \in \mathbb{N}$ be such that A is finite and $d > \max A$. Let $C = A + d * B = \{a + db : a \in A, b \in B\}$. Then*

(1) *$f_C(X) = f_A(X) f_B(X^d)$.*

(2) *For any $n \in \mathbb{N}$, there exist unique integers $q, e \in \mathbb{N}$ such that $n = dq + e$ and $0 \leq e < d$, and we have $r(C, n) = r(A, e) r(B, q) + r(A, d + e) r(B, q - 1)$.*

(3) *For any $x \in \mathbb{N}$, we have $\rho(C, x) \leq \alpha(A, d - 1) \rho(B, [x/d])$*

Proof. (1) We have $C = \bigcup_{a \in A}(a + d * B)$, where the sets $a + d * B = \{a\} + d * B$ are two by two disjoint. Indeed, if for $a, a' \in A$, with $a \leq a'$, there is some $c \in (a + d * B) \cap (a' + d * B)$, then $c = a + db = a' + db'$, with $b, b' \in B$, and therefore $0 \leq a' - a = d(b - b') \leq a' \leq \max A < d$, which is only possible if $b - b' = 0$, i.e. $a = a'$. It follows that $f_C(X) = \sum_{a \in A} \sum_{b \in B} X^{a+db} = \sum_{a \in A} X^a f_B(X^d) = f_A(X) f_B(X^d)$.

(2) Squaring the relation in (1), we get $g_C(X) = g_A(X) g_E(X^d)$, i.e. $\sum_{n=0}^{\infty} r(C, n) X^n = \left(\sum_{n=0}^{\infty} r(A, n) X^n\right) \left(\sum_{n=0}^{\infty} r(B, n) X^{dn}\right)$. Identifying the coefficients of X^n on both sides, for $n \in \mathbb{N}$, we get $r(C, n) = \sum_{j,k} r(A, j) r(B, k)$, where the summation is over all $(j, k) \in \mathbb{N} \times \mathbb{N}$ such that $j + dk = n$. The latter sum can be restricted to $0 \leq j \leq 2m$, where $m = \max A$, since $r(A, j) = 0$ for $j > 2m$. Now, for a given $n \in \mathbb{N}$, the existence and uniqueness of q and e result from the Euclidean division of n by d. Since $n = dq + e$, with $0 \leq e < d$, the condition $j + dk = n$ amounts to $j = d(q - k) + e$, with the restriction $0 \leq d(q - k) + e \leq 2m < 2d$, so that $0 \leq q - k < 2$, i.e. either $k = q$, $j = e$ or $k = q - 1$, $j = d + e$. Therefore $r(C, n) = r(A, e) r(B, q) + r(A, d + e) r(B, q - 1)$.

(3) Let $x = du + v$, where $u, v \in \mathbb{N}$ satisfy $0 \leq v < d$ and are uniquely determined by Euclidean division. Then for $0 \leq n \leq x$, similarly expressed by $n = dq + e$, with $q, e \in \mathbb{N}$ and $0 \leq e < d$, we have $0 \leq q \leq u$, so that $r(B, q - 1)$ and $r(B, q)$ are $\leq \rho(B, u)$. Therefore, by (2), $r(C, n) = r(A, e) r(B, q) + r(A, d + e) r(B, q - 1) \leq (r(A, e) + r(A, d + e)) \rho(B, u)$, for all $0 \leq n \leq x$. Hence $\rho(C, x) \leq \max\{r(A, e) + r(A, d + e) : 0 \leq e \leq d - 1\} \rho(B, u) = \alpha(A, d - 1) \rho(B, u)$, which is the stated inequality, since $u = [x/d]$. $\square$

Lemma 8. *Let* $x, y \in \mathbb{N}$, $A \in \mathcal{B}(x)$ *and* $B \in \mathcal{B}(y)$, *and let* $C = A + (x + 1) * B$. *Then*

(1) $C \in \mathcal{B}(xy + x + y)$.

(2) $\rho(C, xy + x + y) \leq \alpha(A, x) \rho(B, y)$.

(3) $s(C) \leq \alpha(A, x) s(B)$.

Proof.

(1) Let $d = x + 1$ and $z = xy + x + y$. Then $C = A + d * B$ and $z = dy + x$, with $0 \leq x < d$; furthermore $0 \leq \max A \leq x < d$. So, by Lemma 7, for any $0 \leq n \leq z$, if $n = dq + e$ with $0 \leq e < d$, then $0 \leq q \leq y$ and $r(C, n) = r(A, e) r(B, q) + r(A, d + e) r(B, q - 1)$. Since $A \in \mathcal{B}(x)$ and $e \leq d - 1 = x$, then $r(A, e) \geq 1$ and since $B \in \mathcal{B}(y)$ and $q \leq y$, then $r(B, q) \geq 1$. Therefore $r(C, n) \geq r(A, e) r(B, q) \geq 1$, for all $0 \leq n \leq z$, i.e. $\mathbb{N}[z] \subset C + C$. Moreover, $C = A + d * B \subset \mathbb{N}[x] + (x + 1) * \mathbb{N}[y] \subset \mathbb{N}[z]$. Hence $C \in \mathcal{B}(z)$.

(2) By Lemma 7, and since $d - 1 = x$ and $[z/d] = y$, we have $\rho(C, z) \leq \alpha(A, x) \rho(B, y)$.

(3) Since $C \subset \mathbb{N}[z]$, we have $s(C) = \rho(C, 2z)$. But, by Lemma 7, we have $\rho(C, 2z) \leq \alpha(A, x) \rho(B, [2z/d]) \leq \alpha(A, x) s(B)$. Hence $s(C) \leq \alpha(A, x) s(B)$. $\square$

Theorem 2. *For any* $x, y \in \mathbb{N}$, *we have*

(1) $\rho(xy + x + y) \leq \alpha(x) \rho(y)$.

(2) $\tau(xy + x + y) \leq \alpha(x) \tau(y)$.

Proof. Let $z = xy + x + y$. By Lemma 8, for any $A \in \mathcal{B}(x)$ and any $B \in \mathcal{B}(y)$, there exists $C \in \mathcal{B}(z)$ satisfying $\rho(z) \leq \rho(C, z) \leq \alpha(A, x) \rho(B, y)$ and $\tau(z) \leq s(C) \leq \alpha(A, x) s(B)$. Hence $\{\rho(z) \leq \min\{\alpha(A, x) : A \in \mathcal{B}(x)\} \cdot \min\{\rho(B, y) : B \in \mathcal{B}(y)\} = \alpha(x) \cdot \rho(y)$ and similarly $\tau(z) \leq \min\{\alpha(A, x) : A \in \mathcal{B}(x)\} \cdot \min\{s(B) : B \in \mathcal{E}(y)\} = \alpha(x) \cdot \tau(y)$. $\square$

Corollary. *For any* $x, y \in \mathbb{N}$, *we have*

(1) $\rho(xy + x + y) \leq 2\tau(x)\rho(y)$, *and also* $\rho(xy + x + y) \leq [(x + 3)/2]\rho(y)$.

(2) $\tau(xy + x + y) \leq 2\tau(x)\tau(y)$, *and also* $\tau(xy + x + y) \leq [(x + 3)/2]\tau(y)$.

Proof. The inequalities follow immediately from Theorem 2 and Lemma 6. $\square$

Proposition. *For any* $x \in \mathbb{N}^*$, *if* $\rho(x + 1) > \rho(x)$, *then* $\tau(x) > \rho(x)$.

Proof. Assume that $\rho(x + 1) > \rho(x)$. If $B \in \mathcal{B}(x)$ is such that $\rho(B, x) = \rho(x)$, then $B \in \mathcal{B}(x + 1)$; for otherwise, we would have $r(B, x + 1) = 0$ and therefore $C = B \cup \{x + 1\}$ would lie in $\mathcal{B}(x + 1)$ and satisfy $r(C, x + 1) = 2$, so that $\rho(x + 1) \leq \rho(C, x + 1) = \max(\rho(B, x), r(C, x + 1)) = \rho(B, x) = \rho(x) \leq \rho(x + 1)$, in contradiction with the assumption. It follows that if $B \in \mathcal{B}(x)$ is such that $\rho(B, x) = \rho(x)$, then $s(B) > \rho(x)$; indeed, B being in $\mathcal{B}(x + 1)$, we have $s(B) \geq \rho(B, x + 1) \geq \rho(x + 1) > \rho(x)$, by the assumption. Therefore, for all $B \in \mathcal{B}(x)$, we have $s(B) > \rho(x)$; indeed, this was proved if B satisfies the condition $\rho(B, x) = \rho(x)$, and if it does not satisfy this condition we would have $s(B) \geq \rho(B, x) > \rho(x)$. We thus conclude that $\tau(x) = \min\{s(B) : B \in \mathcal{B}(x)\} > \rho(x)$. $\square$

§4 NUMERICAL RESULTS

For $x \in \mathbb{N}$, a basis $B \in \mathcal{B}(x)$ is called ρ-optimal (resp. τ-optimal) if $\rho(B, x) = \rho(x)$ (resp. $s(B) = \tau(x)$); also, if $B \in \mathcal{B}(\#x)$ is such that $s(B) = \sigma(x)$, it is called a σ-optimal basis of cardinality x. The set of ρ-optimal (resp. τ-optimal) bases in $\mathcal{B}(x)$ is written $\mathcal{O}(\rho, x)$ (resp. $\mathcal{O}(\tau, x)$), and the set of σ-optimal bases of cardinality x is written $\mathcal{O}(\sigma, x)$. To compute $\rho(x)$, $\tau(x)$ or $\sigma(x)$, we have to determine at least one corresponding optimal basis. But the number of bases grows so rapidly with x that exhaustive searches are impossible for large x. For instance, $|\mathcal{B}(15)| = 8134$, while $|\mathcal{O}(\rho, 15)| = 155$ and $|\mathcal{O}(\tau, 15)| = 102$; also, $|\mathcal{B}(18)| = 63910$. On the other hand, $|\mathcal{B}(\#10)| = 47098$. As to examples of such bases, we mention $B = \{0, 1, 2, 5, 8, 11\}$, which is in $\mathcal{O}(\rho, 13) \cap \mathcal{O}(\tau, 13) \cap \mathcal{O}(\sigma, 6)$, and also $B = \{0, 1, 2, 3, 4, 5, 7, 9, 11, 15, 21, 26, 34, 35, 39, 46, 54, 62, 72, 79, 89, 94, 101, 110, 128, 137, 150, 153, 166, 182, 193, 206, 218\}$, which is in $\mathcal{O}(\rho, 223) \cap \mathcal{O}(\tau, 223) \cap \mathcal{O}(\sigma, 33)$.

Here are some values of the three main functions studied above that were obtained by computer calculations, using the software Maple.

The ρ function.

$\rho(0) = 1$
$\rho(x) = 2$ for $1 \leq x \leq 5$
$\rho(x) = 3$ for $6 \leq x \leq 12$
$\rho(x) = 4$ for $13 \leq x \leq 55$
$\rho(x) = 5$ for $56 \leq x \leq 69$
$\rho(x) = 6$ for $70 \leq x \leq 233$; and $\rho(234) \geq 6$.

The τ function.

$\tau(0) = 1$
$\tau(x) = 2$ for $1 \leq x \leq 4$
$\tau(x) = 3$ for $5 \leq x \leq 10$

$\tau(x) = 4$ for $11 \leq x \leq 45$
$\tau(x) = 5$ for $46 \leq x \leq 59$
$\tau(x) = 6$ for $60 \leq x \leq 223$; and $\tau(224) \geq 6$.

The σ function.

$\sigma(1) = 1$
$\sigma(x) = 2$ for $2 \leq x \leq 3$
$\sigma(x) = 3$ for $4 \leq x \leq 5$
$\sigma(x) = 4$ for $6 \leq x \leq 12$
$\sigma(x) = 5$ for $13 \leq x \leq 14$
$\sigma(x) = 6$ for $15 \leq x \leq 33$; and $\sigma(34) \geq 6$.

Theorem 3. *We have $\Lambda \geq 6$; i.e. for any $B \in \mathcal{B}(\mathbb{N})$, we have $s(B) \geq 6$.*

Proof. This results from Lemma 4, the fact that the function ρ, or τ or σ, is increasing, and from its highest calculated value, listed above. $\square$

REFERENCES

1. G. A. Dirac, *Note on a problem in additive number theory*, J. London Math. Soc. **26** (1951), 312-313: MR 13,326b.
2. M. Dowd, *Questions related to the Erdős-Turán conjecture*, SIAM J. Discrete Math. **1** (1988), 142-150; MR 89h:11006.
3. P. Erdős and P. Turán, *On a problem of Sidon in additive number theory, and on some related problems*, J. London Math. Soc. **16** (1941), 212-215; MR 3,270e.
4. P. Erdős and W. H. J. Fuchs, *On a problem of additive number theory*, J. London Math. Soc. **31** (1956), 67-73; MR 17,586d.
5. G. Grekos, L. Haddad, C. Helou, J. Pihko, *On the Erdős-Turán conjecture*, to appear in J. Number Theory.
6. M. B. Nathanson, *Unique representation bases for the integers*, arXiv:math.NT/0202137 **v1** (February 14, 2002), 10 pages.
7. M. B. Nathanson, *Generalized additive bases, König's lemma, and the Erdős-Turán conjecture*, preprint (February 21, 2003), 8 pages.
8. I. Z. Ruzsa, *A just basis*, Monatsh. Math. **109** (1990), 145-151; MR 91e:11016.

PENN STATE UNIV., 25 YEARSLEY MILL RD, MEDIA, PA 19063, USA; E-MAIL: CXH22@PSU.EDU

The boundary structure of the sumset in $\mathbf{Z}^2$ *

Shu-Ping Sandie Han

Department of Mathematics
New York City College of Technology
City University of New York
shan@citytech.cuny.edu

Abstract

Let A be a finite subset of $\mathbf{Z}^2$. Let h be a positive integer. Let hA be a sumset defined by

$$\{h_1 a_1 + \cdots + h_k a_k \mid a_i \in A, \ \sum_{i=1}^{k} h_i = h\}$$

where $k = |A|$. It is found that the distribution of the elements of hA in the boundary region of the convex hull of hA exhibited a repeating pattern. In other words, if each side of the boundary region of the convex hull of hA is partitioned into h cells, for h sufficiently large, there exists a constant C and there exist a consecutive $h - C$ congruent parallelograms such that the elements of hA in each parallelogram can be translated by a constant vector to obtain elements of hA in the next parallelogram. By counting the number of parallelograms and the cardinality of hA in each parallelogram, it can be found that the cardinality of hA in the boundary region is a linear function of h.

1 Introduction

Many studies have been done on the structure and the cardinality of sum of sets. In particular, let h be a positive integer, and let A be a finite subset of $\mathbf{Z}^n$, the structure and the cardinality of the h-fold sumset of A, denoted by hA, for sufficiently large h can be approximated by studying the convex hull of hA.

---

*supported by PSC CUNY Grant

It was found by Nathanson that when A is a set of integers, the structure of the h-fold sumset of A consists of an interval of consecutive integers and the cardinality of hA is a linear function of h. It was found by Khovanskii that when A is a finite set of lattice points in $\mathbf{Z}^n$, the structure of the h-fold sumset of A consists of a polytope such that all of the lattice points in the polytope are contained in hA. In other words, Khovanskii showed that there exists a positive real number ρ such that all lattice points in the convex hull of hA, whose distance from the boundary is greater than or equal to ρ, belong to hA. By using the volume of the polytope to approximate the cardinality of hA, Khovanskii showed that the cardinality of hA is a function of h^n.

The author examines the structure of sumset from a different perspective. Both Nathanson and Khovanskii studied the "core" structure of hA. In other words, Nathanson and Khovanskii studied the distribution of those elements of hA that constitute the interior lattice points of the convex hull of hA. The author will study instead the "boundary" structure of hA. The focus is on the distribution of those elements of hA that is less than a given distance away from the boundary of the convex hull of hA. As shown by Khovanskii, the cardinality of the elements of hA in the "core" structure is a function of h^n. It is conjectured that the cardinality of these "boundary" elements of hA is h^{n-1}. This paper will examine the case where A is a finite subset of $\mathbf{Z}^2$. The author will show that there is a repeating and consistent pattern in the "boundary" structure of hA and that the cardinality of these "boundary" elements is a linear function of h.

2   Notation and example

Let $\triangle_{hA}$ denote the convex hull of hA. When we speak of the boundary or the interior of hA, we mean the boundary or the interior of $\triangle_{hA}$. Let x, y, a, b be elements in $\mathbf{R}^2$. Let $l(x, y)$ denote the line segment connecting the two points x and y. Let $d(a, b)$ denote the distance between the two points a and b. Let the set

$$\{a \in \mathbf{R}^2 \mid d(a, l(x, y)) = \rho\}$$

denote the set of all elements a such that a is exactly ρ distance away from the line segment $l(x, y)$. Similarly, the set

$$\{a \in \mathbf{R}^2 \mid d(a, l(x, y)) < \rho\}$$

denotes the set of all elements a such that a is less than ρ distance away from the line segment $l(x, y)$.

If $l(x, y)$ is a boundary line of $\triangle_{hA}$, the set

$$\{a \in \triangle_{hA} \mid d(a, l(x, y)) < \rho\}$$

is also referred to as the "boundary region." The elements of hA in the boundary region are called the "boundary elements" of hA.

The following simple example in $\mathbf{Z}^2$ will illustrate the regularity in the distribution of the boundary elements of hA. In fact, the example is so simple that there is a regularity in the distribution of the elements of hA throughout the interior of $\triangle_{hA}$.

Suppose A is a set of lattice points in $\mathbf{Z}^2$ such that A contains only three elements. Let $A = \{0, a_1, a_2\}$, where $0 = (0,0)$ and $a_1 \neq k a_2$ for any real number k. Let h be a positive integer, then

$$hA = \{k_1 a_1 + k_2 a_2 \mid k_1 = 0, 1, \ldots, h \text{ and } k_2 = 0, 1, \ldots, h - k_1\}$$

The cardinality of hA is

$$|hA| = \frac{(h+2)(h+1)}{2} = \frac{h^2}{2} + \frac{3h}{2} + 1$$

The boundary of $\triangle_{hA}$ consists of three line segments: $l(0, ha_1)$, $l(0, ha_2)$, and $l(ha_1, ha_2)$. Moreover, the boundary of $\triangle_{hA}$ consists of these elements of hA:

$$\{k_1 a_1 \mid k_1 = 0, 1, \ldots, h-1\} \cup \{k_2 a_2 \mid k_2 = 1, \ldots, h-1\} \cup \{k_1 a_1 + k_2 a_2 \mid k_1 = 0, 1, \ldots, h, \ k_2 = h - k_1\}$$

Thus,

$$|hA \cap boundary| = 3h$$

A more interesting problem is to be able to examine not just those elements that lie on the boundary of $\triangle_{hA}$, but also those elements of hA that fall within a given distance from the boundary. Define the following boundary regions with thickness ρ:

$$
\begin{aligned}
F_1 &= F_1(h, \rho) = \{a \in \triangle_{hA} \mid d(a, l(0, ha_1)) < \rho\} \\
F_2 &= F_2(h, \rho) = \{a \in \triangle_{hA} \mid d(a, l(0, ha_2)) < \rho\} \\
F_3 &= F_3(h, \rho) = \{a \in \triangle_{hA} \mid d(a, l(ha_1, ha_2)) < \rho\}
\end{aligned}
$$

Let m_1, m_2, m_3 be three positive integers defined as follows:

$$
\begin{aligned}
m_1 &= \max\{n \in \mathbf{Z}^+ \mid d(na_2, l(0, ha_1)) < \rho\} \\
m_2 &= \max\{n \in \mathbf{Z}^+ \mid d(na_1, l(0, ha_2)) < \rho\} \\
m_3 &= \max\{n \in \mathbf{Z}^+ \mid d(ha_1 - na_1, l(ha_1, ha_2)) < \rho\}
\end{aligned}
$$

For h sufficiently large, the distance ρ from the boundary is small in comparison to the convex hull of hA, therefore, $F_1 \cap F_2 \cap F_3 = \emptyset$. The distribution of the elements of hA in the boundary region F_i can be described as follows:

$$
\begin{aligned}
F_1 \cap hA &= \bigcup_{k_1=0}^{h-m_1} \{k_1 a_1 + k_2 a_2 \mid k_2 = 0, 1, \ldots, m_1\} \cup \\
&\qquad \bigcup_{k_1=h-m_1+1}^{h} \{k_1 a_1 + k_2 a_2 \mid k_2 = 0, 1, \ldots, h - k_1\}
\end{aligned}
$$

$$F_2 \cap hA \;=\; \bigcup_{k_2=0}^{h-m_2} \{k_1 a_1 + k_2 a_2 \mid k_1 = 0, 1, \ldots, m_2\} \;\cup$$

$$\bigcup_{k_2=h-m_2+1}^{h} \{k_1 a_1 + k_2 a_2 \mid k_1 = 0, 1, \ldots, h-k_2\}$$

$$F_3 \cap hA \;=\; \bigcup_{k_1=0}^{h-m_3-1} \{k_1 a_1 + k_2 a_2 \mid k_2 = h - m_3 - k_1, \ldots, h-k_1\} \;\cup$$

$$\bigcup_{k_1=h-m_3}^{h} \{k_1 a_1 + k_2 a_2 \mid k_2 = 0, 1, \ldots, h-k_1\}$$

Note that for $i = 1, 2$, the set $F_i \cap hA$ consists of a disjoint union of $h - m_i + 1$ sets, where the consecutive sets in $F_1 \cap hA$ differ by a_1 and the consecutive sets in $F_2 \cap hA$ differ by a_2. The union of the remaining sets in $F_i \cap hA$ has a constant cardinality. Thus, for $i = 1, 2$, cardinality of $F_i \cap hA$ is

$$
\begin{aligned}
|F_i \cap hA| &= (h - m_i + 1)(m_i + 1) + \sum_{k=h-m_i+1}^{h} h - k \\
&= (h+1)m_i - \frac{(m_i - 2)(m_i + 1)}{2}
\end{aligned}
$$

which is a linear function of h.

Similarly, $F_3 \cap hA$ consists of a disjoint union of $h - m_3$ sets where the consecutive sets differ by $a_1 - a_2$. The union of the remaining sets also has a constant cardinality. Thus, the cardinality of $F_3 \cap hA$ is

$$
\begin{aligned}
|F_3 \cap hA| &= (h - m_3)m_3 + \sum_{k=h-m_3}^{h} h - k + 1 \\
&= (h+1)m_3 - \frac{(m_3 - 2)(m_3 + 1)}{2}
\end{aligned}
$$

which is again a linear function of h.

Furthermore,

$$|hA \cap (F_i \cap F_j)| = (m_i + 1)(m_j + 1)$$

The cardinality of hA in the boundary region can be computed:

$$
\begin{aligned}
|hA \cap (F_1 \cup F_2 \cup F_3)| &= |hA \cap F_1| + |hA \cap F_2| + |hA \cap F_3| \\
&\quad - |hA \cap (F_1 \cap F_2)| - |hA \cap (F_2 \cap F_3)| - |hA \cap (F_1 \cap F_3)| \\
&= \sum_{i=1}^{3}(h+1)m_i - \frac{(m_i - 2)(m_i + 1)}{2} - \sum_{i<j}(m_i + 1)(m_j + 1) \\
&= Ch - D
\end{aligned}
$$

Since m_1, m_2, m_3 are constants, then C and D are also constants where

$$C = m_1 + m_2 + m_3$$
$$D = \sum_{i=1}^{3} \frac{(m_i - 2)(m_i + 1)}{2} - m_i + \sum_{i<j} m_i m_j$$

Thus, the cardinality of hA in the boundary region is a linear function of h.

3   The distribution of hA in the boundary region

Let A be a finite set of lattice points in $\mathbf{Z}^2$. Let h be a positive integer.

Definition 3.1 *An element of A is a vertex of A if it is the vertex of the polytope formed by the convex hull of A. Similarly, an element of A is an interior point of A if it is the interior point of the polytope formed by the convex hull of A.*

Let V_A denote the set of vertices of A.

Definition 3.2 *Let a and b be two vertices of a polytope. We say that a and b are the adjacent vertices if the line segment connecting a and b is an edge of the polytope.*

Without loss of generality, we can assume that $0 \in A$. If there is an element $a \in A$ such that a is an interior point of A, then we can assume that 0 is an interior point of A by considering the set $A - a$. If A does not contain any element that is an interior point of A, then we can assume that 0 is a vertex of A by considering the set $A - a$ where a is a vertex of A.

Label the nonzero vertices of A in the following way,

$$V_A \backslash \{0\} = \{a_i\}_{i=1}^{l}$$

so that consecutive vertices are the adjacent vertices. If $0 \notin V_A$, then a_1 is adjacent to a_l. If $0 \in V_A$, then 0 is adjacent to both a_1 and a_l. However, a_1 and a_l are not adjacent to each other. This particular case will be excluded from the proofs of Lemmas 3.1 and 3.2, and will be discussed separately in the proof of Theorem 4.2. Let $\triangle(0, a_i, a_{i+1})$ be the convex hull formed by the three elements $\{0, a_i, a_{i+1}\}$. Partition the elements of A according to $\triangle(0, a_i, a_{i+1})$ where a_i, a_{i+1} are adjacent vertices:

$$A_i = A \cap \triangle(0, a_i, a_{i+1}) \qquad \text{for } i = 1, \ldots, l-1$$
$$A_l = A \cap \triangle(0, a_l, a_1)$$

Lemma 3.1 *Let a_i, $a_{i+1} \in \mathbf{Z}^2$. Suppose $z \in \triangle(0, a_i, a_{i+1}) \cap \mathbf{Z}^2$. Then there exist nonnegative rational numbers q' and q'' such that $0 \le q' + q'' \le 1$, and*

$$z = q'a_i + q''a_{i+1}$$

Proof.
Let $a_i = (u_1, u_2)$, $a_{i+1} = (v_1, v_2)$, and $z = (z_1, z_2)$, where $u_1, u_2, v_1, v_2, z_1, z_2$ are all integers. The solutions q' and q'' to

$$z = a_i q' + a_{i+1} q''$$

is the solution to the system of equations

$$
\begin{aligned}
z_1 &= u_1 q' + v_1 q'' \\
z_2 &= u_2 q' + v_2 q''
\end{aligned}
$$

The solution is rational. Moreover, since $z \in \triangle(0, a_i, a_{i+1})$, z is contained in the set defined by

$$\{\lambda_1 a_i + \lambda_2 a_{i+1} \mid \lambda_1, \lambda_2 \in \mathbf{R}^+ \cup \{0\},\ 0 \le \lambda_1 + \lambda_2 \le 1\}$$

thus prove the lemma. $\square$.

By Lemma 3.1, for all $a \in A_i \subseteq A$, there exist nonnegative rational numbers q'_a and q''_a such that $0 \le q'_a + q''_a \le 1$ and

$$a = q'_a a_i + q''_a a_{i+1}$$

Let

$$m_a = \min\{m \in \mathbf{Z}^+ \cup \{0\} \mid mq'_a \in \mathbf{Z}^+ \cup \{0\},\ \text{and}\ mq''_a \in \mathbf{Z}^+ \cup \{0\}\}$$

Thus, there is a linear expression

$$m_a a = m'_a a_i + m''_a a_{i+1} \tag{1}$$

where $m'_a = m_a q'_a$ and $m''_a = m_a q''_a$ are nonnegative integer coefficients and

$$m_a \ge m_a(q' + q'') \ge m'_a + m''_a$$

Since A is finite, let

$$
\begin{aligned}
m^* &= \max_{a \in A}\{m_a\} \\
k^* &= \max_{i=1,\dots,l} \mid A_i \mid \\
N &= \max_{a \in A} \parallel a \parallel
\end{aligned}
$$

Let

$$\epsilon = m^* k^* N$$

For $i = 1, \ldots, l$, define a subset of $\triangle_{hA}$ such that

$$K_i(h) = \triangle(0, ha_1, \ldots, ha_{i-1}, ha_{i+1}, \ldots, ha_l)$$

is the convex hull formed by the elements of $V_{hA} \cup \{0\} \setminus \{ha_i\}$. Define a subset of $\triangle(0, ha_i, ha_{i+1})$ that intersects the boundary region of $\triangle_{hA}$ as follows:

$$
\begin{aligned}
\triangle_i(h) &= \{x \in \triangle(0, ha_i, ha_{i+1}) \mid d(x, K_i \cup K_{i+1}) \geq \epsilon\} \quad \text{for } i = 1, \ldots, l-1 \\
\triangle_l(h) &= \{x \in \triangle(0, ha_l, ha_1) \mid d(x, K_l \cup K_1) \geq \epsilon\}
\end{aligned}
$$

The following lemma considers a special property of an element in the boundary region of hA.

Lemma 3.2 *Let A be a finite subset of $\mathbf{Z}^2$ containing 0. Let h be a positive integer. Suppose a_i and a_{i+1} are adjacent vertices. For $i = 1, \ldots, l$, if $w \in \triangle_i(h) \cap hA$, then $w - a_i \in (h-1)A$ and $w - a_{i+1} \in (h-1)A$.*

Proof.
Define a subset $A_i' \subseteq A_i \subseteq A$, where

$$A_i' = A_i \setminus l(0, a_i)$$

which does not contain those elements of A_i that lie on the line segment $l(0, a_i)$.

As an element of hA, w can be represented in terms of the elements of A:

$$w = \sum_{\substack{a \in A \\ a \notin A_i'}} h_a a + \sum_{a \in A_i'} h_a a$$

where $\sum_{a \in A} h_a = h$. Let

$$w' = \sum_{\substack{a \in A \\ a \notin A_i'}} h_a a$$

Then $w' \in K_i(h)$. Since $w \in \triangle_i(h)$, by the definition of $\triangle_i(h)$,

$$|w - w'| = \left| \sum_{a \in A_i'} h_a a \right| \geq \epsilon = m^* k^* N$$

But

$$\sum_{a \in A_i'} h_a N \geq \sum_{a \in A_i'} h_a |a| \geq \left| \sum_{a \in A_i'} h_a a \right| \geq \epsilon = m^* k^* N$$

This implies that

$$\sum_{a \in A_i'} h_a \geq m^* k^*$$

Since $k^* \geq |A_i'|$ for all i, there exists an $a \in A_i'$ such that $h_a \geq m^*$. Thus, by equation (1),

$$h_a a = (h_a - m_a)a + m_a' a_i + m_a'' a_{i+1}$$

where $m_a \geq m_a' + m_a''$ and m_a' and m_a'' are both nonnegative integers. Furthermore, $m_a'' \neq 0$, because $a \in A_i'$. This proves that $w - a_{i+1} \in (h-1)A$.

Similarly, let

$$A_i'' = A_i \backslash l(0, a_{i+1})$$

It can be proven that $w - a_i \in (h-1)A$. $\square$

Let ρ be a nonnegative real number. Define

$$l_i(h, \rho) = \{x \in \triangle(0, ha_i, ha_{i+1}) \mid d(x, l(ha_i, ha_{i+1})) = \rho\}$$

to be the line that intersects $\triangle(0, ha_i, ha_{i+1})$, parallel to $l(ha_i, ha_{i+1})$ and is equal to ρ distance away from $l(ha_i, ha_{i+1})$. Define the boundary region of $\triangle(0, ha_i, ha_{i+1})$ as:

$$F_i(h, \rho) = \{x \in \triangle(0, ha_i, ha_{i+1}) \mid d(x, l(ha_i, ha_{i+1})) < \rho\}$$

which is the region that contains all those elements of $\triangle(0, ha_i, ha_{i+1})$ that are less than ρ distance away from the boundary line $l(ha_i, ha_{i+1})$.

For small h, $F_i(h, \rho) \cap \triangle_i(h) = \triangle_i(h)$. For h sufficiently large, $F_i(h, \rho) \cap \triangle_i(h) \not\subset \triangle_i(h)$. In other words, there exists an element $x \in \triangle_i(h)$ such that $x \notin F_i(h, \rho)$.

Lemma 3.3 *Let h be a positive integer. For h sufficiently large, there exist positive integers M_i and N_i independent of h such that for all real numbers r such that $M_i \leq r \leq h - N_i$,*

$$\{ra_i + ta_{i+1} \mid t \in \mathbf{R}, \ h - \delta_i - r \leq t \leq h - r\} \subset \triangle_i(h)$$

where

$$\delta_i = \{\delta \in \mathbf{R}^+ \mid (h - \delta)a_{i+1} \in l_i(h, \rho)\}$$

Proof.
Define for all positive integers h,

$$\lambda_1(h) = \min\{\lambda \in \mathbf{R} \mid \lambda a_i + (h - \lambda - \delta_i)a_{i+1} \in \triangle_i(h)\}$$

Let h' be a positive integer, and let $u(h') = \lambda_1(h')a_i + (h' - \lambda_1(h') - \delta_i)a_{i+1}$, then $u(h')$ is an intersection of $l_i(h', \rho)$ with the boundary of $\triangle_i(h')$. Since the line $l_i(h', \rho)$ is parallel to the line $l(h'a_i, h'a_{i+1})$, and the boundary of $\triangle_i(h')$ is parallel to the boundary of $K_i(h')$, the four lines form a parallelogram. Furthermore, $u(h')$ and $h'a_{i+1}$ are the opposite vertices of the parallelogram. Let h'' be a positive integer different from h'. Define $u(h'') = \lambda_1(h'')a_i + (h'' - \lambda_1(h'') - \delta_i)a_{i+1}$. Then $u(h'')$ and $h''a_{i+1}$ are the opposite vertices of the parallelogram formed by the lines $l_i(h'', \rho)$, $l(h''a_i, h''a_{i+1})$, and the boundary lines of $\triangle_i(h'')$, and $K_i(h'')$. Since

$l_i(h, \rho)$ is ρ distance away from $l(ha_i, ha_{i+1})$ and the boundary of $\triangle_i(h)$ is ϵ distance away from the boundary of $K_i(h)$ for all positive integers h, the two parallelograms are congruent. Thus, the corresponding diagonals of the parallelograms, $u(h') - h'a_{i+1}$ and $u(h'') - h''a_{i+1}$, are equal. Hence, for all positive integers h' and h'',

$$u(h'') - h''a_{i+1} = u(h') - h'a_{i+1}$$
$$\lambda_1(h'')a_i + (h'' - \lambda_1(h'') - \delta_i)a_{i+1} - h''a_{i+1} = \lambda_1(h')a_i + (h' - \lambda_1(h') - \delta_i)a_{i+1} - h'a_{i+1}$$
$$\lambda_1(h'')(a_i - a_{i+1}) - \delta_i a_{i+1} = \lambda_1(h')(a_i - a_{i+1}) - \delta_i a_{i+1}$$

This implies
$$\lambda_1(h') = \lambda_1(h'') = \lambda_1$$

is a constant independent of h. Let $M_i = \lceil \lambda_1 \rceil$ the least integer greater than λ_1, then M_i is a constant independent of h.

Similarly, for all positive integers h, define

$$\lambda_2(h) = \min\{\lambda \in \mathbf{R} \mid (h - \lambda)a_i + (\lambda - \delta_i)a_{i+1} \in \triangle_i(h)\}$$

It can be shown again that for two different positive integers h' and h'', the parallelogram with opposite vertices $(h' - \lambda_2(h'))a_i + (\lambda_2(h') - \delta_i)a_{i+1}$ and $h'a_i$ is congruent to the parallelogram with opposite vertices $(h'' - \lambda_2(h''))a_i + (\lambda_2(h'') - \delta_i)a_{i+1}$ and $h''a_i$. Thus,

$$(h'' - \lambda_2(h''))a_i + (\lambda_2(h'') - \delta_i)a_{i+1} - h''a_i = (h' - \lambda_2(h'))a_i + (\lambda_2(h') - \delta_i)a_{i+1} - h'a_i$$
$$\lambda_2(h'')(a_{i+1} - a_i) - \delta_i a_{i+1} = \lambda_2(h')(a_{i+1} - a_i) - \delta_i a_{i+1}$$

This implies
$$\lambda_2(h') = \lambda_2(h'') = \lambda_2$$

is a constant independent of h. Let $N_i = \lfloor \lambda_2 \rfloor$ the greatest integer less than λ_2. So N_i is a constant independent of h.

Moreover, for all real numbers r such that $M_i \le r \le h - N_i$,

$$\{ra_i + ta_i \mid t \in \mathbf{R}, \ h - \delta_i - r \le t \le h - r\} \subset \triangle_i(h)$$

$\square$

Let $b_{i,n}(h) \in l(ha_i, ha_{i+1})$ be determined by n and h as follows:

$$b_{i,n}(h) = na_i + (h - n)a_{i+1}$$

For each nonnegative integer n, such that $0 \le n \le h - 1$, define a subset of $\mathbf{R}^2$:

$$B_{i,n}(h) = \{b_{i,n}(h) - ra_{i+1} + ta_2(a_i - a_{i+1}) \mid 0 \le r < \delta_i, \ 0 \le t < 1\} \tag{2}$$

For $M_i \le n \le h - N_i - 1$,

$$B_{i,n}(h) \subset F_i(h, \rho) \cap \triangle_i(h)$$

The following lemma shows that for all nonnegative integers, m and n such that $0 \leq m, n \leq h - 2$, the elements of $B_{i,m}(h)$ are the elements of $B_{i,n}(h)$ translated by special elements of hA. This will play a role later in the article when we consider the boundary elements of hA.

Lemma 3.4 *Let h be a positive integer whose value is fixed. For all nonnegative integers n such that $0 \leq n \leq h - 2$, $B_{i,n+1}(h) = B_{i,n}(h) + a_i - a_{i+1}$.*

Proof.
Let $x = b_{i,n}(h) - ra_{i+1} + t(a_i - a_{i+1}) \in B_{i,n}(h)$ for some r and t such that $0 \leq r < \delta_i$, and $0 \leq t < 1$. Then

$$
\begin{aligned}
x &= b_{i,n}(h) - ra_{i+1} + t(a_i - a_{i+1}) \\
&= na_i + (h - n)a_{i+1} - ra_{i+1} + t(a_{i+1} - a_i) \\
x + a_i - a_{i+1} &= (n+1)a_i + (h - (n+1))a_{i+1} - ra_{i+1} + t(a_{i+1} - a_i) \\
&= b_{i,n+1}(h) - ra_{i+1} + t(a_{i+1} - a_i)
\end{aligned}
$$

By definition, $x + a_i - a_{i+1} \in B_{i,n+1}(h)$. Thus, $B_{i,n}(h) + a_i - a_{i+1} \subset B_{i,n+1}(h)$.
 Conversely, if $x \in B_{i,n+1}(h)$, then

$$
\begin{aligned}
x &= b_{i,n+1}(h) - ra_{i+1} + t(a_i - a_{i+1}) \\
&= (n+1)a_i + (h - (n+1))a_{i+1} - ra_{i+1} + t(a_i - a_{i+1}) \\
&= na_i + (h - n)a_{i+1} - ra_{i+1} + t(a_i - a_{i+1}) + a_i - a_{i+1} \\
&= b_{i,n}(h) - ra_{i+1} + t(a_i - a_{i+1}) + a_i - a_{i+1}
\end{aligned}
$$

Let $y = b_{i,n}(h) - ra_{i+1} + t(a_i - a_{i+1}) \in B_{i,n}(h)$, so $x = y + a_i - a_{i+1} \in B_{i,n}(h) + a_i - a_{i+1}$.
Thus, $B_{i,n+1}(h) \subset B_{i,n}(h) + a_i - a_{i+1}$. Therefore $B_{i,n+1}(h) = B_{i,n}(h) + a_i - a_{i+1}$. $\square$

The following lemma shows that for h sufficiently large, the set $B_{i,n}(h+1)$ is a translation of the set $B_{i,n}(h)$.

Lemma 3.5 *For all nonnegative integers n such that $0 \leq n \leq h - 1$, $B_{i,n}(h+1) = B_{i,n}(h) + a_{i+1}$.*

Proof.
Again, let $x \in B_{i,n}(h)$, then for some real numbers r and t such that $0 \leq r < \delta_i$, $0 \leq t < 1$, we have

$$
\begin{aligned}
x &= b_{i,n}(h) - ra_{i+1} + t(a_i - a_{i+1}) \\
&= na_i + (h - n)a_{i+1} - ra_{i+1} + t(a_i - a_{i+1}) \\
x + a_{i+1} &= na_i + ((h + 1) - n)a_{i+1} - ra_{i+1} + t(a_i - a_{i+1}) \\
&= b_{i,n}(h + 1) - ra_{i+1} + t(a_i - a_{i+1})
\end{aligned}
$$

Thus, $x + a_{i+1} \in B_{i,n}(h+1)$ implying $B_{i,n}(h) + a_{i+1} \subset B_{i,n}(h+1)$.
Conversely, if $x \in B_{i,n}(h+1)$, then

$$
\begin{aligned}
x &= b_{i,n}(h+1) - ra_{i+1} + t(a_i - a_{i+1}) \\
&= na_i + ((h+1) - n)a_{i+1} - ra_{i+1} + t(a_i - a_{i+1}) \\
&= na_i + (h-n)a_{i+1} - ra_{i+1} + t(a_i - a_{i+1}) + a_{i+1} \\
&= b_{i,n}(h) - ra_{i+1} + t(a_i - a_{i+1}) + a_{i+1}
\end{aligned}
$$

There exists an element $y = b_{i,n}(h) - ra_{i+1} + t(a_i - a_{i+1}) \in B_{i,n}(h)$ such that $x = y + a_{i+1}$, thus, $x \in B_{i,n}(h) + a_{i+1}$ implying $B_{i,n}(h+1) \subset B_{i,n}(h) + a_{i+1}$. Hence, $B_{i,n}(h+1) = B_{i,n}(h) + a_{i+1}$.
$\square$

Define

$$
B_{i,F}(h) = \bigcup_{n=0}^{M_i-1} B_{i,n}(h) \tag{3}
$$

$$
B_{i,L}(h) = \left(\bigcup_{n=h-N_i}^{h-1} B_{i,n}(h) \right) \cap F_i(h, \rho) \tag{4}
$$

The set $B_{i,F}(h)$ represents the area of the first cell which is bounded by the lines $l(ha_i, ha_{i+1})$, $l_i(h, \rho)$, $l(0, ha_{i+1})$, and $l_{i,M_i}(h)$. On the other hand, $B_{i,L}(h)$ represents the area of the last cell which is bounded by the lines $l(ha_i, ha_{i+1})$, $l_i(h, \rho)$, $l_{i,N_i}(h)$, and $l(0, ha_i)$.

The next two lemmas consider the first and the last cell separately, but prove the same result: the $h+1$ level cell is the translation of the h level cell, hence the cardinality of the cell is independent of h.

Lemma 3.6 $B_{i,F}(h+1) = B_{i,F}(h) + a_{i+1}$.

Proof.
¿From equation (3),

$$
\begin{aligned}
B_{i,F}(h) + a_{i+1} &= \bigcup_{n=0}^{M_i-1} B_{i,n}(h) + a_{i+1} \\
&= \bigcup_{n=0}^{M_i-1} B_{i,n}(h+1) \\
&= B_{i,F}(h+1)
\end{aligned}
$$

Thus proves the lemma. $\square$

Lemma 3.7 $B_{i,L}(h+1) = B_{i,L}(h) + a_{i+1}$

Proof.
Suppose $x \in B_{i,L}(h)$, then $x \in F_i(h, \rho)$ and $x \in B_{i,n}(h)$ for some nonnegative integer n such that $h - N_i \leq n \leq h - 1$. By Lemma 3.5, it has been shown that $x + a_{i+1} \in B_{i,n}(h+1)$. To complete the proof, we need to show that $x + a_{i+1} \in F_i(h+1, \rho)$. Let w be a point on the line $l(ha_i, ha_{i+1})$, such that $|w - x| = D < \rho$. Since $w + a_{i+1}$ is a point on the line $l((h+1)a_i, (h+1)a_{i+1})$, then

$$d(x + a_{i+1}, l((h+1)a_i, (h+1)a_{i+1})) < d(w + a_{i+1}, x + a_{i+1})| = |w - x| = D < \rho$$

Thus, $x + a_{i+1} \in F_i((h+1), \rho)$, therefore $x + a_{i+1} \in B_{i,L}(h+1)$ implying $B_{i,L}(h) + a_{i+1} \subset B_{i,L}(h+1)$.

Conversely, let $x \in B_{i,L}(h+1)$, then $x \in F_i(h+1, \rho)$ and $x \in B_{i,n}(h+1)$. There exists an element $y \in B_{i,n}(h)$ such that $x = y + a_{i+1}$. Using the similar argument as before, it can be shown that $y \in F_i(h, \rho)$ and $y \in B_{i,L}(h)$, thus implying $x \in B_{i,L}(h) + a_{i+1}$. Hence, prove the lemma. $\square$

Let $S_{i,n}(h)$ denote the subset of hA that belongs in $B_{i,n}(h)$. Let $U_i(h)$ and $V_i(h)$ denote the subset of hA that belong in the first and the last cell respectively, thus, we have the following definitions:

$$
\begin{align}
S_{i,n}(h) &= hA \cap B_{i,n}(h) \quad \text{for } M_i \leq n \leq h - N_i - 1 \tag{5} \\
U_i(h) &= hA \cap B_{i,F}(h) \tag{6} \\
V_i(h) &= hA \cap B_{i,L}(h) \tag{7}
\end{align}
$$

The next two lemmas consider the elements in the boundary region of hA and prove that the cardinalities of hA in the cells $B_{i,n}(h)$ are equal and independent of h.

Lemma 3.8 *For all nonnegative integers n such that $M_i \leq n \leq h - N_i - 2$,*

$$
\begin{align}
S_{i,n+1}(h) &= S_{i,n}(h) + a_i - a_{i+1} \quad \text{and} \\
|S_{i,n+1}(h)| &= |S_{i,n}(h)|.
\end{align}
$$

Proof.
Suppose $x \in S_{i,n}(h)$, then $x \in hA$ and $x \in B_{i,n}(h)$. Lemma 3.4 implies that $x + a_i - a_{i+1} \in B_{i,n+1}(h)$. There left to prove that $x + a_i - a_{i+1} \in hA$. Since $B_{i,n}(h) \subset (F_i(h, \rho) \cap \Delta_i(h))$ for all nonnegative integers n such that $M_i \leq n \leq h - N_i - 1$, so $x \in \Delta_i(h)$. Moreover, $x \in \Delta_i(h) \cap hA$. By Lemma 3.2, both $x - a_i$ and $x - a_{i+1}$ are in $(h-1)A$. Thus,

$$x + a_i - a_{i+1} = (x - a_{i+1}) + a_i \in (h-1)A + A = hA$$

Therefore, $x + a_i - a_{i+1} \in S_{i,n+1}(h)$ and thus $S_{i,n}(h) + a_i - a_{i+1} \subset S_{i,n+1}(h)$

Conversely, if $x \in S_{i,n+1}(h)$, by definition, $x \in hA$ and $x \in B_{i,n+1}(h)$. Lemma 3.4 shows that there exists an element $y \in B_{i,n}(h)$ such that $x = y + a_i - a_{i+1}$. Let $y = x - a_i + a_{i+1}$. Again, $x \in \triangle_i(h)$ since $B_{i,n+1}(h) \subset F_i(h, \rho) \cap \triangle_i(h)$, so by Lemma 3.2, both $x - a_i$ and $x - a_{i+1}$ are in $(h-1)A$. Therefore,

$$y = x - a_i + a_{i+1} = (x - a_i) + a_{i+1} \in (h-1)A + A = hA$$

implying $y \in S_{i,n}(h)$ which implies $x \in B_{i,n}(h) + a_i - a_{i+1}$. Hence, $S_{i,n+1}(h) \subset S_{i,n}(h) + a_i - a_{i+1}$. Hence, $S_{i,n+1}(h) = S_{i,n}(h) + a_i - a_{i+1}$. This also proves that $|S_{i,n+1}(h)| = |S_{i,n}(h)|$. $\square$

Lemma 3.9 *For all nonnegative integers n such that $M_i \leq n \leq h - N_i - 1$,*

$$\begin{aligned} S_{i,n}(h+1) &= S_{i,n}(h) + a_{i+1} \quad and \\ |S_{i,n}(h+1)| &= |S_{i,n}(h)|. \end{aligned}$$

Proof.
Result follows directly from Lemma 3.5. $\square$

Lemma 3.10

$$U_i(h+1) = U_i(h) + a_{i+1} \quad and \quad V_i(h+1) = V_i(h) + a_{i+1}$$

Furthermore,

$$|U_i(h+1)| = |U_i(h)| \quad and \quad |V_i(h+1)| = |V_i(h)|.$$

Proof.
Refer to equations (6) and (7) for the definitions of $U_i(h)$ and $V_i(h)$. The result follows directly from Lemmas 3.6 and 3.7. $\square$

The interesting finding of this research is that for h sufficiently large and for all $M_i \leq n \leq h - N_i - 1$, the cardinality of hA in each set $S_{i,n}(h)$ is a constant independent of h and n. The cardinalities of $U_i(h)$ and $V_i(h)$, the first and the last cell of the partition, are also constants independent of h. Thus the cardinality of hA in the boundary region of $\triangle(0, ha_i, ha_{i+1})$ is

$$|S_{i,n}(h)| \cdot (\text{ number of cells }) + |U_i(h)| + |V_i(h)|$$

The only component of the above expression that depends on h is the number of cells which is determined by M_i and N_i.

4 Main theorems on the distribution and the cardinality of the boundary elements of hA in $\mathbf{Z}^2$

The following theorem proves the special case by considering the pattern of hA in the boundary region of $\triangle(0, ha_i, ha_{i+1})$ where a_i and a_{i+1} are two adjacent vertices of A.

Theorem 4.1 *Let A be a finite subset of $\mathbf{Z}^2$ containing 0. Suppose a_i and a_{i+1} are two adjacent vertices of A. Let h be a positive integer and ρ a nonnegative real number. Define*

$$F_i(h, \rho) = \{x \in \triangle(0, ha_i, ha_{i+1}) \mid d(x, l(ha_i, ha_{i+1})) < \rho\} \tag{8}$$

Then for h sufficiently large

$$\mid F_i(h, \rho) \cap hA \mid \ = \ C_i \cdot h - D_i$$

where C_i and D_i are some constants.

Proof.
Let M_i and N_i be two constants independent of h, determined by the intersection of the line $l_i(h, \rho)$ with the boundary of $\triangle_i(h)$ as in Lemma 3.3. Define $B_{i,n}(h)$, $S_{i,n}(h)$, $U_i(h)$, and $V_i(h)$ the same as in equations (2), (5), (6), and (7).

For h sufficiently large,

$$F_i(h, \rho) \cap hA \ = \ \bigcup_{i=M_i}^{h-N_i-1} S_{i,n}(h) \ \cup \ U_i(h) \ \cup \ V_i(h)$$

where the intersection of any two sets in the union is empty. Thus the cardinality is

$$|F_i(h, \rho) \cap hA| \ = \ \sum_{i=M_i}^{h-N_i-1} |S_{i,n}(h)| \ + \ |U_i(h)| \ + \ |V_i(h)|$$

Let $C_i = |S_{i,n}(h)|$. Then C_i is a constant independent of n and h by Lemmas 3.8 and 3.9. Furthermore, the cardinality of $U_i(h)$ and $V_i(h)$ are constants independent of n and h by Lemma 3.10.

Therefore, the number of elements of hA that are within ρ distance away from the face $l(ha_i, ha_{i+1})$ is:

$$\begin{aligned}
\mid F_i(h, \rho) \cap hA \mid \ &= \ \sum_{i=M_i}^{h-N_i-1} |S_{i,n}(h)| \ + \ |U_i(h)| \ + \ |V_i(h)| \\
&= \ (h - N_i - M_i)C_i + \mid U_i(h) \mid + \mid V_i(h) \mid \\
&= \ C_i \cdot h - D_i
\end{aligned}$$

where $D_i = (M_i + N_i) \cdot C_i \ - \ |U_i(h)| \ - \ |V_i(h)|$. $\square$

By using the result of Theorem 4.1, Theorem 4.2 proves the general case.

Theorem 4.2 *Let A be a finite set of lattice points in $\mathbf{Z}^2$. Let $\partial\triangle_{hA}$ denote the boundary of $\triangle_{hA}$. Let h be a positive integer. Define*

$$F_A(h, \rho) = \{x \in \triangle_{hA} \mid d(x, \partial\triangle_{hA}) < \rho\}$$

to be the region that is less than ρ distance away from the boundary. Then for h sufficiently large

$$\mid hA \cap F_A(h, \rho) \mid = C \cdot h + D$$

where C and D are some constants independent of h.

Proof.

Case 1: A contains an element that is in the interior of the convex hull of A.

Without loss of generality, we can assume that $0 \in A$ and that 0 is in the interior of the convex hull of A. Since we can consider the set $A - a$, where $a \in A$ is an interior element of the convex hull of A.

Let V_A be the set of vertices of A. Let

$$V_A = \{a_i\}_{i=1}^l$$

where $l = |V_A|$. Assume the consecutive vertices are the adjacent vertices, and a_1 and a_l are the adjacent vertices. Define

$$
\begin{aligned}
F_i(h, \rho) &= \{x \in \triangle(0, ha_i, ha_{i+1}) \backslash l(0, ha_{i+1}) \mid d(x, l(ha_i, ha_{i+1})) < \rho\} \quad \text{for } i = 1, \ldots, l-1 \\
F_l(h, \rho) &= \{x \in \triangle(0, ha_l, ha_1) \backslash l(0, ha_1) \mid d(x, l(ha_l, ha_1)) < \rho\}
\end{aligned}
$$

apply Theorem 4.1 to each set $\triangle(0, ha_i, ha_{i+1})$ and $\triangle(0, ha_l, ha_1)$, thus, for all $i = 1, \ldots, l$,

$$|hA \cap F_i(h, \rho)| = C_i h - D_i$$

where C_i and D_i are constants.

Let $T_i(h) = U_i(h) \cap V_{i+1}(h)$.

$$T_i(h) = \{n \in \mathbf{Z}^+ \mid d(na_{i+1}, ha_{i+1}) < \rho\}$$

Since the cardinality of $U_i(h)$ and $V_i(h)$ are independent of h, the cardinality of $T_l(h)$ is also independent of h. Therefore,

$$
\begin{aligned}
|hA \cap F_A(h, \rho)| &= \sum_{i=1}^l |hA \cap F_i(h, \rho)| - \sum |T_i(h)| \\
&= \sum_{i=1}^l C_i h - D_i \\
&= Ch - D
\end{aligned}
$$

where

$$C = \sum_{i=1}^{l} C_i \qquad \text{and} \qquad D = \sum_{i=1}^{l} D_i + \sum |T_i(h)|$$

Case 2: A does not contain an element that is in the interior of the convex hull of A.

Without loss of generality we can assume $0 \in A$ and 0 is a vertex of the convex hull of A. Since we can consider the set $A - a$ where $a \in A$ is a vertex of the convex hull of A. Let V_A be the set of vertices of A. Let

$$V_A = \{a_i\}_{i=1}^{l} \cup \{0\}$$

where $l = |V_A \backslash \{0\}|$. Assume the consecutive vertices are the adjacent vertices, and 0 is adjacent to both a_1 and a_l. Define

$$F_i(h, \rho) = \{x \in \triangle(0, ha_i, ha_{i+1}) \backslash l(0, ha_{i+1}) \mid d(x, l(ha_i, ha_{i+1})) < \rho\} \qquad \text{for } i = 1, \ldots, l-1$$

apply Theorem 4.1 to each set $\triangle(0, ha_i, ha_{i+1})$, thus for all $i = 1, \ldots, l-1$,

$$|hA \cap F_i(h, \rho)| = C_i h - D_i$$

where C_i and D_i are constants.

Define the remaining two boundary regions of $\triangle_{hA}$ in the following way:

$$F_0(h, \rho) = \{x \in \triangle(0, ha_1, ha_2) \mid d(x, l(0, ha_1)) < \rho\}$$
$$F_l(h, \rho) = \{x \in \triangle(0, ha_{l-1}, ha_l) \mid d(x, l(0, ha_l)) < \rho\}$$

To be able to apply Theorem 4.1 to the case of $|hA \cap F_0(h, \rho)|$ and $|hA \cap F_l(h, \rho)|$, we will translate the set A by a_2. Consider the set $A' = A - a_2$ and let $a'_0 = a_2 - a_2 = 0$; $a'_1 = a_1 - a_2$; $a'_2 = a_0 - a_2$; and $a'_3 = a_l - a_2$. Applying Theorem 4.1 to $\{0, a'_1, a'_2\}$ and $\{0, a'_2, a'_3\}$, we have

$$|hA \cap F_0(h, \rho)| = |hA' \cap F_0(h, \rho)|$$

and

$$|hA \cap F_l(h, \rho)| = |hA' \cap F_l(h, \rho)|$$

Putting it all together we have for $i = 0$, and l,

$$|hA \cap F_i(h, \rho)| = |hA' \cap F_i(h, \rho)| = C_i h - D_i$$

Let

$$T_i(h) = hA \cap F_i(h, \rho) \cap F_{i+1}(h, \rho) = U_i(h) \cap V_i(h) \qquad \text{for } i = 0, 1, \ldots, l-1$$
$$T_l(h) = hA \cap F_l(h, \rho) \cap F_0(h, \rho) = U_l(h) + V_0(h)$$

By Lemma 3.10, the elements of $U_i(h+1)$ and $V_i(h+1)$ are simply the translation of the elements of $U_i(h)$ and $V_i(h)$ by a_{i+1}. Thus, the elements of $T_i(h+1)$ are also the translation of the elements of $T_i(h)$ by a_{i+1}. Since the cardinality of $U_i(h)$ and $V_i(h)$ are independent of h, so is the cardinality of $T_i(h)$. The cardinality of hA in the boundary region is:

$$
\begin{aligned}
| \, hA \cap F_A(h,\rho) \, | &= \sum_{i=0}^{l} | \, hA \cap F_i(h,\rho) \, | - \sum_{i=0}^{l} |T_i(h)| \\
&= \sum_{i=0}^{l} C_i \cdot h - D_i - |T_i(h)| \\
&= C \cdot h - D
\end{aligned}
$$

where $C = \sum_{i=0}^{l} C_i$ and $D = \sum_{i=0}^{l} D_i + |T_i(h)|$. $\square$

Define

$$
\triangle_{hA}(\rho) = \{ x \in \triangle_{hA} \mid d(x, \partial\triangle_{hA}) \geq \rho \}
$$

to be the set of elements of $\triangle_{hA}$ that is greater than or equal to ρ distance away from the boundary of $\triangle_{hA}$.

Theorem 4.3 (Khovanskii) *Let A be a finite subset of $\mathbf{Z}^n$ such that A generates $\mathbf{Z}^n$. Then there exists a constant ρ with the following property: for an arbitrary natural number h, every lattice point of the polytope $\triangle_{hA}(\rho)$ belongs to hA.*

Theorem 4.4 *Let A be a finite subset of $\mathbf{Z}^2$ such that A generates $\mathbf{Z}^2$. For all sufficiently large positive integers h, there exists a nonnegative real number ρ independent of h such that*

$$
hA = \left(\triangle_{hA}(\rho) \cap \mathbf{Z}^2 \right) \cup (F_A(h,\rho) \cap hA)
$$

where $|F_A(h,\rho) \cap hA|$ is a linear function of h.

Proof.
Let ρ be the real number in Theorem 4.3. Thus $\triangle_{hA}(\rho) \cap \mathbf{Z}^2 \subseteq hA$. For all h sufficiently large,

$$
\begin{aligned}
hA &= (\triangle_{hA}(\rho) \cap hA) \cup (F_A(h,\rho) \cap hA) \\
&= (\triangle_{hA}(\rho) \cap \mathbf{Z}^2) \cup (F_A(h,\rho) \cap hA)
\end{aligned}
$$

By Theorem 4.2, $|F_A(h,\rho) \cap hA|$ is a linear function of h. $\square$

In conclusion, Theorem 4.4 has presented a general structure for hA where A is a finite subset of $\mathbf{Z}^2$. The general structure is this: There exists a nonnegative real number ρ such that all lattice points whose distance from the boundary of $\triangle_{hA}$ is greater or equal to ρ belongs in the "core" structure of hA, and those lattice points whose distance from the boundary is less than ρ belongs in the boundary region. The distribution of these boundary elements of hA exhibited a regular pattern since they are related by a translation of an element. Khovanskii had shown that the cardinality of those elements of hA in the "core" structure is a function of h^n where n is the dimension of the space, thus in the case of $\mathbf{Z}^2$, the cardinality is h^2. The author had shown explicitly by using simple counting principles that the cardinality of the boundary elements of hA is h^{n-1}, which in the case of $\mathbf{Z}^2$ is a linear function in h.

References

[1] G. Ewald, *Combinatorial Convexity and Algebraic Geometry,* volume 168 of *Graduate Texts in Mathematics.* Springer-Verlag, New York, 1996.

[2] S. Han, C. Kirfel, and M. Nathanson, "Linear forms in finite sets of integers," *The Ramanujan Journal.* 2(1998), pp.271-281.

[3] A.G. Khovanskii, "Newton polyhedron, Hilbert polynomial, and sums of finite sets," *Funksional. Anal. Prilozhen.* **26** (1992), pp. 276-281.

[4] M.B. Nathanson, *Additive Number Theory: Inverse Problems and the Geometry of Sumsets,* volume 165 of *Graduate Texts in Mathematics.* Springer-Verlag, New York, 1996.

ON NTU's IN FUNCTION FIELDS

Howard Kleiman

1. Introduction. Let $g(x,y)$ be a polynomial of degree n in x with coefficients in $\mathbf{Z}[y]$ which is absolutely irreducible. Suppose that the coefficient of x^n is c' while $g(0,y) = c$ with c, c' $\in$ $\mathbf{Z}$. Let b be a root of $g(x,y)$. Then b is an NTU (non-trivial unit) of $\mathbf{F} = \mathbf{Q}(y)(b)/\mathbf{Q}(y)$. We can generalize Hilbert Class Field Theory using NTU units. Also, by elementary algebra, there are but a finite number of solutions of $q(x,y) = 0$ in $\mathbf{Q}$ X $\mathbf{Z}$. We can explicitly obtain them. Furthermore, using a computer algorithm, we can explicitly obtain solutions in $\mathbf{Q}$ X $\mathbf{Z}$ of a large number of minimal equations of $\mathbf{F}$.

2. **THEOREM**. Let $f(x,y) = 0$ be a polynomial of degree n with coefficients in $\mathbf{Z}[y]$ defining the field $\mathbf{L} = \mathbf{Q}(y)(a)/\mathbf{Q}(y)$. Assume that $f(x,y)$ is absolutely irreducible. Suppose that

    (1) $f(0,y)$ factors into $cq(y)$ in $\mathbf{Z}[y]$ where $|c| > 1$, c $\in$ $\mathbf{Z}$, c is not a perfect n-th power in $\mathbf{Z}$, and $q(y)$ is monic and irreducible;

    (2) there exist rational integers y' and y'' such that both $f(0,y')$ amd $f(0,y'')$ are irreducible over $\mathbf{Q}$ with indices relatively prime to c;

    (3) $f(0,y')$ and $f(0,y'')$ have c as their gcd;

219

(4) $f(0,y^*)/c$ $(y^* = y', y'')$ is relatively prime to c.

(5) If $a' = a(y')$ and $a'' = a(y'')$, the class fields of
$L' = Q(a')$ and $L'' = Q(a'')$ have Q as their intersection.

Then L contains an NTU.

Proof. Either a is divisble by an NTU, u, in L whose absolute
norm is c, or else there exists an algebraic integer A, unique to
within an algebraic unit, which divides $a = a(y)$ for every
specialization, y^*, of y into a rational integer. In the latter
case, if conditions (1) - (4) hold for $y = y^*$, the ideal
$(c, a^*) = (A)^*$ is a factor of $(cq(y^*))$, $y^* \in z$, . Therefore, $(A)^*$
is contained in the class field of $L^* = Q(y^*)(a^*)$ $(y^* = y', y'')$.
implying that their respective class fields both contain $Q(A)$ as
a subfield, contradicting (5). Therefore, a is divisible by an
NTU u (not necessarily in L). But, from (1), $q(y)$ is an
irreducible element of $Q[y]$. Therefore, $(a, q(y))$ is a prime
ideal of degree 1, that is, a principal ideal in o_L . Therefore,
$(a, q(y)) = (a)$. (a) is representable by the ideal number ua
where u is a unit in L. u is either an element of Q or an NTU.
But its absolute norm is $\pm c$ which isn't a perfect n-th power. It
follows that u is an NTU.

Continued Fractions and Quadratic Irrationals

Joseph Lewittes[*]

Department of Mathematics and Computer Science

Lehman College - City University of New York

Bronx, NY 10468

lewittes@alpha.lehman.cuny.edu

Most books on number theory contain a chapter or two on continued fractions, which really are indispensable in a number of areas. Nevertheless, they have not achieved a mainstream popularity and are often omitted in courses on number theory. Of course there are reasons for this; their basic construction strikes one as rather bizarre and they are notoriously impossible to manipulate with respect to the usual operations of arithmetic. Furthermore, they have no satisfactory generalization to fit into a more comprehensive framework. But, they are surprising and interesting. One of the key roles played by continued fractions is in the construction of units in real quadratic fields. In studying this topic we have found that the entire theory of units in such fields can in fact be derived via continued fractions. Also, the periods of the continued fractions associated with quadratic irrationals exhibit a certain pattern or structure when classified by discriminant, which has not been previously noted.

In presenting these results I am also attempting an introductory exposition of continued fractions. This paper is meant to be self-contained in the sense that all basic terms will be explained; specific facts quoted from the literature can simply be accepted, for the moment, without impeding the reader's progress. Our focus is narrow, concentrating on the continued fractions themselves, omitting any discussion of their intimate connections with ideal theory and binary quadratic forms. The following theorem may be considered an ultimate goal; eventually its statement will become meaningful, numerical examples will be given and perhaps the reader will be intrigued.

[*]Partially Supported by a PSC-CUNY Research Grant.

221

<u>Theorem</u> Let d be a discriminant and $\varepsilon = \dfrac{t + u\sqrt{d}}{2}$, the fundamental unit. Let x be a reduced quadratic irrational with discriminant d, having the purely periodic continued fraction $\left[\overline{b_0, b_1, \ldots, b_{k-1}}\right]$. Let e be the number of terms b_i that are even. Then the parity of e depends on d only, as follows:

If d is odd, then e is even. If d is even, then $e \equiv u$ (mod 2).

We formally designate as theorems those statements that contain new results. A standard reference on continued fractions in Perron's book [5]; this is the third edition. The classical results that we cite are all found there in the first three chapters. A recent book in English is [6]. For more on orders in quadratic fields and binary quadratic forms see [2], chapter 2 and [1], a more leisurely approach. But these do not integrate continued fractions into their discussion. While [4] does - see, in particular, the preface - it treats only the case of field discriminants.

1. Finite Continued Fractions

Continued fractions may be motivated by the Euclidean Algorithm, which we recall with a simple example. To find the greatest common divisor of 26 and 7 (denoted gcd(26,7) or just (26,7)) one performs successive integer divisions yielding quotient and remainder to obtain:

$$26 = \underline{3} \times 7 + 5$$
$$7 = \underline{1} \times 5 + 2$$
$$5 = \underline{2} \times 2 + 1$$
$$2 = \underline{2} \times 1$$

The last non-zero remainder, 1, is (26,7); the quotients generated, 3, 1, 2, 2, underlined above, seem to be irrelevant. But, waste not want not! Dividing both sides of each equation, save the last, by the divisor, they become:

$$\frac{26}{7} = 3 + \frac{1}{\frac{7}{5}}, \qquad \frac{7}{5} = 1 + \frac{1}{\frac{5}{2}}, \qquad \frac{5}{2} = 2 + \frac{1}{2}$$

By successive substitutions one then obtains

$$\frac{26}{7} = \underline{3} + \cfrac{1}{\underline{1} + \cfrac{1}{\underline{2} + \cfrac{1}{\underline{2}}}} \tag{1}$$

and $\dfrac{26}{7}$ is now presented as a complicated compound fraction, but using only the underlined quotients, and a 1 for each numerator. Such an expression is called a (finite) continued fraction. In general, if $b_0, b_1, \ldots, b_{n-1}$ are any numbers then

$$b_0 + \cfrac{1}{b_1 + \cfrac{1}{b_2 + \ddots \cfrac{1}{b_{n-2} + \cfrac{1}{b_{n-1}}}}} \tag{2}$$

is called a continued fraction. Note that $n \geq 1$ and if $n=1$ we have only b_0 - but it still qualifies technically as a continued fraction. We refer to $b_0, b_1, \ldots, b_{n-1}$ as the terms of the continued fraction (the n terms are indexed from 0 to $n-1$) though in the literature they are called the 'partial quotients.' (2) is too bulky so we write it in condensed form $[b_0, b_1, \ldots, b_{n-1}]$. As long as $b_1, \ldots, b_{n-1}$ are positive real numbers, which we assume from now on, though b_0 may be ≤ 0, (2) has a well-defined numerical value, again designated by the symbol $[b_0, b_1, \ldots, b_{n-1}]$. Thus (1) shows $\dfrac{26}{7} = [3,1,2,2]$. Since $2 = 1 + \frac{1}{1}$,

$\dfrac{26}{7} = [3,1,2,1,1]$ also.

What worked for $\dfrac{26}{7}$ shows that any rational number x can be expressed as a continued fraction. Write $x = \dfrac{c}{d}$, c, d integers, $d>0$, not necessarily relatively prime, and apply the successive divisions of the Euclidean Algorithm starting with division of c by d. If n lines are produced with the quotients being $b_0, b_1, \ldots, b_{n-1}$, then $x = [b_0, b_1, \ldots, b_{n-1}]$. Note that if x is not an integer then $n>1$ and the last term $b_{n-1} > 1$ so that also $x = [b_0, b_1, \ldots, b_{n-1} - 1, 1]$. If x is an integer then $n=1$, the one line of the algorithm reads $c = b_0 d$, with $b_0 = x$, so $x = [b_0] = [b_0 - 1, 1]$. In particular, every rational x has two continued fraction representations, one having an even number of terms and the other an odd number. This observation plays an important role later on.

The reader will have noticed that we've been tacitly assuming that the terms b_i are integers. This indeed is the case, but later on we will want to use the option of having the last term b_{n-1} not an integer. So this should be kept in mind, but from now on, unless otherwise noted, the terms of the continued fraction will be integers only.

Actually, the two ways shown to represent x as a continued fraction are the only ways. This can be seen as follows. First note that if $b_0 + \dfrac{1}{(b_1 + \cdots)}$ is any continued fraction with three or more terms then its value is not an integer, since $(b_1 + \cdots) > 1$; also $b_0 + \dfrac{1}{b_1}$ is not an integer if $b_1 > 1$. Thus the only possible continued fractions for an integer x are those two given above. Suppose now x not an integer and $\left[c_0, c_1, \ldots, c_{n-1}\right]$ is any continued fraction whose value is x. Then $x = c_0 + \dfrac{1}{y}$, $y = \left[c_1, \ldots, c_{n-1}\right]$ and $y > 1$ since x not an integer. Thus $c_0 = \lfloor x \rfloor$. Here, for any real number t, $\lfloor t \rfloor$ denotes the greatest integer $\leq t$, the unique integer k, such that $k \leq t < k+1$. We use this notation rather than the customary $[t]$ to minimize confusion with the square brackets used for continued fractions. If also $x = \left[a_0, a_1, \ldots, a_{k-1}\right]$ we have $a_0 = \lfloor x \rfloor = c_0$, hence $\left[a_1, \ldots, a_{k-1}\right] = \left[c_1, \ldots, c_{n-1}\right]$. If this common value is an integer we have only the two possibilities described above, while if not, then $a_1 = c_1$ and continue on the same way.

So far we've discussed how a rational number can be represented by a continued fraction. We now turn around and ask, given $\left[b_0, b_1, \ldots, b_{n-1}\right]$ how should it be evaluated? Consider, for example, $[5,4,3,6] = 5 + \dfrac{1}{4 + \dfrac{1}{3 + \dfrac{1}{6}}}$. The natural inclination is to do it from the bottom up, $3 + \dfrac{1}{6} = \dfrac{19}{6}$, then $4 + \dfrac{6}{19} = \dfrac{82}{19}$ and finally $5 + \dfrac{19}{82} = \dfrac{429}{82}$. This is fine, but eventually continued fractions will become 'infinitely long' by allowing the number of terms to increase, and from this perspective it is better to evaluate from the top down. So first calculate $[5] = \dfrac{5}{1}$, $[5,4] = 5 + \dfrac{1}{4} = \dfrac{21}{4}$, $[5,4,3] = 5 + \dfrac{1}{4 + \dfrac{1}{3}} = \dfrac{68}{13}$, then finally,

$[5,4,3,6] = \dfrac{429}{82}$. But how are all these related to each other and the final result? To answer this, consider, for the moment $b_0, b_1, \ldots, b_{n-1}$ as variables. Then $[b_0] = \dfrac{b_0}{1}$,

$[b_0, b_1] = b_0 + \dfrac{1}{b_1} = \dfrac{b_1 b_0 + 1}{b_1}$. If we set $A_0 = b_0$, $B_0 = 1$, $A_1 = b_1 b_0 + 1$, $B_1 = b_1$ then

$[b_0] = \dfrac{A_0}{B_0}$, $[b_0, b_1] = \dfrac{A_1}{B_1}$ and $[b_0, b_1, b_2] = b_0 + \dfrac{1}{b_1 + \dfrac{1}{b_2}} = \dfrac{b_2 b_1 b_0 + b_2 + b_0}{b_2 b_1 + 1} = \dfrac{b_2 A_1 + A_0}{b_2 B_1 + B_0}$,

which suggests defining $A_2 = b_2 A_1 + A_0$, $B_2 = b_2 B_1 + B_0$ so that $[b_0, b_1, b_2] = \dfrac{A_2}{B_2}$. This scheme may then be continued. Formally we define, given the sequence $b_0, b_1, b_2, \ldots$,

$$A_{-1} = 1, \quad B_{-1} = 0, \quad A_0 = b_0, \quad B_0 = 1 \tag{3}$$

and inductively for $i \geq 1$

$$A_i = b_i A_{i-1} + A_{i-2}, \quad B_i = b_i B_{i-1} + B_{i-2}. \tag{4}$$

The A_{-1}, B_{-1} are a useful convenience. They, along with B_0, are constants while for $i \geq 0$, A_i, B_i are polynomials in $b_0, b_1, \ldots, b_i$, but B_i is independent of b_0. A suitable induction argument then shows that for all $n \geq 1$

$$[b_0, b_1, \ldots, b_{n-1}] = \dfrac{A_{n-1}}{B_{n-1}}. \tag{5}$$

This is a formal algebraic identity which then remains true if one substitutes for b_0 any number and for the b_i, $i > 0$, any positive numbers ; so $B_{n-1} \neq 0$. Here number may mean integer or real but we are interested only in integers. Returning to the original question as to how to evaluate the continued fraction $[b_0, b_1, \ldots, b_{n-1}]$ the answer is to calculate $A_0, B_0, \ldots, A_{n-1}, B_{n-1}$ successively, and then the value is $\dfrac{A_{n-1}}{B_{n-1}}$. This is done best with a table which we present for our example $[5, 4, 3, 6]$.

i	-1	0	1	2	3
b		5	4	3	6
A	1	5	21	68	429
B	0	1	4	13	82

(6)

First the two rows labeled i and b are filled in; i acts as column index. Then the rows labeled **A** and **B** are filled in left to right, according (3), (4). Then for any $i \geq 0$, the value of $[b_0, b_1, \ldots, b_i]$ is read off from the table as $\dfrac{A_i}{B_i}$.

2. Matrices and the Group G

The definition (3), (4) can be written $\begin{pmatrix} A_i & A_{i-1} \\ B_i & B_{i-1} \end{pmatrix} = \begin{pmatrix} A_{i-1} & A_{i-2} \\ B_{i-1} & B_{i-2} \end{pmatrix} \begin{pmatrix} b_i & 1 \\ 1 & 0 \end{pmatrix}$ for $i \geq 1$.

If $i \geq 2$, then the first matrix on the right is similarly $\begin{pmatrix} A_{i-2} & A_{i-3} \\ B_{i-2} & B_{i-3} \end{pmatrix} \begin{pmatrix} b_{i-1} & 1 \\ 1 & 0 \end{pmatrix}$ and iterating this one obtains finally:

$$\begin{pmatrix} A_i & A_{i-1} \\ B_i & B_{i-1} \end{pmatrix} = \begin{pmatrix} b_0 & 1 \\ 1 & 0 \end{pmatrix} \begin{pmatrix} b_1 & 1 \\ 1 & 0 \end{pmatrix} \cdots \begin{pmatrix} b_i & 1 \\ 1 & 0 \end{pmatrix}. \tag{7}$$

We introduce the abbreviation

$$M[b_0, b_1, \ldots, b_{n-1}] = \begin{pmatrix} b_0 & 1 \\ 1 & 0 \end{pmatrix} \begin{pmatrix} b_1 & 1 \\ 1 & 0 \end{pmatrix} \cdots \begin{pmatrix} b_{n-1} & 1 \\ 1 & 0 \end{pmatrix}. \tag{8}$$

Thus we have the following continued fraction - matrix connection:

$$\begin{pmatrix} A_{n-1} & A_{n-2} \\ B_{n-1} & B_{n-2} \end{pmatrix} = M[b_0, b_1, \ldots, b_{n-1}]. \tag{9}$$

Since the matrix $\begin{pmatrix} b & 1 \\ 1 & 0 \end{pmatrix}$ has determinant -1 we immediately obtain

$$A_{n-1} B_{n-2} - A_{n-2} B_{n-1} = (-1)^n. \tag{10}$$

Here is a nice immediate application of this matrix viewpoint. Since $\begin{pmatrix} b & 1 \\ 1 & 0 \end{pmatrix}$ is symmetric, transposing both sides of (9) yields $\begin{pmatrix} A_{n-1} & B_{n-1} \\ A_{n-2} & B_{n-2} \end{pmatrix} = M[b_{n-1}, \ldots, b_1, b_0]$, so the value of the continued fraction obtained by taking the terms in reverse order is $\dfrac{A_{n-1}}{A_{n-2}}$.

Thus, noting (6), $[6, 3, 4, 5] = \dfrac{429}{68}$.

Basic to all that follows is the set G of all 2x2 matrices $M = \begin{pmatrix} a & b \\ c & d \end{pmatrix}$ with a, b, c,

d integers and det $M = ad - bc = \pm 1$, which ensures that $M^{-1} \in G$ and G is a group with respect to matrix multiplication. Note that any two entries in the same row or column of M must be relatively prime and if a 0 occurs the other entries in its row and column must be ± 1.

A notation that will be useful is to define, given a rational x, $CF_+[x]$ is the continued fraction for x with an even number of terms and $CF_-[x]$ is that with an odd number. For example, from (6), $CF_+\left[\dfrac{429}{82}\right] = [5,4,3,6]$ and $CF_-\left[\dfrac{429}{82}\right] = [5,4,3,5,1]$. If $\sigma = \pm 1$ we write CF_σ for CF_+ or CF_- according as $\sigma = 1$ or -1. The following result is a lemma that is implicit in the literature but not in the form we need it.

<u>Lemma</u>   Let $M = \begin{pmatrix} a & b \\ c & d \end{pmatrix} \in G$, $c > d > 0$. Let $\sigma = \det M$,

$CF_\sigma\left[\dfrac{a}{c}\right] = [b_0, b_1, \ldots, b_{n-1}]$. Then $M = M[b_0, b_1, \ldots, b_{n-1}]$ and this representation - as an M

matrix - is unique.

<u>Proof</u>   Let $N = M[b_0, b_1, \ldots, b_{n-1}]$. We have to show $M=N$. By construction, $(-1)^n = \det N$ and $(-1)^n = \sigma = \det M$. Writing, as usual,

$N = \begin{pmatrix} A_{n-1} & A_{n-2} \\ B_{n-1} & B_{n-2} \end{pmatrix}$ we have $\dfrac{A_{n-1}}{B_{n-1}} = [b_0, b_1, \ldots, b_{n-1}] = \dfrac{a}{c}$. Both fractions are reduced and

$c > 0$, by hypothesis, and $B_{n-1} > 0$ follows from (3), (4). Thus $A_{n-1} = a$, $B_{n-1} = c$. Note

that $n \geq 2$ since $c > d > 0$ makes $c > 1$, $\dfrac{a}{c}$ not an integer. $\det M = \det N$ now implies

$ad - bc = aB_{n-2} - cA_{n-2}$ or $a(d - B_{n-2}) = c(b - A_{n-2})$. So c divides $a(d - B_{n-2})$, $(c,a) = 1$,

hence c divides $d - B_{n-2}$. Now $1 \leq d < c$ and also $1 \leq B_{n-2} \leq B_{n-1} = c$, so $|d - B_{n-2}| < c$,

and being divisible by c this forces $d = B_{n-2}$ and then $b = A_{n-2}$, hence $M = N$. For the uniqueness, if also $M = M[c_0, c_1, \ldots, c_{k-1}]$ then we would have $\dfrac{a}{c} = \dfrac{A'_{k-1}}{B'_{k-1}}$, where A'_{k-1},

B'_{k-1} are the quantities associated with $c_0, c_1, \ldots, c_{k-1}$. But this implies

$\dfrac{a}{c} = [b_0, b_1, \ldots, b_{n-1}] = [c_0, c_1, \ldots, c_{k-1}]$ and $(-1)^n = \det M = (-1)^k$ so $k \equiv n \pmod 2$. But $\dfrac{a}{c}$

has only two continued fractions, one with an even number of terms and the other with an

odd number, so $k = n$ and $c_i = b_i$ for all i.

Now we are interested in the set G^+ of all matrices in G that can be expressed as

a product of matrices $\begin{pmatrix} b & 1 \\ 1 & 0 \end{pmatrix}$ where b is a positive integer. Thus

$$G^+ = \{ M \in G \mid M = M[b_0, b_1, \ldots, b_{n-1}], \ n \geq 1, \ b_0, b_1, \ldots, b_{n-1}, \text{ all positive integers} \}. \quad (11)$$

G^+ is a subset -- not subgroup - of G and is clearly closed under multiplication: M,

$N \in G^+$ implies $MN \in G^+$. The identity matrix $I = \begin{pmatrix} 1 & 0 \\ 0 & 1 \end{pmatrix} \notin G^+$. Our first theorem

gives another characterization of G^+.

<u>Theorem</u> $M = \begin{pmatrix} a & b \\ c & d \end{pmatrix} \in G^+$ if and only if

$$a \geq b \geq d, \ a \geq c \geq d, \ d \geq 0. \quad (12)$$

If $M \in G^+$, let $\sigma = \det M$, $CF_\sigma\left[\dfrac{a}{c}\right] = [b_0, b_1, \ldots, b_{n-1}]$. Then $M = M[b_0, b_1, \ldots, b_{n-1}]$ is

the unique representation of M as a product of matrices of type $\begin{pmatrix} g & 1 \\ 1 & 0 \end{pmatrix}$, g a positive

integer.

<u>Proof</u>   Let, for the moment, G' be the set of matrices M satisfying the

inequalities (12). Given $M_i = \begin{pmatrix} a_i & b_i \\ c_i & d_i \end{pmatrix}$, $i = 1, 2$, in G' set

$$\begin{pmatrix} A & B \\ C & D \end{pmatrix} = M_1 M_2 = \begin{pmatrix} a_1 & b_1 \\ c_1 & d_1 \end{pmatrix}\begin{pmatrix} a_2 & b_2 \\ c_2 & d_2 \end{pmatrix} = \begin{pmatrix} a_1 a_2 + b_1 c_2 & a_1 b_2 + b_1 d_2 \\ c_1 a_2 + d_1 c_2 & c_1 b_2 + d_1 d_2 \end{pmatrix}.$$ Clearly $D \geq 0$.

$C \geq D$ is equivalent to $c_1(a_2 - b_2) + d_1(c_2 - d_2) \geq 0$. Since $c_1 \geq d_1 \geq 0$ and $a_2 \geq b_2$,

$c_2 \geq d_2$ it follows that $C \geq D$. In the same way one shows the other inequalities are satisfied to assure $\begin{pmatrix} A & B \\ C & D \end{pmatrix} \in G'$. Thus $M_1 M_2 \in G'$ and G' is closed under multiplication. Since $\begin{pmatrix} g & 1 \\ 1 & 0 \end{pmatrix}$, $g \geq 1$, is in G' it follows, recalling the definition of G^+, that $G^+ \subset G'$. We now show the reverse inclusion along with the uniqueness. Assume $M = \begin{pmatrix} a & b \\ c & d \end{pmatrix} \in G'$. If $c > d > 0$ then $a > c$, $\dfrac{a}{c} > 1$, not an integer, so M has the unique representation stated, via the lemma, and $b_0 = \left\lfloor \dfrac{a}{c} \right\rfloor \geq 1$, so $M \in G^+$. If $c > d > 0$ fails, then, since $M \in G'$, either $1 = c > d = 0$ or $1 = c = d > 0$. In the former case we must have $b = 1$, $M = \begin{pmatrix} a & 1 \\ 1 & 0 \end{pmatrix}$ with $a \geq 1$. Then $\det M = -1$, $CF_-\left[\dfrac{a}{1}\right] = [a]$ so $M = \mathrm{M}[a] \in G^+$ and the uniqueness is clear. If $c = d = 1$, $M = \begin{pmatrix} a & b \\ 1 & 1 \end{pmatrix}$,

$\det M = a - b = \pm 1$. But $a \geq b$ then forces $a = b + 1$, $\det M = 1$. Thus

$M = \begin{pmatrix} b+1 & b \\ 1 & 1 \end{pmatrix} = \begin{pmatrix} b & 1 \\ 1 & 1 \end{pmatrix}\begin{pmatrix} 1 & 1 \\ 1 & 0 \end{pmatrix} = \mathrm{M}[b,1]$ and $CF_+\left[\dfrac{a}{c}\right] = CF_+\left[\dfrac{b+1}{1}\right] = [b,1]$. Thus again $M \in G^+$ and the uniqueness is clear. This completes the proof of the theorem.

An example: $\begin{pmatrix} 22 & 135 \\ 7 & 43 \end{pmatrix}$, $\begin{pmatrix} 7 & 43 \\ 22 & 135 \end{pmatrix}$, $\begin{pmatrix} 43 & 135 \\ 7 & 22 \end{pmatrix}$ are in G but not G^+, while

$M = \begin{pmatrix} 135 & 22 \\ 43 & 7 \end{pmatrix} \in G^+$. Since $\det M = -1$ we calculate $CF_-\left[\dfrac{135}{43}\right] = [3,7,6]$ and so

$M = \begin{pmatrix} 3 & 1 \\ 1 & 0 \end{pmatrix}\begin{pmatrix} 7 & 1 \\ 1 & 0 \end{pmatrix}\begin{pmatrix} 6 & 1 \\ 1 & 0 \end{pmatrix}$. On the other hand $N = \begin{pmatrix} 135 & 113 \\ 43 & 36 \end{pmatrix} \in G^+$, $\det N = 1$, we

calculate $CF_+\left[\dfrac{135}{43}\right] = [3,7,5,1]$ and so $N = \begin{pmatrix} 3 & 1 \\ 1 & 0 \end{pmatrix}\begin{pmatrix} 7 & 1 \\ 1 & 0 \end{pmatrix}\begin{pmatrix} 5 & 1 \\ 1 & 0 \end{pmatrix}\begin{pmatrix} 1 & 1 \\ 1 & 0 \end{pmatrix}$.

3. Infinite Continued Fractions

We've seen that the process for obtaining the continued fraction of a rational number can be expressed in terms of the greatest integer function. As such, it can also be applied to an irrational number x. Set $x_0 = x$, $b_0 = \lfloor x_0 \rfloor$. Then $b_0 < x_0 < b_0 + 1$ so setting

$x_1 = \dfrac{1}{x_0 - b_0} > 1$ we have $x_0 = b_0 + \dfrac{1}{x_1}$. Again set $b_1 = \lfloor x_1 \rfloor$, $x_2 = \dfrac{1}{x_1 - b_1} > 1$ and

continuing this way we generate a sequence $x_0, x_1, x_2 \ldots$ of real numbers and integers

$b_0, b_1, b_2, \ldots$ such that for each $i \geq 1$, $x_{i-1} = b_{i-1} + \dfrac{1}{x_i}$, $x_i > 1$ and

$$x_0 = b_0 + \cfrac{1}{b_1 + \cfrac{1}{b_2 + \cfrac{\cdots}{b_{i-1} + \cfrac{1}{x_i}}}} = [b_0, b_1, \ldots, b_{i-1}, x_i]. \tag{13}$$

Since x_0 is irrational so is each x_i and the subsequent $x_{i+1} = \dfrac{1}{x_i - b_i}$ is always well-defined. (13) expresses x_0 as a finite continued fraction but the last term x_i is not an integer. Nevertheless, as noted earlier, the formalism of (3), (4) continues and one has for all $i \geq 1$

$$x_0 = [b_0, b_1, \ldots, b_{i-1}, x_i] = \frac{x_i A_{i-1} + A_{i-2}}{x_i B_{i-1} + B_{i-2}} \tag{14}$$

where the A_{i-1}, A_{i-2}, B_{i-1}, B_{i-2} depend only on the integers $b_0, b_1, \ldots, b_{i-1}$.

We now define an infinite continued fraction to be the formal infinite expression

$b_0 + \cfrac{1}{b_1 + \cfrac{1}{b_2 + \cdots}} = [b_0, b_1, b_2, \ldots]$ where b_0 is any integer and $b_1, b_2, \ldots$ an infinite sequence

of positive integers. Thus we have now associated with every irrational x an infinite continued fraction, denoted $CF[x]$. Consider a numerical example. Take $x_0 = x = -\sqrt{3}$.

Then $b_0 = -2$, $x_1 = \dfrac{1}{x_0 - b_0} = \dfrac{1}{2 - \sqrt{3}}$. At this point it is best to proceed with elementary algebra and rationalize the denominator: $x_1 = \dfrac{1}{2-\sqrt{3}}\dfrac{2+\sqrt{3}}{2+\sqrt{3}} = \sqrt{3}+2$, $b_1 = 3$. Leaving

the simple steps to the reader we then have $x_2 = \dfrac{\sqrt{3}+1}{2}$, $b_2 = 1$, $x_3 = \sqrt{3}+1$, $b_3 = 2$,

$x_4 = \dfrac{\sqrt{3}+1}{2}$, $b_4 = 1$. But wait - $x_4 = x_2$, $b_4 = b_2$ so then $x_5 = x_3$, $b_5 = b_3$ and everything

repeats after two steps. Doing it directly on a calculator, without rationalizing

denominators, I find $x_1 = 3.732050807$, $x_2 = 1.366025404$, $x_3 = 2.732050806$,

$x_4 = 1.366025407$, so $x_2 \neq x_4$! So we've found $CF\left[-\sqrt{3}\right] = \left[-2,3,1,2,1,2,...\right] =$

$\left[-2,3,\overline{1,2}\right]$, where the bar over the block 1,2 indicates its infinite repetition. In general, a

continued fraction is called periodic if it has the form $\left[c_0,c_1,...,c_{h-1},\overline{b_0,b_1,...,b_{k-1}}\right]$. The

block $b_0,b_1,...,b_{k-1}$ is called the period and $k \geq 1$ is the length of the period. $c_0,c_1,...,c_{h-1}$

is the pre-period and may be absent, in which case the continued fraction is called purely

periodic. Note that for $CF[x] = \left[\overline{b_0,b_1,...,b_{k-1}}\right]$ purely periodic, the b_i, x_i have period k:

$b_{i+k} = b_i$, $x_{i+k} = x_i$ for all $i \geq 0$ and $b_i = b_j$, $x_i = x_j$ whenever $i \equiv j(\mathrm{mod}\,k)$. It is tacitly

assumed that the period is minimal, does not consist of a repeated shorter block, and

starts 'as soon as possible.' Thus, though $CF\left[-\sqrt{3}\right]$ can be written $\left[-2,3,1,\overline{2,1,2,1}\right]$ the

period is 1,2 and the pre-period -2,3. Because of periodicity we can describe all the

terms. Euler found, for e the base of the natural logarithm,

$CF[e] = \left[2,1,2,1,1,4,1,1,6,1,1,8,1...\right]$ and though not periodic there is a pattern after the initial

$b_0 = 2$ and the terms are 'known.' On the other hand $CF[\pi] = \left[3,7,15,1,292,1,1,1,2...\right]$ and

no one has found a pattern to predict the terms. Apparently $CF[x]$ in not known in its

entirety for any algebraic number of degree >2.

So far infinite continued fractions are only formal objects - like infinite series

before one has defined the notions of convergence and sum. Considering again (14)

along with (10) we have $x_0 - \dfrac{A_{i-1}}{B_{i-1}} = \dfrac{A_{i-1}x_i + A_{i-2}}{B_{i-1}x_i + B_{i-2}} - \dfrac{A_{i-1}}{B_{i-1}} = \dfrac{\pm 1}{B_{i-1}\left(B_{i-1}x_i + B_{i-2}\right)}$. Thus

$$\left| x_0 - \frac{A_{i-1}}{B_{i-1}} \right| = \frac{1}{B_{i-1}\left(B_{i-1}x_i + B_{i-2}\right)} < \frac{1}{B_{i-1}^2}, \text{ since } x_i > 1 \text{ for } i \geq 1. \text{ But } B_0 = 1 \leq B_1 < B_2 < \cdots$$

shows $B_{i-1} \to \infty$, $\lim_{i \to \infty} \dfrac{A_{i-1}}{B_{i-1}} = x_0$. Thus the 'partial' finite continued fractions

$$\left[b_0, b_1, \ldots, b_{i-1}\right] = \frac{A_{i-1}}{B_{i-1}}$$ converge to a limit, the limit being x_0. Moreover it can be shown

that given any sequence of integers $b_0, b_1, \ldots$ with b_i positive for $i \geq 1$, then the finite

continued fractions $\left[b_0, b_1, \ldots, b_{i-1}\right]$ tend to a limit as $i \to \infty$, which is then defined to be

the value of the infinite continued fraction. In fact, the function that assigns to each

infinite continued fraction its value is a bijection onto the set of all irrational numbers; if

the value of $\left[b_0, b_1, \ldots\right]$ is x then $CF[x] = \left[b_0, b_1, \ldots\right]$. The rationals $\dfrac{A_{i-1}}{B_{i-1}}$ are called the

convergents of x, and give 'best' rational approximations; their study is the topic of

diophantine approximation.

It was not just luck that $CF\left[-\sqrt{3}\right]$ turned out to be periodic. The classical

theorem of Lagrange asserts that $CF[x]$ is periodic if and only if x is a quadratic

irrational - the root of a quadratic equation with rational coefficients. Furthermore,

Galois showed that $CF[x]$ is purely periodic if and only if x is reduced. Reduced means

$x > 1$ and $-1 < x' < 0$, where x' is the conjugate of x, the other root of the equation

satisfied by x. For example, from $CF\left[-\sqrt{3}\right]$ we had $x_2 = \dfrac{\sqrt{3}+1}{2} = [\overline{1,2}]$ and $x_2 > 1$,

$-1 < x_2' = \dfrac{-\sqrt{3}+1}{2} < 0$.

Let I denote the set of irrational numbers, I_2 the quadratic irrationals and R the

reduced ones. The x_i generated by $CF[x]$ are called the 'complete quotients' of x. The

construction shows that if $CF[x] = \left[b_0, b_1, b_2, \ldots\right]$ then $CF[x_i] = \left[b_i, b_{i+1}, b_{i+2}, \ldots\right]$, which

might be called a 'tail' of $CF[x]$. Recall the group G; $M = \begin{pmatrix} a & b \\ c & d \end{pmatrix} \in G$ acts on I by sending x into $M(x) = \dfrac{ax+b}{cx+d}$. The map $G \times I \to I$ that sends (M,x) to $M(x)$ is a

standard group action. Under it I breaks up into G-orbits or equivalence classes; for

$x, y \in I$ we say $y \sim x$ if $y = M(x)$ for some $M \in G$. Thus (14) shows $x_i \sim x$ and Serret

proved the beautiful result that describes the G-equivalence class of x in terms of

continued fractions: $y \sim x$ if and only if for some integers $i, j \geq 0$, $y_j = x_i$, that is y, x

have identical 'tails' from some point on.

From now on we shall be concerned only with quadratic irrationals and our goal

is to show how the periodic structure of $CF[x]$ reflects arithmetic properties. Numbers

$x \in I_2$ are classified by their discriminant. This word requires careful definition since it

is used in different ways in various contexts. x is the root of a quadratic equation with

rational coefficients. By multiplying, dividing the equation by integers one can arrange it

to be in the form $aX^2 + bX + c = 0$ with a, b, c integers, $a>0$ and $\gcd(a,b,c)=1$; X is

simply an indeterminate. This equation is now uniquely determined and we call it the

standard equation for x. Then $d = b^2 - 4ac$ is called the discriminant of x, denoted

d=disc(x). For example, with $x = -\sqrt{3}$, the standard equation is $X^2 - 3 = 0$ and d=12.

Note that any integer d arriving as disc(x) is positive, since x is real, not a (perfect)

square, since x is irrational and $d \equiv b^2 \equiv 0$ or $1 \pmod 4$. We call any positive integer d

not a square and $\equiv 0$ or $1 \pmod 4$ a discriminant. In fact, d=disc(x) where x is a root of

$$X^2 - \left(\frac{d}{4}\right) = 0,\ \text{if } d \equiv 0 \pmod 4,\ \text{and a root of } X^2 - X - \left(\frac{d-1}{4}\right) = 0,\ \text{if } d \equiv 1 \pmod 4.$$

The first few discriminants are 5, 8, 12, 13, 17, 20, 21. Note, for later use, that if the

discriminant $d = s^2 D$, s, D integers and $D \equiv 0$ or $1 \pmod 4$, then D is also a discriminant.

Consider $y = \dfrac{1}{x} = \begin{pmatrix} 0 & 1 \\ 1 & 0 \end{pmatrix}(x)$. If $aX^2 + bX + c = 0$ is the standard equation for x

and d=disc(x), then $\pm(cX^2 + bX + a) = 0$ is the equation for y, so disc(y)=d. if

$z = x - m = \begin{pmatrix} 1 & -m \\ 0 & 1 \end{pmatrix}(x)$, m an integer, then $x = z + m$ satisfies

$a(z+m)^2 + b(z+m) + c = 0$ from which one deduces that the standard equation for z is

$AX^2 + BX + C = 0$ where $A=a$, $B=2am+b$, $C=am^2+bm+c$ whence

$\mathrm{disc}(z) = B^2 - 4AC = b^2 - 4ac = d$. Easier to see is that $-x = \begin{pmatrix} -1 & 0 \\ 0 & 1 \end{pmatrix}(x)$ has $\mathrm{disc}(-x)=d$.

Since $\begin{pmatrix} 1 & b \\ 0 & 1 \end{pmatrix}\begin{pmatrix} 0 & 1 \\ 1 & 0 \end{pmatrix} = \begin{pmatrix} b & 1 \\ 1 & 0 \end{pmatrix}$ and the results of Section 2 enable one to show that any

$M \in G$ may be expressed as a product of matrices of the type $\begin{pmatrix} b & 1 \\ 1 & 0 \end{pmatrix}, \begin{pmatrix} -1 & 0 \\ 0 & 1 \end{pmatrix}, \begin{pmatrix} 0 & 1 \\ 1 & 0 \end{pmatrix}$ it

follows that if $y = M(x)$, $M \in G$, then $\mathrm{disc}(y) = \mathrm{disc}(x)$. Thus $\mathrm{disc}(x)$ is an invariant of

the G-equivalence class of x. We also see directly from $x = x_0 = b_0 + \dfrac{1}{x_1}$, that $\mathrm{disc}(x) =$

$\mathrm{disc}(x_1)$ and in general all the complete quotients x_i of x have the same discriminant.

Thus we've shown that for $x, y \in I_2$, $y \sim x$ implies $\mathrm{disc}(y) = \mathrm{disc}(x)$. However,

the converse is false, as we shall see. Given a discriminant d, let $I_2(d)$ be the set of

$x \in I_2$ having $\mathrm{disc}(x) = d$ and $P(d)$ the set of reduced x having $\mathrm{disc}(x) = d$. By what we

have just shown, $I_2(d)$ is a union of G-orbits. By Lagrange's theorem every $x \in I_2$ has

some x_i which is purely periodic, hence in R by Galois, thus every $x \in I_2(d)$ is G-

equivalent to some $y \in R(d)$. In other words every G-orbit in $I_2(d)$ meets $R(d)$. The

number of G-orbits in $I_2(d)$ is called the class number of d, denoted $h = h(d)$ and we

will show $h(d)$ is finite by showing $R(d)$ is finite. But more than showing $R(d)$ finite,

we want to give an easy algorithm for listing the members of $R(d)$.

      __Theorem__   Let d be a discriminant. The set $R(d)$ is finite and its members

may be enumerated as follows.

Let g be that one of $\lfloor \sqrt{d} \rfloor$, $\lfloor \sqrt{d} \rfloor - 1$ that is $\equiv d \pmod 2$. For

$p = g, g-2, g-4, \ldots$ and >0 (so the last p is 2 or 1 according as d is even or odd)

determine all ordered triples (p, q, q^*) of positive integers satisfying

q, q^* are even, $d = p^2 + qq^*$, $g - p + 2 \le q \le g + p$ and $\gcd(\tfrac{1}{2}q, p, \tfrac{1}{2}q^*) = 1$.   (15)

Assign to $(p,q,q^{\cdot})$ the number $x = \dfrac{\sqrt{d}+p}{q}$. Then $x \in R(d)$ and in this way

each member of $R(d)$ is listed exactly once. By this arrangement the standard equation

for x is $\left(\tfrac{1}{2}q\right)X^2 - pX - \left(\tfrac{1}{2}q^{\cdot}\right) = 0$.

<u>Proof</u>   Suppose $x \in R(d)$ and has standard equation $aX^2 + bX + c = 0$.

By the quadratic formula the roots are $\dfrac{-b+\sqrt{d}}{2a}$, $\dfrac{-b-\sqrt{d}}{2a}$ and the former is the larger

root - by our standardization $a>0$. But x reduced says $x>1$, $-1<x'<0$ so x is the larger

root, $x = \dfrac{-b+\sqrt{d}}{2a}$. Also $0 > xx' = \dfrac{c}{a}$ so $c<0$ and $-\dfrac{b}{a} = x+x' > 1-1 = 0$ so $b<0$.

Setting $p = -b$, $q = 2a$, $q^{\cdot} = -2c$ we have p,q,q^{*} positive integers, q,q^{*} even and

$d = p^2 + qq^{\cdot}$. With this notation $x = \dfrac{\sqrt{d}+p}{q}$ (we write the $\sqrt{d}$ first to emphasize the

role of the discriminant) and the standard equation is $\left(\tfrac{1}{2}q\right)X^2 - pX - \left(\tfrac{1}{2}q^{\cdot}\right) = 0$. We now

reverse direction and ask, given p,q,q^{*} positive integers, with q,q^{*} even, $d = p^2 + qq^{\cdot}$

and $\gcd\left(\tfrac{1}{2}q, p, \tfrac{1}{2}q^{\cdot}\right) = 1$ will $x = \dfrac{\sqrt{d}+p}{q} \in R(d)^{\cdot}$? Note that x is indeed a root of the

standard equation $\left(\tfrac{1}{2}q\right)X^2 - pX - \left(\tfrac{1}{2}q^{\cdot}\right) = 0$ and has discriminant d, so we only need to

check if x is reduced. Now $x>1$ is equivalent to $q < \sqrt{d} + p$, $x' = \dfrac{-\sqrt{d}+p}{q} < 0$ is

equivalent to $p < \sqrt{d}$ and $x' > -1$ is equivalent to $\sqrt{d} - p < q$. Combining these we see

that

$$0 < \sqrt{d} - p < q < \sqrt{d} + p \qquad (16)$$

must be satisfied. We now want to eliminate the $\sqrt{d}$. Since $d = p^2 + qq^{\cdot}$, q,q^{*} even, we

have $p \equiv d \pmod 2$ and $0 < p < \sqrt{d}$, so that p is $\le$ the largest integer that is $< \sqrt{d}$ and

$\equiv d \pmod 2$, which is precisely g as defined in the theorem. Thus the possible p are as

described. By definition of g, $\sqrt{d} = g + \vartheta$ with $0 < \vartheta < 2$ and (16) is equivalent to

$g - p + 9 < q < g + p + 9$. Noting that g-p, q, g+p are all even this is equivalent to $g - p + 2 \le q \le g + p$.

Our description shows that the conditions of (15) are precisely what is needed to guarantee $x = \dfrac{\sqrt{d} + p}{q} \in R(d)$. Since there are only finitely many values of p and for each p only finitely many values of q, q^* this shows $R(d)$ finite and the proof is complete.

Note that $qq^* = d - p^2 = \left(\sqrt{d} - p\right)\left(\sqrt{d} + p\right)$ along with (16) shows that also $\sqrt{d} - p < q^* < \sqrt{d} + p$ so our conditions are symmetric with respect to q, q^*. Along with $x = \dfrac{\sqrt{d} + p}{q}$ in $R(d)$ we also have $\dfrac{\sqrt{d} + p}{q^*} = -\dfrac{1}{x'} \in R(d)$. If we set $x^* = -\dfrac{1}{x'}$, then $x \to x^*$ is an involution of $R(d)$.

Note from (15) that q=2 can occur only when g=p. Since $d = g^2 + 2q^*$ determines the even integer q^* uniquely and the gcd condition is obviously satisfied, we see that $\dfrac{\sqrt{d} + g}{2}$, which we denote as z or $z(d)$ is the unique member of $R(d)$ with q=2.

Suppose in listing members of $R(d)$ according to the theorem one has p, q, q^* satisfying all conditions of (15) except $\gcd\left(\frac{1}{2}q, p, \frac{1}{2}q^*\right) = s > 1$. Then $q = sQ$, $p = sP$, $q^* = sQ^*$, with Q, Q^* even and $D = P^2 + QQ^*$, $d = s^2 D$ so D is a discriminant and P, Q, Q^* satisfy (15) for D. The inequalities in (15) are not so obvious though, but the proof showed they are equivalent to (16) and the terms there are homogeneous in s, so we obtain $0 < \sqrt{D} - P < Q < \sqrt{D} + P$. Thus $\dfrac{\sqrt{d} + p}{q} = \dfrac{\sqrt{D} + P}{Q} \in R(D)$. As a consequence, if d cannot be written as $s^2 D$, $s > 1$, $D \equiv 0$ or $1 \pmod 4$ the gcd condition of (15) needn't be checked since it always holds. Such d are called fundamental (or field) discriminants; the first few are 5, 8, 12, 13, 17, 21.

It's worth remarking that when presented with an $x \in I_2$, disc(x) is often not obvious. For example, $x = \dfrac{\sqrt{7}+2}{4} > 1$ and $-1 < x' = \dfrac{-\sqrt{7}+2}{4} < 0$, so $x \in R$. What is disc(x)? Find the standard equation: $(4x-2)^2 = 7$, $16x^2 - 16x - 3 = 0$, $d = 448$,

$x = \dfrac{\sqrt{448}+16}{32} \in R(448)$. From now on we always write a reduced x in the form

described in the theorem, $x = \dfrac{\sqrt{d}+p}{q}$, d=disc(x), and call this the standard form of x.

For d a discriminant we denote $|R(d)|$, the number of members of $R(d)$, by $r(d)$. Since $z(d) \in R(d)$, $r(d) \geq 1$. We consider some numerical examples.

$\underline{d=37}$ is a prime, so the gcd condition need not be checked. $\lfloor \sqrt{37} \rfloor = 6$, g=5 and the range of p is p=5, 3, 1.

$\underline{p=5}$ Solve $qq^* = 37 - 5^2 = 12$ with q, q^* even and $2 \leq q \leq 10$. There are two solutions, $12 = 2*6 = 6*2$.

$\underline{p=3}$ Solve $qq^* = 37 - 3^2 = 28$ with q, q^* even and $4 \leq q \leq 8$. There are no solutions.

$\underline{p=1}$ Solve $qq^* = 37 - 1^2 = 36$ with q, q^* even and $6 \leq q \leq 6$. There is only one solution, $36 = 6*6$. Thus $r(37) = 3$,

$R(37) = \left\{ z = \dfrac{\sqrt{37}+5}{2}, x = \dfrac{\sqrt{37}+5}{6}, y = \dfrac{\sqrt{37}+1}{6} \right\}$. Note that $x = z^*$ and $y = y^*$.

$\underline{d=72}$ (not a fundamental discriminant) $\lfloor \sqrt{72} \rfloor = 8$, $g = 8$, p=8, 6, 4, 2.

$\underline{p=8}$ $qq^* = 72 - 8^2 = 8$, $2 \leq q \leq 16$, has solutions, $8 = 2*4 = 4*2$, $\gcd\left(\tfrac{1}{2}(2), 8, \tfrac{1}{2}(4)\right) = 1$.

$\underline{p=6}$ $qq^{\bullet} = 72 - 6^2 = 36$, $4 \le q \le 14$, has one solution, $36 = 6*6$, but $\gcd\left(\tfrac{1}{2}(6),6,\tfrac{1}{2}(6)\right) = 3$; reject.

$\underline{p=4}$ $qq^{\bullet} = 72 - 4^2 = 56$, $6 \le q \le 12$, no solutions.

$\underline{p=2}$ $qq^{\bullet} = 72 - 2^2 = 68$, $8 \le q \le 10$, no solutions. Thus $r(72) = 2$,

$$R(72) = \left\{ z = \frac{\sqrt{72} + 8}{2}, x = \frac{\sqrt{72} + 8}{4} \right\}.$$

The function $r(d)$ is extremely erratic. The reader may wish to calculate $r(281)$ and $r(293)$. For larger d one has $r(1109) = 11$, $r(1129) = 51$, $r(1181) = 9$, $r(1201) = 53$; all these discriminants are prime. These results were not obtained by hand but by a computer program implementing the algorithm of the theorem. I want to thank my Lehman College student Mr. E. Moss for his assistance with the computer work.

If $x \in R(d)$, $CF[x] = \left[\overline{b_0, b_1, \ldots, b_{k-1}}\right]$ is purely periodic, $x_0 = x$, $x_1, \ldots x_{-1}$ are all $\sim x$ and in $R(d)$. Suppose also $y \in R(d)$, $CF[y] = \left[\overline{c_0, c_1, \ldots, c_{m-1}}\right]$. If $y \sim x$, by Serret's theorem $y_j = x_i$, for some $j, i \ge 0$. But then for any $n \ge 0$, $y_{j+n} = x_{i+n}$ so choose n so that $j + n = sm$, a multiple of m. By periodicity $y_{sm} = y_0 = y$, so $y = x_{i+n}$, and reducing $i + n \bmod k$, say $i + n \equiv \ell(\bmod k)$, $0 \le \ell \le k - 1$, $x_{i+n} = x_\ell$, or $y = x_\ell$ is one of the k distinct complete quotients of x. Thus from $CF[x] = \left[\overline{b_0, b_1, \ldots, b_{k-1}}\right]$ we deduce that the G-orbit of x meets $R(d)$ in the set $\{x_0 = x, x_1, \ldots x_{k-1}\}$. We refer to this as the cycle of x, since it comes with a specific ordering determined by $CF[x]$. If one had started with some x_i, $0 \le i \le k - 1$, $CF[x_i] = \left[\overline{b_i, b_{i+1}, \ldots, b_{k-1}, b_0, \ldots, b_{i-1}}\right]$ and the cycle of x_i is $\{x_i, x_{i+1}, \ldots, x_{k-1}, x_0 = x, \ldots, x_{i-1}\}$. Thus the ordering is determined uniquely up to a cyclic permutation. If one then partitions $R(d)$ into cycles, the number of cycles is precisely $h(d)$, the class number of d, defined earlier as the number of G-orbits of $I_2(d)$.

For example, starting with $z = z_0 = \dfrac{\sqrt{37}+5}{2}$ obtained above one finds

$CF[z_0] = [\overline{5,1,1}]$, $z_1 = x$, $z_2 = y$, in the notation of that example, $\{z, x, y\}$ is the only cycle and $h(37) = 1$. Observe that the period 5, 1, 1 after the first term is a palindrome, 1, 1, which reads the same backwards as forwards. This phenomenon always occurs for $CF[z]$ for any d. $CF[y] = [\overline{1,5,1}]$ has the whole period as a palindrome, and $y = \dfrac{\sqrt{37}+1}{6}$, $37 = 1^2 + 6^2$. Again this will always happen when $q = q^*$, $d = p^2 + q^2$. In the literature these observations are usually made in connection with the study of $CF[\sqrt{n}]$, n not a square. We have examined these special symmetries in a recent paper [3].

The function $h(d)$ is also quite erratic. Something is known concerning its divisibility by powers of 2 from Gauss' genus theory of binary quadratic forms. Gauss also conjectured that $h(d) = 1$ occurs infinitely often and this still remains an open question. For the d mentioned earlier, $h(1109) = 1$, $h(1129) = 9$, $h(1181) = 1$, $h(1201) = 1$.

4. Quadratic Fields   Significant aspects of purely periodic continued fractions can only be understood in the context of algebraic number theory. Actually, all that is needed are some basic notions of that theory as they pertain to real quadratic fields. We shall take these for granted, and recall a few points to fix the ideas.

Let K be a real quadratic field, O_K the ring of algebraic integers in K. If $x, y \in K$ are linearly independent over the rationals they form a basis for K/Q; every $\lambda \in K$ has a unique representation as $\lambda = rx + sy$, $r, s \in Q$. If we restrict ourselves to those λ for which $r, s \in Z$ the resulting set L is a subgroup of the additive group of K and is a free abelian group of rank 2. We write $L = \langle x, y \rangle$ to indicate the construction of L via the basis x, y for L/Z. L is called a (full) module in K. An order in K is a full module that is also a subring of K and contains the ring Z. (This has nothing to do with the notion of order in the sense of 'less than' or 'greater than.') If L is a full module the set

$\{\xi \in K | \xi L \subset L\}$ is called the coefficient ring L. Here ξL is the set of all elements $\xi \lambda$, λ ranging over L. The coefficient ring of L is in fact an order.

Every K is obtained uniquely (up to isomorphism) as $K = Q(\sqrt{m})$ where m is an integer>1 and square free. Set

$$\omega = \begin{cases} \dfrac{1+\sqrt{m}}{2} & , \text{ if } m \equiv 1(\text{mod } 4) \\ \sqrt{m} & , \text{ if } m \equiv 2,3(\text{mod } 4) \end{cases} \tag{17}$$

Then 1, ω is a basis for K/Q and $O_K = \langle 1, \omega \rangle$. Every order O is contained in O_K and is uniquely determined by its index $f = [O_K : O]$ and then $O = \langle 1, f\omega \rangle$. For $x \in K$, irrational, $x' \neq x$ and $x' \in K$. Defining $x' = x$ for $x \in Q$, the map $x \to x'$ is the non-trivial automorphism of K/Q; it and the identity map constitute the Galois Group of K/Q. The discriminant of O, disc(O) is defined to be $\begin{vmatrix} x & y \\ x' & y' \end{vmatrix}^2 = (xy' - x'y)^2$, where x, y is a basis for O. It is independent of the choice of basis. Using the basis 1, $f\omega$ one obtains disc$(O) = f^2(\omega - \omega')^2$. By (17) then

$$disc(O) = \begin{cases} f^2 m & , \text{ if } m \equiv 1(\text{mod } 4) \\ 4f^2 m & , \text{ if } m \equiv 2,3(\text{mod } 4). \end{cases} \tag{18}$$

Thus disc$(O)=d$, a positive integer, not a perfect square; m is the square free part of d. For $f=1$, $O = O_K$, d is m or $4m$, according as $m \equiv 1(\text{mod } 4)$ or not. This also is called the discriminant of K, denoted d_K. The numbers d_K are the fundamental or field discriminants mentioned earlier.

It should now be clear that if we allow O to range over all orders of all quadratic fields the resulting set of integers $\{disc(O)\}$ is precisely the set of positive integers not perfect squares that are $\equiv 0$ or $1(\text{mod } 4)$, i.e. the set of numbers called 'discriminants' in the previous section arising as disc$(x)= b^2 - 4ac$. Also, given d the numbers m, f are uniquely recoverable, showing that there is a unique O with disc$(O)=d$. We denote by O_d the order with discriminant d. For example, what is O_{12000}?

Write $12000 = 2^4 \cdot 5^2 \cdot 2 \cdot 3 \cdot 5$, so $m=30$ is the square free part. $30 \equiv 2 \pmod 4$ so write $12000 = 4(2 \cdot 5)^2 \cdot 30$ showing $f=10$. Thus $O_{12000} = \langle 1, 10\sqrt{30} \rangle$ contained in $Q(\sqrt{30})$.

Since O is determined by its discriminant (we find) it is preferable to express the elements of O_d 'canonically' in terms of d, rather than use the somewhat arbitrary basis $1, f\omega$. We claim

$$\lambda \in O_d \text{ if and only if } \lambda = \frac{r + s\sqrt{d}}{2}, \ r,s \in Z \text{ and } r \equiv sd \pmod 2 \tag{19}$$

Note the congruence condition, so this is not a basis representation.

To prove the claim, suppose $m \equiv 1 \pmod 4$. Then $d = f^2 m \equiv f^2 \equiv f \pmod 2$ and

$$f\omega = \frac{f + f\sqrt{m}}{2} = \frac{f + \sqrt{d}}{2}. \text{ Hence } \lambda \in O_d \text{ iff } \lambda = t + sf\omega = \frac{(2t + sf) + s\sqrt{d}}{2} \text{ with } t,s \in Z,$$

or iff $\lambda = \frac{r + s\sqrt{d}}{2}$ with $r,s \in Z$ and $r \equiv sf \equiv sd \pmod 2$. The cases $m \equiv 2,3 \pmod 4$ are

proved similarly. Note that with $\lambda = \frac{r + s\sqrt{d}}{2} \in O_d$ then $N\lambda$, the norm of λ,

$$= \lambda\lambda' = \frac{r^2 - ds^2}{4} \text{ and this is an integer. Also we see directly that } \lambda \in O_d \text{ iff } \lambda' \in O_d.$$

The units in K are the algebraic integers τ for which $\dfrac{1}{\tau}$ is also an algebraic

integer. Thus the units are the invertible elements in O_K and form a group U_K, a

subgroup of K^*, the multiplicative group of non-zero elements of K. $\tau \in O_K$ is a unit iff

$N\tau = \tau\tau' = \pm 1$ which is iff its standard equation has the form $X^2 + bX \pm 1 = 0$.

If O_d is an order in K and $\tau \in O_d \cap U_K$ then $\dfrac{1}{\tau} = \pm\tau'$ shows that τ is an invertible

element of O_d. Thus it follows that $O_d \cap U_K$ is the group of invertible elements in O_d, denoted U_d, called the unit group for discriminant d. By (19) we have that

$\tau = \dfrac{r + s\sqrt{d}}{2} \in U_d$ iff r,s satisfy the Pell equation $r^2 - ds^2 = \pm 4$. Now it is not at all

obvious that this equation has non-trivial solutions, i.e. other than $r = \pm 2$, $s = 0$, so not at

all obvious that there are any units τ other than the trivial ones ± 1, which are the units

for the rational field Q with the ring of integers Z. What we want to show here is that the

existence of non-trivial units follows as a direct consequence of results about continued

fractions, and the two theories shed light on each other.

Let $x \in K$ be irrational, L the full module $\langle x,1 \rangle$. $x, 1$ is a basis for K/Q so

multiplication by $\lambda \in K$ is represented by

$$\lambda \cdot x = \alpha x + \beta \cdot 1 = \alpha x + \beta \qquad (20)$$

$$\lambda = \lambda \cdot 1 = \gamma x + \delta \cdot 1 = \gamma x + \delta$$

with α, β, γ, δ rational. As long as $\lambda \neq 0$, the matrix $M = \begin{pmatrix} \alpha & \beta \\ \gamma & \delta \end{pmatrix}$ is invertible;

$M \in GL(2,Q)$. We denote $M = \rho(\lambda)$ and then the map $\rho: K^\times \to GL(2,Q)$ is an injective

homomorphism - the proof makes use of the commutativity of $K^\times$. Recall that

$N\lambda = \det \rho(\lambda)$. ρ really should be denoted ρ_x, to show its dependence on the basis $x,1$.

Let $aX^2 + bX + c = 0$ be the standard equation for x and $d = \mathrm{disc}(x)$. Let O be the

coefficient ring of L; by (20), $\lambda \in O$ iff $\alpha,\beta,\gamma,\delta \in Z$. Thus for $\lambda \in O$, $\lambda = \gamma x + \delta$ and

$(\gamma x + \delta)x = \alpha x + \beta$, or $\gamma x^2 + (\delta - \alpha)x - \beta = 0$. Comparing with the standard equation this

forces $\gamma = na$ for some integer n, $\lambda = n(ax) + \delta \in \langle ax,1 \rangle$. Conversely, one sees that

$\lambda \in \langle ax,1 \rangle$ implies $\lambda \in O$. Thus $O = \langle ax,1 \rangle$ and

$\mathrm{disc}(O) = \begin{vmatrix} ax & 1 \\ ax' & 1 \end{vmatrix}^2 = a^2(x - x')^2 = a^2 \left(\dfrac{\pm\sqrt{d}}{a} \right)^2 = d$. This proves the crucial connection

$$\text{if } \mathrm{disc}(x) = d \text{ then } O_d \text{ is the coefficient ring of } \langle x,1 \rangle. \qquad (21)$$

Let Ω be the set of matrices $M = \begin{pmatrix} \alpha & \beta \\ \gamma & \delta \end{pmatrix} \in GL(2,Q)$ having $\alpha,\beta,\gamma,\delta \in Z$. For a

non-zero integer n let $\Omega_n = \{ M \in \Omega \mid \det(M) = n \}$; so our group $G = \Omega_1 \cup \Omega_{-1}$. Suppose now $\lambda \in O$, as above, $N\lambda = n$ and $\rho(\lambda) = M = \begin{pmatrix} \alpha & \beta \\ \gamma & \delta \end{pmatrix}$. Then $M \in \Omega_n$ and

$$x = \frac{\lambda x}{x} = \frac{\alpha x + \beta}{\gamma x + \delta} = M(x) \, ; x \text{ is fixed under the action of } M \text{ as a Mobius transformation.}$$

Conversely, suppose $M = \begin{pmatrix} \alpha & \beta \\ \gamma & \delta \end{pmatrix} \in \Omega_n$ and $x = M(x) = \dfrac{\alpha x + \beta}{\gamma x + \delta}$. Set $\lambda = \gamma x + \delta$, then

$\lambda x = (\gamma x + \delta)x = \alpha x + \beta$ so $\lambda \in O$, $M = \rho(\lambda)$. If we make the definition $\Lambda_n = \{\lambda \in O \mid$

$N\lambda = n\}$, and for any set of matrices $S \subset GL(2, Q)$, $S_x = \{M \in S \mid M(x) = x\}$ then we have

shown that ρ maps Λ_n one-one onto $\Omega_{n,x}$. In particular with $U = U_d = \Lambda_1 \cup \Lambda_{-1}$, ρ

maps U bijectively onto $\Omega_{1,x} \cup \Omega_{-1,x} = G_x$. Since ρ is also a homomorphism we have

that $\rho : U \to G_x$ is an isomorphism. The inverse isomorphism we denote by $\varphi : G_x \to U$;

for $M = \begin{pmatrix} \alpha & \beta \\ \gamma & \delta \end{pmatrix} \in G_x$, $\varphi(M) = \gamma x + \delta \in U$. Thus if the group structure of G_x is

determined then so is that of U.

So far the only property of x that was used is that disc$(x) = d$, so we are free to

choose any x subject to this condition. We choose $x \in R(d)$, $CF[x] = \overline{[b_0, b_1, \ldots, b_{k-1}]}$.

For $i \geq 1$ set $M_{i-1} = M[b_0, b_1, \ldots, b_{i-1}] = \begin{pmatrix} A_{i-1} & A_{i-2} \\ B_{i-1} & B_{i-2} \end{pmatrix}$. By (14), $x = M_{i-1}(x_i)$. With $i = k$,

set $M = M_{k-1} = M[b_0, b_1, \ldots, b_{k-1}]$ and $x_k = x_0 = x$ we have $x = M(x)$, $M \in G_x$. Since

$M \in G^+$, it has infinite order and this proves the existence of non-trivial units. Let

$\Gamma = \{\pm I\} \times \{M''\}_{n \in Z} = \{\pm M''\}$. Clearly $\Gamma \subset G_x$ and we now show that actually $\Gamma = G_x$.

Suppose $V \in G_x$, $V \neq \pm I$. Let $\varphi(V) = \xi \neq 1$. Then $\varphi(\pm V) = \pm \xi$, $\varphi(\pm V^{-1}) = \pm \dfrac{1}{\xi}$ and one

of these is > 1. Let W be that one of $\pm V$, $\pm V^{-1}$ for which $\varphi(W) > 1$. It will suffice now

to show $W \in \Gamma$. Suppose $W = \begin{pmatrix} \alpha & \beta \\ \gamma & \delta \end{pmatrix}$; set $\eta = \varphi(W) = \gamma x + \delta$, $\eta > 1$ and $\eta \in U$. Thus

$\eta' = \pm \dfrac{1}{\eta}$, $|\eta'| < 1$, then $0 < \eta - \eta' = \gamma(x - x')$ shows $\gamma > 0$ (recall x here is reduced).

Also $\delta = \eta' - \gamma x' = \eta' + |\gamma x'| > \eta' > -1$ shows $\delta \geq 0$. Also $-1 < x'$ implies

$-\gamma + \delta < \gamma x' + \delta = \eta' < 1$, hence $-\gamma + \delta \leq 0$, $\delta \leq \gamma$. Altogether then $0 \leq \delta \leq \gamma$.

We now consider various cases.

If $0 < \delta < \gamma$ then by the lemma of section 2, setting $\sigma = \det W$,

$$CF_\sigma\left[\frac{\alpha}{\gamma}\right] = [c_0, c_1, \ldots, c_{n-1}], \text{ we have } W = M[c_0, c_1, \ldots, c_{n-1}].$$ Again using (14), the value of

the continued fraction $[c_0, c_1, \ldots, c_{n-1}, x] = W(x)$ and $W(x) = x$, since $W \in G_x$. Thus

$x = [c_0, c_1, \ldots, c_{n-1}, x]$, which says $[\overline{b_0, b_1, \ldots, b_{k-1}}] = [c_0, c_1, \ldots, c_{n-1}, \overline{b_0, b_1, \ldots, b_{k-1}}]$. Since the

continued fraction of an irrational is unique this forces $c_0, c_1, \ldots, c_{n-1}$ to be a j-fold

repetition of the period block $b_0, b_1, \ldots, b_{k-1}$, for some $j \geq 1$. Thus $W = M^j \in \Gamma$.

Remaining are the cases $0 = \delta < \gamma = 1$ and $0 < \delta = \gamma = 1$. Rather than consider

these directly it is easier to consider separately the two cases (a) $N\eta = -1$ (b) $N\eta = 1$. For

(a) $N\eta = -1$, $\eta' = -\dfrac{1}{\eta} < 0$. Our previous analysis showed $-\gamma + \delta < \eta'$ hence now

$-\gamma + \delta < 0$, $\delta < \gamma$ so we must have $\delta = 0$, $\gamma = 1$, $W = \begin{pmatrix} \alpha & \beta \\ 1 & 0 \end{pmatrix}$. But $\eta = \varphi(W)$,

$\rho(\eta) = W$, $\det W = N\eta = -1$ so $\beta = 1$. Thus $W = \begin{pmatrix} \alpha & 1 \\ 1 & 0 \end{pmatrix}$, $x = W(x) = \alpha + \dfrac{1}{x}$, hence

$\alpha = \lfloor x \rfloor$, $CF[x] = [\overline{\alpha}]$. Hence $k = 1$, $b_0 = \alpha$, $W = M$.

For (b), $\eta' = \dfrac{1}{\eta} > 0$. We saw previously that $\delta > \eta'$, hence now $\delta > 0$, so we

must have $\delta = \gamma = 1$, $W = \begin{pmatrix} \alpha & \beta \\ 1 & 1 \end{pmatrix}$. But $\det W = N\eta = 1$ then implies $\alpha - \beta = 1$,

$W = \begin{pmatrix} \beta + 1 & \beta \\ 1 & 1 \end{pmatrix}$, so $x = W(x) = \beta + \dfrac{x}{x+1} = \beta + \dfrac{1}{1 + \frac{1}{x}}$. Thus $\beta = \lfloor x \rfloor$ and

$CF[x] = [\beta, 1, \beta, 1, \ldots]$. If $\beta > 1$ then $CF[x] = [\overline{\beta, 1}]$, so $k = 2$; $b_0 = \beta$, $b_1 = 1$,

$$M = \mathrm{M}[\beta,1] = \begin{pmatrix} \beta & 1 \\ 1 & 0 \end{pmatrix}\begin{pmatrix} 1 & 1 \\ 1 & 0 \end{pmatrix} = \begin{pmatrix} \beta+1 & \beta \\ 1 & 1 \end{pmatrix} = W. \text{ If } \beta = 1,\ CF[x] = [\bar{1}],\ k = 1,\ b_0 = 1,$$

$$M = \mathrm{M}[1] = \begin{pmatrix} 1 & 1 \\ 1 & 0 \end{pmatrix} \text{ and } W = \begin{pmatrix} 2 & 1 \\ 1 & 1 \end{pmatrix} = M^2. \text{ This completes the proof that } G_x = \Gamma.$$

Now $U = \varphi(G_x) = \{\pm\varepsilon^n\}_{n\in Z}$ where $\varepsilon = \varphi(M) > 1$. ε clearly is the smallest unit >1 and is called the fundamental unit for the discriminant d. We now write using (19), and keep this notation throughout, $\varepsilon = \dfrac{t+u\sqrt{d}}{2}$, $t,u \in Z$, $t \equiv ud \pmod 2$.

To summarize: to find ε choose any $x \in R(d)$ - there is always at least

$z = \dfrac{\sqrt{d}+g}{2}$ at hand - determine $CF[x] = [\overline{b_0,b_1,\ldots,b_{k-1}}]$ and set $M = \mathrm{M}[b_0,b_1,\ldots,b_{k-1}]$

$= \begin{pmatrix} A_{k-1} & A_{k-2} \\ B_{k-1} & B_{k-2} \end{pmatrix}$. Then $\varepsilon = \varphi(M) = B_{k-1}x + B_{k-2}$. For example, with $d=37$, $z = \dfrac{\sqrt{37}+5}{2}$,

we had (in the previous section) $CF[z] = [\overline{5,1,1}]$. Thus $M = \begin{pmatrix} 11 & 6 \\ 2 & 1 \end{pmatrix}$,

$\varepsilon = 2z+1 = \dfrac{12+2\sqrt{37}}{2}$, $t=12$, $u=2$. Returning to the general case, $\varepsilon = \varphi(M)$ is

equivalent to $M = \rho(\varepsilon)$ which means that once ε is known, M can be calculated as the

matrix representing multiplication by ε on the basis $x,1$. Write $x = \dfrac{\sqrt{d}+p}{q}$ in standard

form. Then straightforward computation shows

$$\varepsilon \cdot x = \left(\frac{t+pu}{2}\right)x + \left(\frac{1}{2}q^*u\right)1$$

$$\varepsilon \cdot 1 = \left(\frac{1}{2}qu\right)x + \left(\frac{t-pu}{2}\right)1$$

provided one recalls $d = p^2 + qq^*$. Thus

$$\begin{pmatrix} A_{k-1} & A_{k-2} \\ B_{k-1} & b_{k-2} \end{pmatrix} = M = \rho(\varepsilon) = \begin{pmatrix} \dfrac{t+pu}{2} & \dfrac{1}{2}q^*u \\ \dfrac{1}{2}qu & \dfrac{t-pu}{2} \end{pmatrix} \tag{22}$$

gives two independent presentations of M; the one on the left depends only on

$CF[x] = \left[\overline{b_0, b_1, \ldots, b_{k-1}}\right]$ and the one on the right depends on knowing ε. This gives a new

way of calculating $CF[x]$, purely rationally, once ε is known. Namely, since $M \in G^+$,

by our theorem of section 2 it has a unique representation as $M[b_0, b_1, \ldots, b_{k-1}]$ found by

computing $CF_\sigma\left[\dfrac{(t + pu)/2}{qu/2}\right]$ where $\sigma = \det M = N\varepsilon$. Thus the period of $CF[x]$ is found

by computing the continued fraction of a rational number. We state this as a theorem.

$\underline{\underline{\text{Theorem}}}$ If $\varepsilon = \dfrac{t + u\sqrt{d}}{2}$ is known, then for $x = \dfrac{\sqrt{d} + p}{q} \in R(d)$, $CF[x]$ may

be found as follows. Let $\sigma = N\varepsilon = \pm 1$, and compute $CF_\sigma\left[\dfrac{(t + pu)/2}{qu/2}\right] = [b_0, b_1, \ldots b_{k-1}]$.

Then $CF[x] = \left[\overline{b_0, b_1, \ldots, b_{k-1}}\right]$.

Note from (22) and (10) that $(-1)^k = \det M = N\varepsilon$ so that the parity of k, the

length of the period of $CF[x]$, depends only on d, not the specific $x \in R(d)$. k is odd for

all $x \in R(d)$ iff $N\varepsilon = -1$ and is even for all $x \in R(d)$ iff $N\varepsilon = 1$.

Here is a numerical example. For $d=1009$, $z = \dfrac{\sqrt{1009} + 31}{2}$ has, by usual method

of computation, $CF[z] = \left[\overline{31, 2, 1, 1, 1, 1, 2}\right]$, $k=7$. Then by the method described in the table

(6) one finds $M = \begin{pmatrix} A_6 & A_5 \\ B_6 & B_5 \end{pmatrix} = \begin{pmatrix} 1067 & 408 \\ 34 & 13 \end{pmatrix}$, so $\varepsilon = 34z + 13 = \dfrac{1080 + 34\sqrt{1009}}{2}$, $t=1080$,

$u=34$, $N\varepsilon=-1$. Consider now $x = \dfrac{\sqrt{1009} + 25}{12} \in R(1009)$, $p=25$, $q=12$, $q'=32$. Now, by

the theorem, we calculate $CF_-\left[\dfrac{965}{204}\right] = [4, 1, 2, 1, 2, 2, 3, 1, 1]$ and so

$CF[x] = \left[\overline{4, 1, 2, 1, 2, 2, 3, 1, 1}\right]$.

For certain d, ε is known a priori. Suppose $d = n^2 + 4$, $n \geq 1$. Then $\dfrac{n + \sqrt{d}}{2}$ is a

unit in O_d and is >1. It must be the fundamental unit ε, for otherwise it is ε^j for some

$j>1$ and then the coefficient of $\sqrt{d}$ would be greater than 1. Thus $\varepsilon = \dfrac{n+\sqrt{d}}{2}$, $t=n$, $u=1$

and $N\varepsilon = -1$. Note that in this case $z(d) = \varepsilon$ and since $\varepsilon^2 - n\varepsilon - 1 = 0$, $\varepsilon = n + \dfrac{1}{\varepsilon}$ shows

$CF[z] = [\overline{n}]$. This is, in fact, the only time the period has length 1. By the theorem we

have that for $x = \dfrac{\sqrt{d}+p}{q} \in R(d)$, $CF[x]$ is found from $CF_{-}\left[\dfrac{(n+p)/2}{q/2}\right]$. For example

$904 = 30^2 + 4$, $x = \dfrac{\sqrt{904}+26}{6}$ is reduced and $CF_{-}\left[\dfrac{28}{3}\right] = [9,2,1]$ so $CF[x] = [\overline{9,2,1}]$.

Other cases where ε is immediately known are:

$$d = n^2 + 1, n \text{ even } \geq 4, \ \varepsilon = \frac{2n+2\sqrt{d}}{2}, \ N\varepsilon = -1,$$

$$d = n^2 - 4, n \geq 4, \ \varepsilon = \frac{n+\sqrt{d}}{2}, \ N\varepsilon = 1,$$

$$d = n^2 - 1, n \text{ odd } \geq 5, \ \varepsilon = \frac{2n+2\sqrt{d}}{2}, \ N\varepsilon = 1.$$

It is interesting to note that our determination of G_x gives a new criterion for

$x \in R$: An irrational number x is a reduced quadratic irrational if and only if x is positive

and $S(x) = x$ for some matrix $S \in G^+$.

We have already seen that the condition is necessary. On the other hand, suppose

$S(x) = x$, $S \in G^+$. Then by the Theorem of Section 2, $S = M[b_0, b_1, \ldots, b_{n-1}]$, for a

uniquely determined sequence of positive integers. There is a unique, largest integer

$j \geq 1$ such that $b_0, b_1, \ldots, b_{n-1}$ is a j-fold repetition of the initial k terms $b_0, b_1, \ldots, b_{k-1}$,

$n = jk$. Let $y = [\overline{b_0, b_1, \ldots, b_{k-1}}]$. Then, $y \in R$ and G_y is generated by

$M = M[b_0, b_1, \ldots, b_{k-1}]$. But $S = M^j$ so S fixes y and y'. Since S has only two fixed

points $x = y$ or y'. But $x>0$, so $x = y$, x is reduced.

5. G mod 2   Let $F = Z \bmod 2$, the field with two elements 0,1. We use the usual integer symbols, so one should note, from the context, when the arithmetic is being done mod 2. In this section all congruences $a \equiv b$ are to be understood mod 2 unless indicated otherwise. The map $n \to n \bmod 2$ of Z onto F induces a homomorphism

$$\Psi : G \to GL(2, F) : \Psi\left(\begin{pmatrix} a & b \\ c & d \end{pmatrix}\right) = \begin{pmatrix} a \bmod 2 & b \bmod 2 \\ c \bmod 2 & d \bmod 2 \end{pmatrix}.$$ For $M \in G$ we denote $\Psi(M)$ by

$\overline{M}$, and $GL(2, F)$ by $\overline{G}$. One easily sees that $\overline{G}$ is a group of order 6 its elements being

$$I = \begin{pmatrix} 1 & 0 \\ 0 & 1 \end{pmatrix}, \text{ the identity element, } U = \begin{pmatrix} 1 & 1 \\ 1 & 0 \end{pmatrix}, U^2 = \begin{pmatrix} 0 & 1 \\ 1 & 1 \end{pmatrix}, R = \begin{pmatrix} 0 & 1 \\ 1 & 0 \end{pmatrix}, UR = \begin{pmatrix} 1 & 1 \\ 0 & 1 \end{pmatrix},$$

$U^2 R = \begin{pmatrix} 1 & 0 \\ 1 & 1 \end{pmatrix}$. U, U^2 have order 3, $R, UR, U^2 R$ have order 2 and $UR = RU^2$. Thus

$\overline{G} \cong S_3$, the symmetric group of degree 3. U, U^2 correspond to 3-cycles and $R, UR, U^2 R$ to transpositions. For $M \in G$, $S \in \overline{G}$ we also say $M \equiv S$ rather than $\overline{M} = S$. For

example $\begin{pmatrix} 7 & 5 \\ 11 & 8 \end{pmatrix} \equiv U$, $\begin{pmatrix} 21 & 8 \\ 8 & 3 \end{pmatrix} \equiv I$. $\Sigma_1 = \{I\}$, $\Sigma_2 = \{U, U^2\}$, $\Sigma_3 = \{R, UR, U^2 R\}$ are the conjugacy classes of $\overline{G}$.

Theorem Let d be a discriminant, $\varepsilon = \dfrac{t + u\sqrt{d}}{2}$, $t \equiv ud$, the fundamental unit. Let

$x \in R(d)$, $CF[x] = \left[\overline{b_0, b_1, \dots, b_{k-1}}\right]$, $M = \mathrm{M}[b_0, b_1, \dots, b_{k-1}]$. Then the conjugacy class of $\overline{M} \in \overline{G}$ depends only on d, as follows:

(a) $d \equiv 1 \pmod 8$ implies $\overline{M} = I$.

(b) $d \equiv 5 \pmod 8$ implies $\overline{M} = I$ if u is even, and $\overline{M} \in \Sigma_2$ if u is odd.

(c) d even implies $\overline{M} = I$ if u is even, and $\overline{M} \in \Sigma_3$ if u is odd.

Proof As usual, write $x = \dfrac{\sqrt{d} + p}{q}$ and $M = \begin{pmatrix} \dfrac{t + pu}{2} & \dfrac{1}{2} q^* u \\ \dfrac{1}{2} qu & \dfrac{t - pu}{2} \end{pmatrix}$ as in (22). If u is

even, then both off-diagonal elements are 0 mod 2 and the only matrix in $\overline{G}$ with this property is I; thus $\overline{M} = I$ whenever u is even. If $d \equiv 1 \pmod 8$, $t^2 - du^2 = \pm 4$ implies $t^2 - u^2 \equiv 4 \pmod 8$. But if t, u are odd, $t^2 - u^2 \equiv 1 - 1 \equiv 0 \pmod 8$, hence u, and t, must be even. This proves (a) and the case u even in (b),(c). Now assume u odd. If

$d \equiv 5 \pmod 8$, $d = p^2 + qq^{\bullet}$, p odd, so $qq^{\bullet} \equiv 5 - 1 \equiv 4 \pmod 8$. Thus $\dfrac{1}{2} q$, $\dfrac{1}{2} q^{\bullet}$ are both

odd and the off-diagonal elements of M are odd. Also, $t \equiv du \equiv 1$ so trace$(M) = t$, is odd.

So one of the diagonal entries of M is even and one is odd. Thus $M \equiv \begin{pmatrix} 1 & 1 \\ 1 & 0 \end{pmatrix} \equiv U$ or

$M \equiv \begin{pmatrix} 0 & 1 \\ 1 & 1 \end{pmatrix} \equiv U^2$, i.e. $\overline{M} \in \Sigma_2$. If d is even, then so is p, and $\gcd\left(\dfrac{1}{2}q, p, \dfrac{1}{2}q^{\bullet}\right) = 1$

implies at least one of $\dfrac{1}{2}q, \dfrac{1}{2}q^{\bullet}$ is odd. Now trace$(M) = t \equiv du \equiv 0$ so the diagonal entries

are both even or both odd. But this characterizes $\overline{M} \in \Sigma_3$ and the proof is done.

Note that we have $M \equiv I$ if and only if u is even, Now $\overline{G}$ being isomorphic to S_3 has a homomorphism $\overline{G} \to \{\pm 1\}$ defined by assigning to each $S \in \overline{G}$ its sign as a permutation. Thus $S \to -1$ if $S \in \Sigma_3$ and $S \to 1$ otherwise. Composing with $\Psi : G \to \overline{G}$ we obtain a homomorphism, or character, χ, of G to $\{\pm 1\}$. $\chi(M) = -1$ for $\overline{M} \in \Sigma_3$ and $\chi(M) = 1$ for $\overline{M} \in \Sigma_1 \cup \Sigma_2$.

Keeping with the notation of the theorem, we have

$M = \begin{pmatrix} b_0 & 1 \\ 1 & 0 \end{pmatrix}\begin{pmatrix} b_1 & 1 \\ 1 & 0 \end{pmatrix} \cdots \begin{pmatrix} b_{k-1} & 1 \\ 1 & 0 \end{pmatrix}$. But $\begin{pmatrix} b & 1 \\ 1 & 0 \end{pmatrix} \equiv R$ or U according as b is even or odd, so

$\chi\left(\begin{pmatrix} b_i & 1 \\ 1 & 0 \end{pmatrix}\right) = -1$ or 1 according as b_i is even or odd. Since χ is a homomorphism

$\chi(M) = (-1)^e$ where is $e = e(x)$ is the number of terms b_i in the period of $CF[x]$ that are even. Combining with the theorem we thus have: e odd iff $\chi(M) = -1$ iff $\overline{M} \in \Sigma_3$ iff d even and u odd. Thus d odd implies e even and d even implies e odd or even according as u is odd or even. But this is precisely the theorem stated in the Introduction, as promised.

Note that for d odd the result in terms of e is weaker than the result of the theorem of this section, since e does not differentiate between Σ_1 and Σ_2. For d odd we always have e even and $M \equiv I$ if u even and $M \equiv U$ or U^2 if u odd.

We conclude with some numerical examples.

(1) At the end of the previous section we had for $d = 1009 \equiv 1(\mathrm{mod}\,8)$,

$$\varepsilon = \frac{1080 + 34\sqrt{1009}}{2}, \quad N\varepsilon = -1, \quad u = 34 \text{ even. We found}$$

(a) $CF\left[\dfrac{\sqrt{1009}+31}{2}\right] = \left[\overline{31,2,1,1,1,1,2}\right]$, $k{=}7$, $e{=}2$, $M \equiv URU^4R \equiv I$.

(b) $CF\left[\dfrac{\sqrt{1009}+25}{12}\right] = \left[\overline{4,1,2,1,2,2,3,1,1}\right]$, $k{=}9$, $e{=}4$, $M \equiv RURUR^2U^3 \equiv I$, all as

predicted.

(2) For $d = 904 = 30^2 + 4$, even, we saw $CF\left[\dfrac{\sqrt{904}+30}{2}\right] = \left[\overline{30}\right]$, $k{=}1$, $e{=}1$,

$$\varepsilon = \frac{30 + \sqrt{904}}{2}, \quad N\varepsilon = -1, \quad u = 1, \text{ odd. Here } M \equiv R. \text{ Also}$$

$$CF\left[\frac{\sqrt{904}+26}{6}\right] = \left[\overline{9,2,1}\right], k{=}3, e{=}1, M \equiv URU \equiv R.$$

(3) Consider $x = \left[\overline{10,2,1}\right]$; what can be said about $d{=}\mathrm{disc}(x)$? Here $e{=}2$ does not tell us too much but $M \equiv R^2U \equiv U$, so we know $d \equiv 5(\mathrm{mod}\,8)$ and u is odd. Also $k{=}3$, so $N\varepsilon = -1$. Calculating $M = \begin{pmatrix} 31 & 21 \\ 3 & 2 \end{pmatrix}$, one can now find x by solving the quadratic equation $M(x) = x$. But it's easier to proceed using (22) which gives

$$\begin{pmatrix} \dfrac{t+pu}{2} & \dfrac{1}{2}q^{\cdot}u \\ \dfrac{1}{2}qu & \dfrac{t-pu}{2} \end{pmatrix} = \begin{pmatrix} 31 & 21 \\ 3 & 2 \end{pmatrix}.$$ Since $u = \gcd\left(\dfrac{1}{2}qu, pu = \left(\dfrac{t+pu}{2} - \dfrac{t-pu}{2}\right), \dfrac{1}{2}q^{\cdot}u\right) =$

$\gcd(3,29,21) = 1$ we have $q=6$, $q^{*}=42$, $p=29$, $d = p^2 + qq^{\cdot} = 1093 \equiv 5 \pmod 8$ and

$x = \dfrac{\sqrt{1093}+29}{6}$. $t=31+2=33$ and $\varepsilon = \dfrac{33+\sqrt{1093}}{2}$.

(4) In the same way consider $x = \left[\overline{2,2,2,1,1,1,1,1}\right]$; $k=8$, $N\varepsilon = 1$. Since $e=3$ we have d even

and u odd; $M \equiv R^3 U^5 \equiv UR$. We find $M = \begin{pmatrix} 121 & 75 \\ 50 & 31 \end{pmatrix}$ so $u = \gcd(50,90,75) = 5$,

$q=20$, $q^{*}=30$, $p=18$, $d = 18^2 + 20 \times 30 = 924$, $x = \dfrac{\sqrt{924}+18}{20}$. Finally

$t=121+31=152$, $\varepsilon = \dfrac{152 + 5\sqrt{924}}{2}$.

252

References

1. Adams,W.W. and Goldstein,L.J., Introduction to Number Theory, Prentice-Hall, 1976.

2. Borevich,Z.I. and Shafarevich,I.R., Number Theory, Academic Press, 1966.

3. Lewittes, J., "Quadratic Irrationals and Continued Fractions" in Number Theory - New York Seminar 1991-1995,Chudnovsky, D.V. and Chudnovsky G.V. and Nathanson, M.B. (Eds.), Springer, New York,1996.

4. Ono,T., An Introduction to Algebraic Number Theory, Plenum Press, 1990.

5. Perron, O., Die Lehre von der Kettenbrüchen (Band I), B.G. Teubner, 1954.

6. Rockett,A. and Szusz, P., Continued Fractions, World Scientific, 1992.

The inverse problem for representation functions of additive bases[*]

Melvyn B. Nathanson[†]

Department of Mathematics

Lehman College (CUNY)

Bronx, New York 10468

Email: nathansn@alpha.lehman.cuny.edu

Abstract

Let A be a set of integers. For every integer n, let $r_{A,2}(n)$ denote the number of representations of n in the form $n = a_1 + a_2$, where $a_1, a_2 \in A$ and $a_1 \leq a_2$. The function $r_{A,2} : \mathbf{Z} \to \mathbf{N}_0 \cup \{\infty\}$ is the *representation function of order 2 for A*. The set A is called an *asymptotic basis of order 2* if $r_{A,2}^{-1}(0)$ is finite, that is, if every integer with at most a finite number of exceptions can be represented as the sum of two not necessarily distinct elements of A. It is proved that every function is a representation function, that is, if $f : \mathbf{Z} \to \mathbf{N}_0 \cup \{\infty\}$ is any function such that $f^{-1}(0)$ is finite, then there exists a set A of integers such that $f(n) = r_{A,2}(n)$ for all $n \in \mathbf{Z}$. Moreover, the set A can be constructed so that $\text{card}\{a \in A : |a| \leq x\} \gg x^{1/3}$.

1 Representation functions

Let $\mathbf{N}$, $\mathbf{N}_0$, and $\mathbf{Z}$ denote the positive integers, nonnegative integers, and integers, respectively. Let A and B be sets of integers. We define the *sumset*

$$A + B = \{a + b : a \in A \text{ and } b \in B\},$$

and, in particular,

$$2A = A + A = \{a_1 + a_2 : a_1, a_2 \in A\}$$

and

$$A + b = A + \{b\} = \{a + b : a \in A\}.$$

---

[*]2000 Mathematics Subject Classification: 11B13, 11B34, 11B05. Key words and phrases. Additive bases, sumsets, representation functions, density, Erdős-Turán conjecture, Sidon set.

[†]This work was supported in part by grants from the NSA Mathematical Sciences Program and the PSC-CUNY Research Award Program.

The *restricted sumsets* are

$$A \hat{+} B = \{a + b : a \in A, b \in B, \text{ and } a \neq b\}$$

and

$$2 \wedge A = A \hat{+} A = \{a_1 + a_2 : a_1, a_2 \in A \text{ and } a_1 \neq a_2\}.$$

Similarly, we define the *difference set*

$$A - B = \{a - b : a \in A \text{ and } b \in B\}$$

and

$$-A = \{0\} - A = \{-a : -a \in A\}.$$

We introduce the *counting function*

$$A(y, x) = \sum_{\substack{a \in A \\ y \leq a \leq x}} 1.$$

Thus, $A(-x, x)$ counts the number of elements $a \in A$ such that $|a| \leq x$.

For functions f and g, we write $f \gg g$ if there exist numbers c_0 and x_0 such that $|f(x)| \geq c_0 |g(x)|$ for all $x \geq x_0$, and $f \ll g$ if $|f(x)| \leq c_0 |g(x)|$ for all $x \geq x_0$.

In this paper we study representation functions of sets of integers. For any set $A \subseteq \mathbf{Z}$, the *representation function* $r_{A,2}(n)$ counts the number of ways to write n in the form $n = a_1 + a_2$, where $a_1, a_2 \in A$ and $a_1 \leq a_2$. The set A is called an *asymptotic basis of order 2* if all but finitely many integers can be represented as the sum of two not necessarily distinct elements of A, or, equivalently, if the function

$$r_{A,2} : \mathbf{Z} \to \mathbf{N}_0 \cup \{\infty\}$$

satisfies

$$\operatorname{card}(r_{A,2}^{-1}(0)) < \infty.$$

Similarly, the *restricted representation function* $\hat{r}_{A,2}(n)$ counts the number of ways to write n in the form $n = a_1 + a_2$, where $a_1, a_2 \in A$ and $a_1 < a_2$. The set A is called a *restricted asymptotic basis of order 2* if all but finitely many integers can be represented as the sum of two distinct elements of A.

Let

$$f : \mathbf{Z} \to \mathbf{N}_0 \cup \{\infty\} \tag{1}$$

be any function such that

$$\operatorname{card}(f^{-1}(0)) < \infty. \tag{2}$$

The *inverse problem for representation functions of order 2* is to find sets A such that $r_{A,2}(n) = f(n)$ for all $n \in \mathbf{Z}$. Nathanson [4] proved that every function f satisfying (1) and (2) is the representation function of an asymptotic basis of order 2, and that such bases A can be arbitrarily *thin* in the sense that the counting functions $A(-x, x)$ tend arbitrarily slowly to infinity. It remained an open problem to construct *thick* asymptotic bases of order 2 for the integers with a prescribed representation function.

In the special case of the function $f(n) = 1$ for all integers n, Nathanson [6] constructed a unique representation basis, that is, a set A of integers with $r_{A,2}(n) = 1$ for all $n \in \mathbf{Z}$, with the additional property that $A(-x, x) \gg \log x$. He posed the problem of constructing a unique representation basis A such that $A(-x, x) \gg x^\alpha$ for some $\alpha > 0$.

In this paper we prove that for *every* function f satisfying (1) and (2) there exist uncountably many asymptotic bases A of order 2 such that $r_{A,2}(n) = f(n)$ for all $n \in \mathbf{Z}$, and $A(-x, x) \gg x^{1/3}$. It is not known if there exists a real number $\delta > 0$ such that one can solve the inverse problem for arbitrary functions f satisfying (1) and (2) with $A(-x, x) \gg x^{1/3+\delta}$.

2   The Erdős-Turán conjecture

The set A of nonnegative integers is an asymptotic basis of order 2 for $\mathbf{N}_0$ if the sumset $2A$ contains all sufficently large integers. If A is a set of nonnegative integers, then

$$0 \leq r_{A,2}(n) < \infty$$

for every $n \in \mathbf{N}_0$. It is not true, however, that if

$$f : \mathbf{N}_0 \to \mathbf{N}_0$$

is a function with

$$\mathrm{card}\left(f^{-1}(0)\right) < \infty,$$

then there must exist a set A of nonnegative integers such that $r_{A,2}(n) = f(n)$ for all $n \in \mathbf{N}_0$. For example, Dirac [1] proved that the representation function of an asymptotic basis of order 2 cannot be eventually constant, and Erdős and Fuchs [3] proved that the mean value $\sum_{n \leq x} r_{A,2}(n)$ of an asymptotic basis of order 2 cannot converge too rapidly to cx for any $c > 0$. A famous conjecture of Erdős and Turán [2] states that the representation function of an asymptotic basis of order 2 must be unbounded. This problem is only a special case of the general *inverse problem for representation functions for bases for the nonnegative integers*: Find necessary and sufficient conditions for a function $f : \mathbf{N}_0 \to \mathbf{N}_0$ satisfying $\mathrm{card}\left(f^{-1}(0)\right) < \infty$ to be the representation function of an asymptotic basis of order 2 for $\mathbf{N}_0$.

It is a remarkable recent discovery that the inverse problem for representation functions for the integers, and, more generally, for arbitrary countably infinite abelian groups and countably infinite abelian semigroups with a group component, is significantly easier than the inverse problem for representation functions for the nonnegative integers and for other countably infinite abelian semigroups (Nathanson [5]).

3   Construction of thick bases for the integers

Let $[x]$ denote the integer part of the real number x.

Lemma 1 *Let* $f : \mathbf{Z} \to \mathbf{N}_0 \cup \{\infty\}$ *be a function such that* $f^{-1}(0)$ *is finite. Let* Δ *denote the cardinality of the set* $f^{-1}(0)$. *Then there exists a sequence* $U = \{u_k\}_{k=1}^\infty$ *of integers such that, for every* $n \in \mathbf{Z}$ *and* $k \in \mathbf{N}$,

$$f(n) = \mathrm{card}\{k \geq 1 : u_k = n\}$$

and

$$|u_k| \leq \left[\frac{k + \Delta}{2}\right].$$

Proof. Every positive integer m can be written uniquely in the form

$$m = s^2 + s + 1 + r,$$

where s is a nonnegative integer and $|r| \leq s$. We construct the sequence

$$V = \{0, -1, 0, 1, -2, -1, 0, 1, 2, -3, -2, -1, 0, 1, 2, 3, \ldots\}$$
$$= \{v_m\}_{m=1}^\infty,$$

where

$$v_{s^2+s+1+r} = r \qquad \text{for } |r| \leq s.$$

For every nonnegative integer k, the first occurrence of $-k$ in this sequence is $v_{k^2+1} = -k$, and the first occurrence of k in this sequence is $v_{(k+1)^2} = k$.

The sequence U will be the unique subsequence of V constructed as follows. Let $n \in \mathbf{Z}$. If $f(n) = \infty$, then U will contain the terms $v_{s^2+s+1+n}$ for every $s \geq |n|$. If $f(n) = \ell < \infty$, then U will contain the ℓ terms $v_{s^2+s+1+n}$ for $s = |n|, |n|+1, \ldots, |n|+\ell-1$ in the subsequence U, but not the terms $v_{s^2+s+1+n}$ for $s \geq |n| + \ell$. Let $m_1 < m_2 < m_3 < \cdots$ be the strictly increasing sequence of positive integers such that $\{v_{m_k}\}_{k=1}^\infty$ is the resulting subsequence of V. Let $U = \{u_k\}_{k=1}^\infty$, where $u_k = v_{m_k}$. Then

$$f(n) = \mathrm{card}\{k \geq 1 : u_k = n\}.$$

Let $\mathrm{card}\left(f^{-1}(0)\right) = \Delta$. The sequence U also has the following property: If $|u_k| = n$, then for every integer $m \notin f^{-1}(0)$ with $|m| < n$ there is a positive integer $j < k$ with $u_j = m$. It follows that

$$\{0, 1, -1, 2, -2, \ldots, n-1, -(n-1)\} \setminus f^{-1}(0) \subseteq \{u_1, u_2, \ldots, u_{k-1}\},$$

and so

$$k - 1 \geq 2(n-1) + 1 - \Delta.$$

This implies that

$$|u_k| = n \leq \frac{k + \Delta}{2}.$$

Since u_k is an integer, we have

$$|u_k| \le \left[\frac{k+\Delta}{2}\right].$$

This completes the proof. $\square$

Lemma 1 is best possible in the sense that for every nonnegative integer Δ there is a function $f : \mathbf{Z} \to \mathbf{N}_0 \cup \{\infty\}$ with card $(f^{-1}(0)) = \Delta$ and a sequence $U = \{u_k\}_{k=1}^\infty$ of integers such that

$$|u_k| = \left[\frac{k+\Delta}{2}\right] \qquad \text{for all } k \ge 1. \tag{3}$$

For example, if $\Delta = 2\delta + 1$ is odd, define the function f by

$$f(n) = \begin{cases} 0 & \text{if } |n| \le \delta \\ 1 & \text{if } |n| \ge \delta + 1 \end{cases}$$

and the sequence U by

$$u_{2i-1} = \delta + i,$$
$$u_{2i} = -(\delta + i)$$

for all $i \ge 1$.

If $\Delta = 2\delta$ is even, define f by

$$f(n) = \begin{cases} 0 & \text{if } -\delta \le n \le \delta - 1 \\ 1 & \text{if } n \ge \delta \text{ or } n \le -\delta - 1 \end{cases}$$

and the sequence U by $u_1 = \delta$ and

$$u_{2i} = \delta + i,$$
$$u_{2i+1} = -(\delta + i)$$

for all $i \ge 1$. In both cases the sequence U satisfies (3).

Theorem 1 *Let $f : \mathbf{Z} \to \mathbf{N}_0 \cup \{\infty\}$ be any function such that*

$$\Delta = \text{card}(f^{-1}(0)) < \infty.$$

Let

$$c = 8 + \left[\frac{\Delta + 1}{2}\right].$$

There exist uncountably many sets A of integers such that

$$r_{A,2}(n) = f(n) \qquad \text{for all } n \in \mathbf{Z}$$

and

$$A(-x, x) \ge \left(\frac{x}{c}\right)^{1/3}.$$

Proof. Let

$$\Delta = \operatorname{card}(f^{-1}(0)).$$

By Lemma 1, there exists a sequence $U = \{u_k\}_{k=1}^{\infty}$ of integers such that

$$f(n) = \operatorname{card}(\{i \in \mathbf{N} : u_i = n\}) \qquad \text{for all integers } n \tag{4}$$

and

$$|u_k| \leq \frac{k + \Delta}{2} \qquad \text{for all } k \geq 1. \tag{5}$$

We shall construct a strictly increasing sequence $\{i_k\}_{k=1}^{\infty}$ of positive integers and an increasing sequence $\{A_k\}_{k=1}^{\infty}$ of finite sets of integers such that, for all positive integers k,

(i)
$$|A_k| = 2k,$$

(ii) There exists a positive number c such that

$$A_k \subseteq [-ck^3, ck^3]$$

(iii)
$$r_{A_k,2}(n) \leq f(n) \qquad \text{for all } n \in \mathbf{Z},$$

(iv) For $j = 1, \ldots, k$,

$$r_{A_k,2}(u_j) \geq \operatorname{card}(\{i \leq i_k : u_i = u_j\}).$$

Let $\{A_k\}_{k=1}^{\infty}$ be a sequence of finite sets satisfying (i)–(iv). We form the infinite set

$$A = \bigcup_{k=1}^{\infty} A_k.$$

Let $x \geq 8c$, and let k be the unique positive integer such that

$$ck^3 \leq x < c(k+1)^3.$$

Conditions (i) and (ii) imply that

$$A(-x, x) \geq |A_k| = 2k > 2 \left(\frac{x}{c}\right)^{1/3} - 2 \geq \left(\frac{x}{c}\right)^{1/3}.$$

Since

$$f(n) = \lim_{k \to \infty} \operatorname{card}(\{i \leq i_k : u_i = n\}),$$

conditions (iii) and (iv) imply that

$$r_{A,2}(n) = \lim_{k \to \infty} r_{A_k,2}(n) = f(n)$$

for all $n \in \mathbf{Z}$.

We construct the sequence $\{A_k\}_{k=1}^\infty$ as follows. Let $i_1 = 1$. The set A_1 will be of the form $A_1 = \{a_1 + u_{i_1}, -a_1\}$, where the integer a_1 is chosen so that $2A_1 \cap f^{-1}(0) = \emptyset$ and $a_1 + u_{i_1} \neq -a_1$. This is equivalent to requiring that

$$2a_1 \notin (f^{-1}(0) - 2u_{i_1}) \cup (-f^{-1}(0)) \cup \{-u_{i_1}\}. \tag{6}$$

This condition excludes at most $1 + 2\Delta$ integers, and so we have at least two choices for the number a_1 such that $|a_1| \leq 1 + \Delta$ and a_1 satisfies (6). Since $|u_{i_1}| = |u_1| \leq (1 + \Delta)/2$ and

$$|a_1 + u_{i_1}| \leq |a_1| + |u_{i_1}| \leq \frac{3(1 + \Delta)}{2},$$

it follows that $A_1 \subseteq [-c, c]$ for any $c \geq 3(1 + \Delta)/2$, and the set A_1 satisfies conditions (i)–(iv).

Let $k \geq 2$ and suppose that we have constructed sets $A_1, \ldots, A_{k-1}$ and integers $i_1 < \cdots < i_{k-1}$ that satisfy conditions (i)–(iv). Let $i_k > i_{k-1}$ be the least integer such that

$$r_{A_{k-1},2}(u_{i_k}) < f(u_{i_k}).$$

Since

$$i_k - 1 \leq \sum_{n \in \{u_1, u_2, \ldots, u_{i_k - 1}\}} r_{A_{k-1},2}(n)$$

$$\leq \sum_{n \in \mathbf{Z}} r_{A_{k-1},2}(n)$$

$$= \binom{2k - 1}{2}$$

$$< 2k^2,$$

it follows that

$$i_k \leq 2k^2.$$

Also, (5) implies that

$$|u_{i_k}| \leq \frac{i_k + \Delta}{2} \leq k^2 + \frac{\Delta}{2}. \tag{7}$$

We want to choose an integer a_k such that the set

$$A_k = A_{k-1} \cup \{a_k + u_{i_k}, -a_k\}$$

satisfies (i)–(iv). We have $|A_k| = 2k$ if

$$a_k + u_{i_k} \neq -a_k$$

and

$$A_{k-1} \cap \{a_k + u_{i_k}, -a_k\} = \emptyset,$$

or, equivalently, if

$$a_k \notin (-A_{k-1}) \cup (A_{k-1} - u_{i_k}) \cup \{-u_{i_k}/2\}. \tag{8}$$

Thus, in order for $A_{k-1} \cup \{a_k + u_{i_k}, -a_k\}$ to satisfy condition (i), we exclude at most $2|A_{k-1}| + 1 = 4k - 3$ integers as possible choices for a_k.

The set A_k will satisfy conditions (iii) and (iv) if

$$2A_k \cap f^{-1}(0) = \emptyset$$

and

$$r_{A_k,2}(n) = \begin{cases} r_{A_{k-1},2}(n) & \text{for all } n \in 2A_{k-1} \setminus \{u_{i_k}\} \\ r_{A_{k-1},2}(n) + 1 & \text{for } n = u_{i_k} \\ 1 & \text{for all } n \in 2A_k \setminus (2A_{k-1} \cup \{u_{i_k}\}). \end{cases}$$

Since the sumset $2A_k$ decomposes into

$$2A_k = 2\left(A_{k-1} \cup \{a_k + u_{i_k}, -a_k\}\right)$$
$$= 2A_{k-1} \cup \left(A_{k-1} + \{a_k + u_{i_k}, -a_k\}\right) \cup \{u_{i_k}, 2a_k + 2u_{i_k}, -2a_k\},$$

it suffices that

$$\left(A_{k-1} + \{a_k + u_{i_k}, -a_k\}\right) \cap 2A_{k-1} = \emptyset, \tag{9}$$
$$\left(A_{k-1} + \{a_k + u_{i_k}, -a_k\}\right) \cap f^{-1}(0) = \emptyset, \tag{10}$$
$$\left(A_{k-1} + a_k + u_{i_k}\right) \cap \left(A_{k-1} - a_k\right) = \emptyset, \tag{11}$$
$$\{2a_k + 2u_{i_k}, -2a_k\} \cap 2A_{k-1} = \emptyset \tag{12}$$
$$\{2a_k + 2u_{i_k}, -2a_k\} \cap f^{-1}(0) = \emptyset \tag{13}$$
$$\{2a_k + 2u_{i_k}, -2a_k\} \cap \left(A_{k-1} + \{a_k + u_{i_k}, -a_k\}\right) = \emptyset. \tag{14}$$

Equation (9) implies that the integer a_k must be chosen so that it cannot be represented either in the form

$$a_k = x_1 + x_2 - x_3 - u_{i_k}$$

or

$$a_k = x_1 - x_2 - x_3,$$

where $x_1, x_2, x_3 \in A_{k-1}$. Since $\text{card}(A_{k-1}) = 2(k-1)$, it follows that the number of integers that cannot be chosen as the integer a_k because of equation (9) is at most $2(2(k-1))^3 = 16(k-1)^3$.

Similarly, the numbers of integers excluded as possible choices for a_k because of equations (10), (11), (12), (13), and (14) are at most $4\Delta(k-1), 4(k-1)^2, 8(k-1)^2, 2\Delta$, and $8(k-1)$, respectively, and so the number of integers that cannot be chosen as a_k is

$$16(k-1)^3 + 12(k-1)^2 + (4\Delta + 8)(k-1) + 2\Delta$$
$$= 16k^3 - 36k^2 + (32 + 4\Delta)k - 2\Delta - 12$$
$$\le (16 + \Delta)k^3 - 4k^2 - 32k(k-1) - 2\Delta - 12.$$

Let
$$c = 8 + \left[\frac{\Delta + 1}{2}\right].$$

The number of integers a with

$$|a| \le ck^3 - k^2 - \left[\frac{\Delta + 1}{2}\right] = \left(8 + \left[\frac{\Delta + 1}{2}\right]\right) k^3 - k^2 - \left[\frac{\Delta + 1}{2}\right] \tag{15}$$

is

$$\left(16 + 2\left[\frac{\Delta + 1}{2}\right]\right) k^3 - 2k^2 - 2\left[\frac{\Delta + 1}{2}\right] + 1$$
$$\ge (16 + \Delta) k^3 - 2k^2 - \Delta.$$

If the integer a satisfies (15), then (7) implies that

$$|a + u_{i_k}| \le |a| + |u_{i_k}| \le ck^3.$$

It follows that there are at least two acceptable choices of the integer a_k such that the set $A_k = A_{k-1} \cup \{a_k + u_{i_k}, -a_k\}$ satisfies conditions (i)–(iv). Since this is true at each step of the induction, there are uncountably many sequences $\{A_k\}_{k=1}^{\infty}$ that satisfy conditions (i)–(iv). This completes the proof. □

We can modify the proof of Theorem 1 to obtain the analogous result for the restricted representation function $\hat{r}_{A,2}(n)$.

Theorem 2 *Let $f : \mathbf{Z} \to \mathbf{N}_0 \cup \{\infty\}$ be any function such that*

$$\operatorname{card}(f^{-1}(0)) < \infty.$$

Then there exist uncountably many sets A of integers such that

$$\hat{r}_{A,2}(n) = f(n) \qquad \text{for all } n \in \mathbf{Z}$$

and

$$A(-x, x) \gg x^{1/3}.$$

4 Representation functions for bases of order h

We can also prove similar results for the representation functions of asymptotic bases and restricted asymptotic bases of order h for all $h \ge 2$.

For any set $A \subseteq \mathbf{Z}$, the *representation function* $r_{A,h}(n)$ counts the number of ways to write n in the form $n = a_1 + a_2 + \cdots + a_h$, where $a_1, a_2, \ldots, a_h \in A$ and $a_1 \le a_2 \le \cdots \le a_h$. The set A is called an *asymptotic basis of order h* if all but finitely many integers can be represented as the sum of h not necessarily distinct elements of A, or, equivalently, if the function

$$r_{A,h} : \mathbf{Z} \to \mathbf{N}_0 \cup \{\infty\}$$

satisfies

$$\text{card}(r_{A,h}^{-1}(0)) < \infty.$$

Similarly, the *restricted representation function* $\hat{r}_{A,h}(n)$ counts the number of ways to write n as a sum of h pairwise distinct elements of A. The set A is called a *restricted asymptotic basis of order h* if all but finitely many integers can be represented as the sum of h pairwise distinct elements of A.

Theorem 3 *Let $f : \mathbf{Z} \to \mathbf{N}_0 \cup \{\infty\}$ be any function such that*

$$\text{card}(f^{-1}(0)) < \infty.$$

There exist uncountably many sets A of integers such that

$$r_{A,h}(n) = f(n) \qquad \text{for all } n \in \mathbf{Z}$$

and

$$A(-x, x) \gg x^{1/(2h-1)},$$

and there exist uncountably many sets A of integers such that

$$\hat{r}_{A,h}(n) = f(n) \qquad \text{for all } n \in \mathbf{Z}$$

and

$$A(-x, x) \gg x^{1/(2h-1)}.$$

References

[1] G. A. Dirac, *Note on a problem in additive number theory*, J. London Math. Soc. **26** (1951), 312–313.

[2] P. Erdős and P. Turán, *On a problem of Sidon in additive number theory and some related questions*, J. London Math. Soc. **16** (1941), 212–215.

[3] P. Erdős and W. H. J. Fuchs, *On a problem of additive number theory*, J. London Math. Soc. **31** (1956), 67–73.

[4] M. B. Nathanson, *Every function is the representation function of an additive basis for the integers*, www.arXiv.org, math.NT/0302091.

[5] ———, *Representation functions of additive bases for abelian semigroups*, Ramanujan J., to appear.

[6] ———, *Unique representation bases for the integers*, Acta Arith., to appear.

On the ubiquity of Sidon sets*

Melvyn B. Nathanson[†]
Department of Mathematics
Lehman College (CUNY)
Bronx, New York 10468
Email: nathansn@alpha.lehman.cuny.edu

Abstract

A Sidon set is a set A of integers such that no integer has two essentially distinct representations as the sum of two elements of A. More generally, for every positive integer g, a $B_2[g]$-set is a set A of integers such that no integer has more than g essentially distinct representations as the sum of two elements of A. It is proved that almost all small subsets of $\{1, 2, \ldots, n\}$ are $B_2[g]$-sets, in the sense that if $B_2[g](k, n)$ denotes the number of $B_2[g]$-sets of cardinality k contained in the interval $\{1, 2, \ldots, n\}$, then $\lim_{n \to \infty} B_2[g](k, n) / \binom{n}{k} = 1$ if $k = o\left(n^{g/(2g+2)}\right)$.

1 Sidon sets

Let A be a nonempty set of positive integers. The *sumset* $2A$ is the set of all integers of the form $a_1 + a_2$, where $a_1, a_2 \in A$. The set A is called a *Sidon set* if every element of $2A$ has a unique representation as the sum of two elements of A, that is, if

$$a_1, a_2, a'_1, a'_2 \in A$$

and

$$a_1 + a_2 = a'_1 + a'_2,$$

and if

$$a_1 \leq a_2 \quad \text{and} \quad a'_1 \leq a'_2,$$

then

$$a_1 = a'_1 \quad \text{and} \quad a_2 = a'_2.$$

---

*2000 Mathematics Subject Classification: 11B13, 11B34, 11B05. Key words and phrases. Sidon sets, sumsets, representation functions.

†Supported in part by grants from the NSA Mathematical Sciences Program and the PSC-CUNY Research Award Program. This paper was written while the author was a visitor at the (alas, now defunct) AT&T Bell Laboratories in Murray Hill, New Jersey, an excellent research institution that split into AT&T Research Labs and Lucent Bell Labs, and provided another instance of a whole being greater than the sum of its parts.

More generally, for positive integers h and g, the *h-fold sumset* hA is the set of all sums of h not necessarily distinct elements of A. The *representation function* $r_{A,h}(m)$ counts the number of representations of m in the form

$$m = a_1 + a_2 + \cdots + a_h,$$

where

$$a_i \in A \qquad \text{for all } i = 1, 2, \ldots, h,$$

and

$$a_1 \leq a_2 \leq \cdots \leq a_h.$$

The set A is called a $B_h[g]$-set if every element of hA has at most g representations as the sum of h elements of A, that is, if

$$r_{A,h}(m) \leq g$$

for every integer m. In particular, a $B_2[1]$-set is a Sidon set, and $B_h[1]$-sets are usually denoted B_h-sets.

Let $h \geq 2$. Let A be a nonempty set of integers, and $a \in A$. Then $r_h(m + a) \geq r_{h-1}(m)$. Therefore, if $r_{A,h-1}(m) > g$ for some $m \in (h-1)A$, then $r_{A,h}(m + a) > g$ for every $a \in A$. It follows that if A is a $B_h[g]$-set, then A is also a $B_{h-1}[g]$-set. In particular, every B_h- set is also a B_{h-1}-set.

Let A be a subset of $\{1, 2, \ldots, n\}$, and let $|A|$ denote the cardinality of A. Then $hA \subseteq \{h, h+1, \ldots, hn\}$. If $|A| = k$, then there are exactly $\binom{k+h-1}{h}$ ordered h-tuples of the form $(a_1, \ldots, a_h)$, where $a_i \in A$ for all $i = 1, \ldots, h$ and $a_1 \leq \cdots \leq a_h$. If A is a $B_h[g]$-set and $|A| = k$, then

$$\frac{k^h}{h!} < \binom{k+h-1}{h} = \sum_{m \in hA} r_{A,h}(m) \leq g|hA| < ghn,$$

and so

$$|A| = k < cn^{1/h}$$

for $c = (h! g h)^{1/h}$. It follows that if A is a "large" subset of $\{1, 2, \ldots, n\}$, then A cannot be a $B_h[g]$-set. In this paper we prove that almost all "small" subsets of $\{1, 2, \ldots, n\}$ are $B_2[g]$-sets and almost all "small" subsets of $\{1, 2, \ldots, n\}$ are B_h-sets.

Notation. If $\{u_n\}_{n=1}^{\infty}$ and $\{v_n\}_{n=1}^{\infty}$ are sequences and $v_n > 0$ for all n, we write $u_n = o(v_n)$ if $\lim_{n \to \infty} u_n/v_n = 0$, and $u_n = O(v_n)$ or $u_n \ll v_n$ if $|u_n| \leq c v_n$ for some $c > 0$ and all $n \geq 1$. The number c in this inequality is called the *implied constant*.

2 Random small $B_2[g]$-sets

We require the following elementary lemma.

Lemma 1 *If $n \geq 1$ and $0 \leq j \leq k \leq n$, then*

$$\frac{\binom{n-j}{k-j}}{\binom{n}{k}} \leq \left(\frac{k}{n}\right)^j.$$

Proof. We have

$$\frac{\binom{n-j}{k-j}}{\binom{n}{k}} = \frac{(n-j)!k!}{(k-j)!n!} = \prod_{i=0}^{j-1} \frac{k-i}{n-i} \leq \left(\frac{k}{n}\right)^j$$

since

$$\frac{k-i}{n-i} \leq \frac{k}{n}$$

for $i = 0, 1, \ldots, n-1$. $\square$

Theorem 1 *For any positive integers g, k, and n, let $B_2[g](k,n)$ denote the number of $B_2[g]$-sets A contained in $\{1, \ldots, n\}$ with $|A| = k$. Then*

$$B_2[g](k,n) > \binom{n}{k}\left(1 - \frac{4k^{2g+2}}{n^g}\right).$$

Proof. Let A be a subset of $\{1, 2, \ldots, n\}$ of cardinality k. If A is not a $B_2[g]$-set, then there is an integer $m \leq 2n$ such that $r_{A,2}(m) > g$, that is, m has at least $g + 1$ representations as the sum of two elements of A. This means that the set A contains $g + 1$ integers $a_1, \ldots, a_{g+1}$ such that

$$1 \leq a_1 < \cdots < a_{g+1} \leq \frac{m}{2},$$

and A also contains the $g + 1$ integers $m - a_i$ for $i = 1, \ldots, g + 1$. If $a_{g+1} < m/2$, then

$$|\{a_i, m - a_i : i = 1, \ldots, g + 1\}| = 2g - 2.$$

If $a_{g+1} = m/2$, then

$$|\{a_i, m - a_i : i = 1, \ldots, g + 1\}| = 2g - 1.$$

Therefore, for each integer m, the number of sets $A \subseteq \{1, \ldots, n\}$ such that $|A| = k$ and $r_{A,2}(m) \geq g + 1$ is at most

$$\binom{\left[\frac{m-1}{2}\right]}{g+1}\binom{n - 2g - 2}{k - 2g - 2} + \binom{\left[\frac{m-1}{2}\right]}{g}\binom{n - 2g - 1}{k - 2g - 1},$$

and so

$$\binom{n}{k} - B_2[g](k,n) \leq \sum_{m \leq 2n} \binom{\left[\frac{m-1}{2}\right]}{g+1}\binom{n - 2g - 2}{k - 2g - 2}$$

$$+ \sum_{m \leq 2n} \binom{\left[\frac{m-1}{2}\right]}{g}\binom{n - 2g - 1}{k - 2g - 1}.$$

Observing that

$$\sum_{m\le 2n}\binom{[\frac{m-1}{2}]}{g+1} < \sum_{m\le 2n}\left(\frac{m}{2}\right)^{g+1} \le 2n^{g+2}$$

and

$$\sum_{m\le 2n}\binom{[\frac{m-1}{2}]}{g} < 2n^{g+1}$$

and applying Lemma 1, we obtain

$$1 - \frac{B_2[g](k,n)}{\binom{n}{k}} \le \sum_{m\le 2n}\left(\frac{\binom{[\frac{m-1}{2}]}{g+1}\binom{n-2g-2}{k-2g-2}}{\binom{n}{k}} + \frac{\binom{[\frac{m-1}{2}]}{g}\binom{n-2g-1}{k-2g-1}}{\binom{n}{k}}\right)$$

$$\le \sum_{m\le 2n}\binom{[\frac{m-1}{2}]}{g+1}\left(\frac{k}{n}\right)^{2g+2} + \sum_{m\le 2n}\binom{[\frac{m-1}{2}]}{g}\left(\frac{k}{n}\right)^{2g+1}$$

$$< 2n^{g+2}\left(\frac{k}{n}\right)^{2g+2} + 2n^{g+1}\left(\frac{k}{n}\right)^{2g+1}$$

$$\le \frac{4k^{2g+2}}{n^g}.$$

This completes the proof. $\square$

Theorem 2 Let $\{k_n\}_{n=1}^{\infty}$ be a sequence of positive integers such that $k_n \le n$ for all n and

$$k_n = o\left(n^{g/(2g+2)}\right).$$

Then

$$\lim_{n\to\infty}\frac{B_2[g](k_n,n)}{\binom{n}{k_n}} = 1.$$

Proof. This follows immediately from Theorem 1. $\square$

Theorem 3 Let $B_2(k,n)$ denote the number of Sidon sets of cardinality k contained in $\{1,\dots,n\}$. If $k_n = o\left(n^{1/4}\right)$, then

$$\lim_{n\to\infty}\frac{B_2(k_n,n)}{\binom{n}{k_n}} = 1.$$

Proof. This follows immediately from Theorem 2 with $g = 1$. $\square$

Theorem 4 Let $\{k_n\}_{n=1}^{\infty}$ be a sequence of positive integers such that $k_n \le n$ for all n and

$$k_n = o\left(n^{g/(2g+3)}\right).$$

Then

$$\lim_{n \to \infty} \frac{\sum_{k \leq k_n} B_2[g](k,n)}{\sum_{k \leq k_n} \binom{n}{k}} = 1.$$

Proof. It suffices to show that

$$\lim_{n \to \infty} \frac{\sum_{k \leq k_n} \left(\binom{n}{k} - B_2[g](k,n)\right)}{\sum_{k \leq k_n} \binom{n}{k}} = 0,$$

where $f(k)$ is defined in the proof of Theorem 1. If $a_1, \ldots, a_\ell, b_1, \ldots, b_\ell$ are positive real numbers and $B = \max(b_1, \ldots, b_\ell)$, then

$$\frac{a_1 + \cdots + a_\ell}{b_1 + \cdots + b_\ell} \leq \frac{a_1 + \cdots + a_\ell}{B} \leq \frac{a_1}{b_1} + \cdots + \frac{a_\ell}{b_\ell}.$$

This implies that

$$\frac{\sum_{k \leq k_n} \left(\binom{n}{k} - B_2[g](k,n)\right)}{\sum_{k \leq k_n} \binom{n}{k}} \leq \sum_{k \leq k_n} \frac{4k^{2g+2}}{n^g}$$

$$\leq \frac{4k_n^{2g+3}}{n^g}$$

$$= 4 \left(\frac{k_n}{n^{g/(2g+3)}}\right)^{2g+3}$$

$$= o(n),$$

and the proof is complete. $\square$

We can restate our results in the language of probability. Let Ω be the probability space consisting of the $\binom{n}{k}$ subsets of $\{1, \ldots, n\}$ of cardinality k, where the probability of choosing $A \in \Omega$ is $1/\binom{n}{k}$. If $P_{h,g}(k,n)$ denotes the probability that a random set $A \in \Omega$ is a $B_h[g]$-set, then Theorem 2 states that

$$\lim_{n \to \infty} P_{2,g}(k_n, n) = 1$$

if $k_n = o\left(n^{g/(2g+2)}\right)$.

Similarly, Theorem 4 states that if $k_n = o\left(n^{g/(2g+3)}\right)$ and if $P_{h,g}(k_n, n)$ denotes the probability that a random set $A \subseteq \{1, \ldots, n\}$ of cardinality $|A| \leq k_n$ is a $B_h[g]$-set, then

$$\lim_{n \to \infty} P_{2,g}(k_n, n) = 1.$$

3 Random small B_h-sets

A set A is called a B_h-set if $r_{A,h}(m) = 1$ for all $m \in hA$. Let $B_h(k,n)$ denote the number of B_h-sets of cardinality k contained in $\{1, \ldots, n\}$. Since every set of integers is a B_1-set, and every B_h set is a B_{h-1}-set, it follows that

$$\binom{n}{k} = B_1(k,n) \geq \cdots \geq B_{h-1}(k,n) \geq B_h(k,n) \geq \cdots$$

We shall prove that almost all "small" subsets of $\{1, \ldots, n\}$ are B_h-sets. The method is similar to that used to prove Theorem 2, but, for $h \geq 3$, we have to consider the possible dependence between different representations of an integer n as the sum of h elements of A. This means the following: Let $(a_1, \ldots, a_h)$ and $(a_1', \ldots, a_h')$ be h-tuples of elements of A such that

$$a_1 + \cdots + a_h = a_1' + \cdots + a_h',$$

$$a_1 \leq \cdots \leq a_h,$$
$$a_1' \leq \cdots \leq a_h',$$

and

$$\{a_1, \ldots, a_h\} \cap \{a_1', \ldots, a_h'\} \neq \emptyset.$$

If $h = 2$, then $a_i = a_i'$ for $i = 1, 2$, but if $h \geq 3$, then it is not necessarily true that $a_i = a_i'$ for all $i = 1, \ldots, h$. For example, in the case $h = 3$ we have $1 + 3 + 4 = 1 + 2 + 5$ but $(1, 3, 4) \neq (1, 2, 5)$. In the case $h = 5$ we have $1 + 1 + 2 + 3 + 3 = 1 + 2 + 2 + 2 + 3$ but $(1, 1, 2, 3, 3) \neq (1, 2, 2, 2, 3)$, even though $\{1, 1, 2, 3, 3\} = \{1, 2, 2, 2, 3\}$.

Because of the lack of independence, we need a careful description of a representation of m as the sum of h not necessarily distinct integers. We introduce the following notation. Let A be a set of positive integers. Corresponding to each representation of m in the form

$$m = a_1 + \cdots + a_h, \tag{1}$$

where $a_1, \ldots, a_h \in A$ and $a_1 \leq \cdots \leq a_h$, there is a unique triple

$$(r, (h_j), (a_j')), \tag{2}$$

where

(i) r is the number of distinct summands in this representation,

(ii) $(h_j) = (h_1, \ldots, h_r)$ is an ordered partition of h into r positive parts, that is, an r-tuple of positive integers such that

$$h = h_1 + \cdots + h_r,$$

(iii) $(a_j') = (a_1', \ldots, a_r')$ is an r-tuple of pairwise distinct elements of A such that

$$1 \leq a_1' < \cdots < a_r' \leq m$$

and

$$\{a_1, \ldots, a_h\} = \{a_1', \ldots, a_r'\},$$

(iv)

$$m = h_1 a_1' + \cdots + h_r a_r',$$

where each integer a_j' occurs exactly h_j times in the representation (1).

There is a one-to-one correspondence between distinct representations of an integer m in the form (1) and triples of the form (2). Moreover, for each r and m, the integer a'_r is completely determined by the ordered partition (h_j) of h and the $(r-1)$-tuple $(a'_1, \ldots, a'_{r-1})$. Therefore, for positive integers m and r, the number of triples of the form (2) does not exceed $\pi_r(h)m^{r-1}$, where $\pi_r(h)$ is number of ordered partitions of h into exactly r positive parts.

Theorem 5 *Let* $h \geq 2$. *For all positive integers* $k \leq n$,

$$B_{h-1}(k,n) - B_h(k,n) \ll \binom{n}{k}\frac{k^{2h}}{n}$$

and

$$B_h(k,n) \gg \binom{n}{k}\left(1 - \frac{k^{2h}}{n}\right).$$

where the implied constants depend only on h.

Proof. Let A be a B_{h-1}-set contained in $\{1, \ldots, n\}$. Then $hA \subseteq \{h, h+1, \ldots, hn\}$. If $m \in hA$ and m has two distinct representations as the sum of h elements of A, then there exist positive integers r_1 and r_2 and triples

$$(r_1, (h_{1,j}), (a'_{1,j})) \quad \text{and} \quad (r_2, (h_{2,j}), (a'_{2,j})) \tag{3}$$

such that, for $i = 1$ and 2, we have

$$m = \sum_{j=1}^{r_i} h_{i,j}a'_{i,j},$$

$$h = \sum_{j=1}^{r_i} h_{i,j},$$

and

$$1 \leq a'_{i,1} < \cdots < a'_{i,r_i} \leq m.$$

The number of pairs of triples of the form (3) for fixed positive integers m, $r_1 \leq h$, and $r_2 \leq h$ is at most

$$\pi_{r_1}(h)m^{r_1-1}\pi_{r_2}(h)m^{r_2-1} \ll m^{r_1+r_2-2},$$

where the implied constant depends only on h. Moreover, since A is a B_{h-1}-set, no number can have two representations as the sum of $h-1$ elements of A. This implies that

$$\{a'_{1,1}, a'_{1,2}, \ldots, a'_{1,r_1}\} \cap \{a'_{2,1}, a'_{2,2}, \ldots, a'_{2,r_2}\} = \emptyset,$$

and so the set

$$\{a'_{i,j} : i = 1, 2 \text{ and } j = 1, \ldots, r_i\}$$

contains exactly $r_1 + r_2$ elements of A. Therefore, given positive integers $r_1 \leq h$, $r_2 \leq h$ and $m \leq hn$, there are

$$\ll m^{r_1+r_2-2} \binom{n - r_1 - r_2}{k - r_1 - r_2}$$

sets A for which m has two representations as the sum of h elements of A, and in which one representation uses r_1 distinct integers and the other representation uses r_2 distinct integers. Summing over $m \leq hn$, we obtain

$$\ll n^{r_1+r_2-1} \binom{n - r_1 - r_2}{k - r_1 - r_2}.$$

Applying Lemma 1, we obtain

$$B_{h-1}(k, n) - B_h(k, n) \ll \binom{n}{k} \sum_{r_1, r_2 \leq h} \frac{n^{r_1+r_2-1} \binom{n-r_1-r_2}{k-r_1-r_2}}{\binom{n}{k}}$$

$$\ll \binom{n}{k} \sum_{r_1, r_2 \leq h} n^{r_1+r_2-1} \left(\frac{k}{n}\right)^{r_1+r_2}$$

$$= \binom{n}{k} \sum_{r_1, r_2 \leq h} \frac{k^{r_1+r_2}}{n}$$

$$\ll \binom{n}{k} \frac{k^{2h}}{n}.$$

It follows that

$$\binom{n}{k} - B_h(k, n) = B_1(k, n) - B_h(k, n)$$

$$= \sum_{j=2}^{h} (B_{j-1}(k, n) - B_j(k, n))$$

$$\ll \sum_{j=2}^{h} \binom{n}{k} \frac{k^{2j}}{n}$$

$$\ll \binom{n}{k} \frac{k^{2h}}{n},$$

and so

$$B_h(k, n) \gg \binom{n}{k} \left(1 - \frac{k^{2h}}{n}\right).$$

This completes the proof. $\square$

Theorem 6 *Let $B_h(k, n)$ denote the number of B_h-sets A contained in $\{1, \ldots, n\}$ with $|A| = k$. Let $\{k_n\}_{n=1}^{\infty}$ be a sequence of positive integers such that*

$$k_n = o\left(n^{1/2h}\right).$$

Then

$$\lim_{n \to \infty} \frac{B_h(k_n, n)}{\binom{n}{k_n}} = 1.$$

Proof. By Theorem 5,

$$1 \geq \frac{B_h(k_n, n)}{\binom{n}{k_n}} \gg 1 - \left(\frac{k_n}{n^{1/2h}} \right)^{2h},$$

and so

$$\lim_{n \to \infty} \frac{B_h(k_n, n)}{\binom{n}{k_n}} = 1.$$

This completes the proof. $\square$

4 Remarks added in proof

A variant of Theorem 3 appears in Nathanson [2, p. 37, Exercise 14]. Godbole, Janson, Locantore, and Rapoport [1] used probabilistic methods to obtain a converse of Theorem 6. They proved that if

$$\lim_{n \to \infty} \frac{k_n}{n^{1/2h}} = \infty,$$

then

$$\lim_{n \to \infty} \frac{B_h(k_n, n)}{\binom{n}{k_n}} = 0.$$

They also analyzed the threshold behavior of $B_h(k, n)$, and proved that if

$$\lim_{n \to \infty} \frac{k_n}{n^{1/2h}} = \Lambda > 0,$$

then

$$\lim_{n \to \infty} \frac{B_h(k_n, n)}{\binom{n}{k_n}} = e^{-\lambda},$$

where $\lambda = \kappa_h \Lambda^{2h}$.

It is natural to conjecture that analogous results hold for the function $B_h[g](k, n)$, namely, if

$$\lim_{n \to \infty} \frac{k_n}{n^{g/(gh+h)}} = 0,$$

then

$$\lim_{n \to \infty} \frac{B_h(k_n, n)}{\binom{n}{k_n}} = 1,$$

and if

$$\lim_{n \to \infty} \frac{k_n}{n^{g/(gh+h)}} = \infty,$$

then

$$\lim_{n \to \infty} \frac{B_h(k_n, n)}{\binom{n}{k_n}} = 0.$$

It should also be possible to describe the threshold behavior of $B_h(k_n, n)/\binom{n}{k_n}$ in the case

$$\lim_{n \to \infty} \frac{B_h(k_n, n)}{\binom{n}{k_n}} = \Lambda > 0.$$

References

[1] A. P. Godbole, S. Janson, N. W. Locantore, Jr., and R. Rapoport, *Random Sidon sequences*, J. Number Theory **75** (1999), 7–22.

[2] M. B. Nathanson, *Additive number theory: Inverse problems and the geometry of sumsets*, Graduate Texts in Mathematics, vol. 165, Springer-Verlag, New York, 1996.